W0042573

PROGRESS IN COLLOID & POLYMER SCIENCE

Editors: F. Kremer (Leipzig) and G. Lagaly (Kiel)

Volume 108 (1998)

The Colloid Science of Lipids

New Paradigms for Self-Assembly
in Science and Technology

Kåre Larsson Festschrift

Guest Editors:

B. Lindman and B. W. Ninham (Lund)

Springer-Verlag Berlin Heidelberg GmbH

IV

ISBN 978-3-662-15606-3
DOI 10.1007/978-3-7985-1655-7
ISSN 0340-255 X

Die Deutsche Bibliothek –
CIP-Einheitsaufnahme

Progress in colloid & polymer science. –
Darmstadt : Steinkopff ; New York :
Springer
   Früher Schriftenreihe
   Reihe Progress in colloid & polymer
   science zu: Colloid & polymer science
Vol. 108. The colloid science of lipids. –
1998
The colloid science of lipids ; new paradigms
for self assembly in science and
technology ; Kåre-Larsson-Festschrift /
guest eds.: B. Lindmann und
B. W. Ninham. – Darmstadt : Steinkopff ;
New York : Springer, 1998
   (Progress in colloid & polymer science ;
   Vol. 108)

ISBN 978-3-7985-1655-7 (eBook)

© Springer-Verlag Berlin Heidelberg 1998
Originally published by Dr. Dietrich
Steinkopff-Verlag GmbH & Co. KG,
Darmstadt in 1998
Softcover reprint of the hardcover 1st edition 1998

Chemistry Editor:
Dr. Maria Magdalene Nabbe;
Production: Holger Frey, Ajit Vaidya.

Typesetting and Copy-Editing:
Macmillan Ltd., Bangalore, India

Prog Colloid Polym Sci (1998) V
© Steinkopff Verlag 1998

Caption. Kåre Larsson on his 60th birthday
In honour of Kåre Larsson's contributions a meeting on the Colloid Science of Lipids was held on June 10, 1997 at the Medical Research Center of Lund University in Malmö. The Editors are grateful for the support of their organising committee Dr. Gunnar Sandberg, Professor Ann-Charlotte Eliasson, and Anita Sandberg of the Department of Food Technology. We are grateful too for the sponsorship of:
Swedish Council for Forestry and Agricultural Research, Swedish Research Council for Engineering Sciences, Nobel Committee for Chemistry, Wenner-Gren Center Foundation for Scientific Research, Lund University, Center for Chemistry and Chemical Engineering, Lund University, GS Development.
Björn Lindman and Barry Ninham (editors).

CAPTION: Larsson's first original research as a 14 year old involved his formulation of an effective skin cream using lipids and attracted national press attention.

Kåre Larsson's first papers were published in 1960, three crystal structure reports on organosilicon compounds. It was probably of no importance, since no one asked for reprints, but he obtained his "licentiat" exam (equivalent to a modern Ph.D). He then moved from Uppsala University to Göteborg University with Einar Stenhagen, professor of medical chemistry and well-known in the lipid field. Kåre's work for his Doctor of Science degree concerned crystal structures of different fatty acid glycerides. At that time the molecular conformations of different polymorphic forms in the solid state of fats were unknown, and many groups were competing in the tricky work of growing good single crystals in order to solve the first structure. During the same month in 1963 two senior crystallographers in the US (Jensen and Mabis) reported the structure of the beta-form of tridecanoin at the same time that Larsson reported the structure of the beta-form of tridodecanoin. The structure determination was a milestone in fat research; the solid state behavior of fats could then be explained. After his Ph. D. (in 1964) he went to Vittorio Luzzati to learn low-angle X-ray methods in order to study glyceride-water phases. In papers published in 1966–67 he reported the structure determinations of all the aqueous phases of fatty acid monoglycerides with chain lengths C_6–C_{18}. For the cubic phases he proposed a reversed structure of water aggregates. He also filed a patent in 1965 where he described how to produce lipid crystal dispersions as gel-like formulations with up to 80 % water from onion-like lipid-water particles (later termed liposomes). This technology has been applied mainly in dermatology, and it is still used in topical treatment of psoriasis and skin infections. In a 1967 paper in Nature Larsson reported a single-crystal X-ray diffraction study of the free rotation of hydrocarbon chains in the solid state. From 1967–1971 he worked in the computing field and led the design

of a complete system for the management of the entire health care system of Stockholm. To demonstrate the system, medical information on patients could be requested and transferred via transatlantic cable to and from New York, much like the internet today .... During his computing interlude he had an assistant to keep some lipid work going. He then returned to a research position and went into a number of different techniques to study lipid structure: surface balance, ESCA on LB films, and Raman spectroscopy. The Raman work using ordinary lipids mixed with perdeuterated lipids was very successful, and the results were applied to study lipid-protein phases. In 1975 Kåre became Professor of Food Technology at Lund University. The following list summarizes his most important contributions since 1975.

– *Fat crystallization* Molecular organization in the liquid state and the metastable phases alfa and beta prime were derived. Application work with industry involved a method to better control phase transition in margarines.

– *Emulsions of fats and oils* The significance of surface-active crystals of food emulsifiers in order to produce stable emulsions was demonstrated. The first edible microemulsion system was described.

– *Structure and functionality of polar lipids* Together with Sten Andersson and Stephen Hyde in 1984 the first complete description of the structure of the two glycerolmonooleate-water cubic phases was reported. It was based on infinite periodic minimal surfaces. The first lipid-protein-water phase diagram reported: lyzosyme/glycerolmonooleate/water. Colloidal cubic particles – cubosomes were first described.

– *Lipid oxidation* The relationship between oxidation reactions and physical state was examined, and a drastic improvement of the antioxidant effect could be achieved by an organized arrangement of ascorbic acid and tocoferol in a microemulsion.

– Gastrointestinal lipid digestion and absorption Association structures between bile acids and lipolysis products were studied; surface protective function of phospholipids from joint work with a group in surgery were reported.

– Structure of native starch and its gelatinization changes The structure of the amylose lipid complex was determined and found to have profound effect on gelatinization and on enzymatic degradation; the molecular packing in starch granules has been related to that in quartz.

– Cereal technology Surface film determination of interfacial disulphide bridge formation between gluten protein molecules and various studies involving all components in wheat led to a complete description on the molecular/colloidal level of breadmaking, which was presented in a monograph written with Ann-Charlotte Eliasson in 1993; "Cereals in breadmaking".

– Proteins at solid surfaces Basic ellipsometry studies of adsorption of blood proteins and milk proteins at metal surfaces showed for the first time formation of protein bilayers; this was in collaboration with Per-Ingvar Brånemark involving his osseointegration of titanium implants.

– Drug delivery A series of patents have been developed based on lipid carriers for drug delivery, for example cubosomes. They have lead to commercial products, such as Elyzol dental gel against periodontal diseases.

Larsson's *main research interest at present* is the significance of periodic curvature along cell membranes and the dynamic properties of these wave shapes, lung surfactants, and the role of lipids generally in physiology.

Scientific Papers and Technological Contributions of Kåre Larsson

1. The crystal structure of octa-(methylsilsesquioxane). Arkiv Kemi 16 (1960) 203
2. A crystal structure investigation of substituted octa-(silsesquioxanes). Arkiv Kemi 16 (1960) 209
3. The crystal structure of octa-(silsesquioxane). Arkiv Kemi 16 (1960) 215
4. On the structure of the crystal form C of 11-bromoundecanoic acid. Acta Chem Scand 16 (1963) 1751
5. The crystal structure of the D-form of 11-bromoundecanoic acid. Acta Chem Scand 17 (1963) 199
6. On the structure of the crystal form E of 11-bromoundecanoic acid. Acta Chem Scand 17 (1963) 215
7. Polymorphism of 11-bromoundecanoic acid. Acta Chem Scand 17 (1963) 221
8. An X-ray diffraction study of crystalline cholesterol in some pathological deposits in man (with H. Bogren). Biochim Biophys Acta 75 (1963) 65
9. Crystalline components of biliary calculi (with H. Bogren). Scand J Clin Lab Invest 15 (1963) 457
10. On the pigment in biliary calculi (with H. Bogren). Scand J Clin Lab Invest 15 (1963) 569
11. The crystal structure of the 1,3-diglyceride of 3-thiadodecanoic acid. Acta Cryst 16 (1963) 741
12. The Crystal Structure of the beta-form of triglycerides. Proc Chem Soc (1963) 87
13. A comparison of the solid state behaviour of stearic acid, 17-bromoheptadecanoic acid and 17-iodoheptadecanoic acid. Acta Chem Scand 18 (1964) 17
14. A set of crystallographic programmes for the Saab D21 computer. 4. Fourier synthesis with location of maxima in the Fourier space. Arkiv Kemi 23 (1964) 17
15 The crystal structure of the β-form of trilaurin. Arkiv Kemi 23 (1964) 1
16. On the crystal structure of 2-monolaurin. Arkiv Kemi 23 (1964) 23
17. On the crystal structure of the forms β1 and β2 of racemic 1-monoglycerides. Arkiv Kemi 23 (1964) 29
18. Solid state behaviour of glycerides. Arkiv Kemi 23 (1964) 35
19. A set of crystallographic programmes for the Saab D21 computer. 1. General data reduction (with S. Abrahamsson). Arkiv Kemi 24 (1965) 383
20. An integrated set of programmes for crystallographic calculations (with S. Abrahamsson, S. Aleby, B. Nilsson, K. Selin and A. Westerdahl). Acta Chem Scand 19 (1965) 758
21. The crystal structure of the beta-form of fatty acids (with E. von Sydow). Acta Chem Scand 20 (1966) 1203
22. A set of crystallographic programmes for the Saab D21 computer. 7. Intensity statistics. Arkiv Kemi 25 (1966) 205
23. The crystal structure of the L-1 monoglyceride of 11-bromoundecanoic acid. Acta Cryst 21 (1966) 267
24. Classification of glyceride crystal forms. Acta Chem Scand 20 (1966) 2255
25. Alternation of melting points in homologous series of long-chain compounds. J. Amer. Oil Chemists' Soc 43 (1966) 559
26. Photoelectron spectroscopy of fatty acid multilayers (with C. Nordling. K. Siegbahn and E. Stenhagen). Acta Chem Scand 20 (1966) 2880
27. Arrangement of rotating molecules in the high-temperature form of normal paraffins. Nature 213 (1967) 383
28. The structure of mesomorphic phases and micelles in aqueous glyceride systems. Z Phys Chem 56 (1967) 173
29. Phase behaviour and rheological properties of aqueous systems of industrially distilled monoglycerides (with N. Krog). Chem Phys Lipids 2 (1968) 129
30. Polymorphism in monomolecular films on water and formation of multimolecular films (with T. Burch and M Lundquist). Chem Phys Lipids 2 (1968) 102
31. Preparation of hydrophilic crytals of water-insoluble lipids. Third Scandinavian Surface Chemistry Symposium, Nordforsk, Stockholm (1967)

32. System Medakod-Samordnad databehandling för hälso-, sjuk- och socialvård (with S. Abrahamsson, S. Bergström and S. Tillman). Läkartidningen 65 (1968) 3196

33. Behaviour of the di-1,1'-monoglyceride of dodecane-1.12-dioic acid in the solid state and in aqueous systems (with L. Sjöström). Arkiv Kemi 30 (1968) 1

34. Structures of emulsifier-water phases. In Surface-active Lipids in Foods, Society of Chemistry and Industry, Monograph 32 (1968)

35. Realtidssystem i drift vid Stockholms läns landsting. Databehandling (1969) 24

36. Some recent studies of structural arrangements of lipids in surface layers and interfaces (with M. Lundquist, S. Ställberg-Stenhagen and E. Stenhagen). J Coll Interface Sci 20 (1969) 268

37. Vattenattraherande fettkristaller – ett nytt ytfenomen. Svensk Naturvetenskap (1969) 133

38. Crystal and molecular structure of 5-(bromomethylene)-10,11-dihydro-5H-dibenzo (a,d) -cycloheptene. Acta Chem Scand 24 (1970) 1503

39. Total regional system for medical care (with S. Abrahamsson, S. Bergström and S. Tillman). Computers and Biomed Res 3 (1970) 30

40. Larodan for cleaning oiled seabirds (with G. Odham). Marine Poll Bull 1 (1970) 122

41. Ett medicinskt informationssystem för en vårdregion (with S. Abrahamsson, S. Bergström, L-E Böttiger, S. O. Lööw and H. Peterson). Läkartidningen 68 (1971) 347

42. Danderyd hospital computer system. 3. Basic software design (with S. Abrahamsson). Computers and Biomed Res 4 (1971) 126

43. Hydrocarbon chain conformation in fats. Chemica Scripta 1 (1971) 21

44. Antimicrobial effect of simple lipids and its relation to surface film behaviour (with B. Noren and G. Odham). Biochem Pharm 21 (1972) 947

45. Molecular arrangement in glycosphingolipids (with S. Abrahamsson, K.-A. Karlsson and I. Pascher). Chem Phys Lipids 8 (1972) 152

46. Molecular arrangements in glycerides. Fette-Seifen-Anstrichmitteln 74 (1972) 136

47. On the structure of isotropic phases in lipid-water systems. Chem Phys Lipids 9 (1972) 181

48. Structural properties of the lipid-water gel phase (with N. Krog). Chem Phys Lipids 10 (1973) 165

49. Conformation-dependent features in the Raman spectra of simple lipids. Chem Phys Lipids 10 (1973) 165

50. Lipid multilayers. Surface Colloid Sci 6 (1973) 261

51. The occurrence of different molecular conformations of diglycerides of the air-water interface. Biochim Biophys Acta 318 (1973) 1

52. Detection of changes in the environment of hydrocarbon chains by Raman spectroscopy and its application to lipid-protein sytems (with R. P. Rand). Biochim Biophys Acta 326 (1973) 245

53. Crystal and liquid-crystal structures of lipids. Food Science series, Marcel Dekker 5 (1976) 39

54. On lipid surface films on the sea (with G. Odham and A. Södergren): Marine Chem 2 (1974) 9

55. Antimicrobial effect of simple lipids and the effect of pH and positive ions (with B. Noren and G. Odham). Antimicrobial Agents and Chemotherapy 8 (1975) 742

56. Phase behaviour of the binary system tetradecyl amine-water (with A. Al-Mamun). Chem Phys Lipids 12 (1974) 176

57. On the effect of urea on human epidermis (with L. Hellgren). Dermatologica 149 (1974) 289

58. Antimicrobial effect of simple lipids and the effect of pH and positive ions (with B. Noren and G. Odham). Antimicrobial Agents and Chemotherapy 8 (1975) 733

59. A Study of the combined Raman- and fluorescence scattering of human blood plasma (with L. Hellgren). Experientia 30 (1974) 481

60. Ny princip för salvbas med kristallina lipider – skyddseffekt vid oljeexposition (with A. Björnberg and L. Hellgren). Opuscula Med 20 (1975) 162

61. The significance of crystalline hydrocarbon chains in aqueous dispersions and emulsions of lipids. Chem Phys Lipids 14 (1975) 233

62. Tilted molecules of dipalmitoylphosphatidylcholine become vertical at the pretransition (with R. P. Rand and D. Chapman). Biophys J 15 (1975) 1117

63. Lecithin-diglyceride interaction in surface films on water and in aqueous systems (with R. Faiman and B. Szalontai). Acta Chem Scand A30 (1976) 281

64. Surface balance study of the interaction between microorganisms and lipid monolayer at the air/water interface (with S. Kjelleberg, B. Norcrans and H. Löfgren). Appl Environ Microbiol 31 (1976) 609

65. Raman spectroscopy studies of interface structure in aqueous dispersions. In "Lipids" vol. 2 Eds. R. Paoletti, G. Jacini and R. Porcellati. Raven Press, NY, 1976

66. Liquid crystals and emulsions (with S. Friberg). In "Advances in Liquid Crystals". Ed. G. H. Brown, Academic Press, NY, 1976

67. Assignment of the C-H stretching vibrational frequencies in the Raman spectra of lipids (with R. Faiman). J Raman Spectroscopy 4 (1976) 387

68. A Raman spectroscopy study of the effect of hydrocarbon chain length and unsaturation on lecithin-cholesterol interaction (with R. Faiman and D. A. Long). J Raman Spectroscopy 5 (1976) 3

69. Liquid-crystalline phases in biological model systems (with I. Lundström). In "Lyotropic Liquid Crystals" Ed. S. Friberg, ACS-Series 152 (1976) 43

70. A new skin protecting ointment against acrylic resins (with P.-O. Glantz and G. Nyquist). Odont Rev 27 (1976) 265

71. Cholesteryl sulphate and phosphate in the solid state and i aqueous systems (with J. Abrahamsson, S. Abrahamsson, B. Hellqvist). Chem Phys Lipids 19 (1977) 213

72. Folded bilayers – An alternative to the rippled lamellar lechitin structure. Chem Phys Lipids 20 (1977) 225

73. Influence of some inflammatory and anti-inflammatory substances on skin biopsies investigated with microcalorimetry and respiratory techniques (with L. Hellgren and J. Vincent). Arch Derm Res 258 (1977) 295

74. A Raman spectroscopy study of micellar structures in ternary systems of water-sodium octanoate-pentanol/decanol (with J. B. Rosenholm and N. Dinh-Nguyen). Colloid&Polymer Sci 255 (1977) 1098

75. Stability of emulsions formed by polar lipids. Prog Chem Fats and Lipids 16 (1978) 163

76. Thermal effects from degranulation of mastcells in cutaneous mastocytosis (with L. Hellgren). Experientia 33 (1977) 97

77. Hemolytic effect of some polar lipids used as food additives (with L. Å. Johansson). Lebensmittel-Wiss u Techn 11 (1978) 206

78. Phase equilibria and structures in the aqueous system of wheat lipids (with T. Carlsson and Y. Miezis). Cereal Cemistry 55 (1978) 168–179

79. Phase behaviour of some aqueous systems involving monoglycerides, cholesterol and bile acids (with K. Gabrielsson and B. Lundberg). J Sci Food Agric 29 (1978) 909

80. A study of the amylose-monoglyceride complex by Raman spectroscopy (with T. L.-G. Carlsson, N. Dinh-Nguyen and N. Krog). Starch 31 (1979) 222–224

81. Lyotropic and multilamellar systems. Ann Phys 3 (1978) 283

82. Anhydrous state of amphiphilic compounds (polymorphism). In "Physicochimie des composés amphiphiles", Col Nat du CNRS No 938 (1978) 17

83. The aqueous system of monogalactosyl diglycerides and digalactosyl diglycerides – Significance to the structure of the thylakoid membrane (with S. Puang-Ngern). Adv Biochem Phys Plant Lipids, Elsevier 1979

84. On the possibilty of dietary fiber formation by interaction in the intestine between starch and lipids (with Y. Miezis). Starch 31 (1979) 301

85. Phase equilibria in a ternary system saponin-sunflower oil monoglycerides-water. Interactions between aliphatic and alicyclic amphiphiles (with P. Barla, H. Ljusberg-Wahren, T. Norin and K. Roberts). J Sci Food Agric 30 (1979) 864

86. An X-ray scattering study of the L2-phase in monoglyceride-water systems. J Colloid Interface Sci 72 (1979) 152

87. The cubic phase of monoglyceride-water systems. Arguments for a structure based upon lamellar units (with G. Lindblom, L. Johansson, S.Forsen and K Fontell). J Am Chem Soc 101 (1979) 5465

88. Phase equilibria in the aqueous system of wheat gluten lipids and in the aqueous salt system of wheat lipids (with Y. Miezis and S. Poovardom). Cereal Chem 56 (1979) 417

89. Inhibition of starch gelatinization by amylose-lipid complex formation. Starch 32 (1980) 125

90. Physical structure and phase properties of aqueous systems of lipids from rye and triticale in relation to wheat lipids (with T. L.-G. Carlson and Y. Miezis). J Dispersion Science and Technology 1 (1980) 197

91. Inverse micellar phases in ternary systems of polar lipids/fat/water and protein emulsification of such phases to w/o/w-microemulsion-emulsions (with E. Pilman and E. Tornberg). J Dispersion Science and Technology 1 (1980) 267

92. Structural relationships between lamellar, cubic and hexagonal structures in monoglyceride-water systems. Possibility of cubic structures in biological systems (with K. Fontell and N. Krog). Chem Phys Lipids 27 (1980) 321

93. A Raman spectroscopy study of mixed bile salt-monoglyceride micelles (with H. Ljusberg-Wahren). Chem Phys Lipids 28 (1981) 25

94. Technical effects in cereal products of lipids. In Cereals for Foods and Beverages, Eds. G. Inglett and L. Munck, Academic Press (1980)

95. Some effects of starch lipids on the thermal and rheological properties of wheat starch (with A. C. Eliasson, T. Carlsson and Y Miezis). Starch 33 (1981) 130

96. Interfacial phenomena – Bioadhesion and biocompatibility. Desalination 35 (1980) 105

97. On the possibility of modifying the gelatinization properties of starch by lipid surface coating (with A. C. Eliasson and Y. Miezis). Starch 33 (1981) 231

98. Microbial adhesion to surfaces with different surface charges (with P. O. Glantz). Acta Odont Scand 39 (1981) 79

99. Aqueous lipid phases of relevance to intestinal fat digestion and absorption (with M. Lindström, H. Ljusberg-Wahren and B. Borgström). Lipids 16 (1981) 749

100. Polymorphism of rapeseed oil with a low content of erucic acid and possibilities to stabilize the beta prime crystal form in fats (with L. Hernquist, B. Herslöf and O. Podlaha). J Sci Food Agric 32 (1981) 1197

101. Ellipsometry studies of adsorbed lipids and milk proteins on metal surfaces (with P. O. Hegg). In: Proceedings in the fouling and cleaning in the food industry, Lund University Press (1981)

102. On surface charge and intraoral adsorption (with P. O. Glantz). In: Intraoral adhesion, IRL Press Ltd., London (1981)

103. Water-in-oil emulsions (with U. Hoppe). J Dispersion Sci Techn 2 (1981) 433

104. Molecular amphiphile bilayers forming a cubic phase in amphiphile-water systems (with G. Lindblom). J Dispersion Sci Techn 3 (1982) 61

105. Some effects of lipids on the structure of foods. Food Microstructure 1 (1982) 55

106. On the crystal structure of the beta prime form of triglycerides and structural changes at the phase transitions liq. → alpha → beta prime → beta (with L. Hernquist). Fat Sci Technol 84 (1982) 349

107. Interfacial tension between an inverse micellar phase of lipid components and aqueous protein solution (with E. Pilman and E. Tornberg). J Dispersion Sci Techn 3 (1982) 335

108. An antimicrobial system for the oral cavity based on an aqueous lipid dispersion (with P.-O. Glantz). J Dispersion Sci Techn 3 (1982) 373

109. Lipidfunktioner i livsmedel. Kemisk Tidskrift 3 (1982) 35

110. Bread baked from wheat/rice mixed flours using liquid-crystalline lipid phases in order to improve bread volume (with D. Rajapaksa and A.-C. Eliasson). J Cereal Sci 1 (1983) 53

111. A cubic protein-monoolein-water phase (with B. Ericsson and K. Fontell). Biochim Biophys Acta 23 (1983) 729

112. On structural relations between lipid mesophases and reverse micellar (L2) solutions (with K. Fontell, L. Hernqvist and J. Sjöblom). J Colloid Interface Sci 93 (1983) 45

113. Two cubic phases in the monoolein-water system. Nature 304 (1983) 664

114. A comparison of the phase behaviour of monoolein isomers in excess water (with M. Herslöf and H. Ljusberg-Wahren). Chem Phys Lipids 33 (1983) 211

115. Digestibility of amylose-lipid complexes in-vitro and in-vivo (with J. Holm, I. Björk, A.-C. Eliasson, N.G. Asp and I Lundqvist). Starch/Stärke 9 (1983) 294

116. The interface zone of inorganic implants in-vivo: titanium implants in bone (with T. Albrektsson, P. I. Brånemark, H. A. Hansson, B. Kasemo, I. Lundström and R. Skalak). Ann Biomed Engin 11 (1983) 1

117. Physical state of lipids and their effect in baking. In: "Lipids in Cereal Technology", Ed. P. Barnes, Academic Press, London, 1983

118. Relation betwen adsorption on a metal surface and monolayer formation at the air/water interface (with T. Arnebrant, T. Nylander, P. A. Cuypers and P. O. Hegg). In "Surfactants in Solution". Eds. K. L. Mittal and B. Lindman, Plenum Publ Co, 1984

119. An electron microscopy study of the L 2-phase (microemulsion) in a ternary system: triglyceride/monoglyceride/water (with T. Gulik-Krzywicki). Chem Phys Lipids 35 (1984) 127

120. Cleaning of polymer and metal surfaces studied by ellipsometry (with K. Bäckström, S. Engström, B. Lindman, T. Arnebrant and T. Nylander). J Colloid Interface Sci 99 (1984) 549

121. The use of freeze-fracture and freeze-etching electron microscopy for phase analysis and structure determination of lipid systems (with T. Gulik-Krzywicki and L. P. Aggerbäck). In: Surfactants in Solution Eds. K. L. Mittal and B. Lindman, Plenum Publ Co (1994)

122. A cubic structure consisting of a lipid bilayer forming an infinite periodic minimal surface of the gyroid type in the glycerolmonooleat-water system (with S. T. Hyde and B. Ericsson). Z Kristallographie 168 (1984) 213

123. Bilayer formation at adsorption of proteins from aqueous solution on metal surfaces (with T. Arnebrant, B. Ivarsson, I. Lundström and T. Nylander). Progr Colloid & Polymer Sci70 (1985) 62

124. Relation between antioxidant effect of tocopherol and emulsion structure (with C. Ruben). J Dispersion Sci Techn. 6 (1985) 213

125. Fat oxidation analysis using a Wilhelmy surface balance (with L. Moberger). J Dispersion Sci Techn 6 (1985) 383

126. Applications in the Food Industry (with N. Krog and T. Riisom). In Encyclopedia of Emulsion Technology, 2 (1985) 321, Marcel Dekker, New York

127. A phase transition model of cooperative phenomena in membranes (with S. Andersson). Acta Chem Scand B40 (1986) 1

128. Effects of anesthetics on a planar to curved lipid bilayer transition. Acta Chem Scand A40 (1986) 313

129. Effect of a glycoprotein monomolecular layer on the integration of titanium implants in bone (with T. Albrektsson, T. Arnebrant, T. Nylander and L. Sennerby). In: Biological and Biomechanical Performance of Biomaterials. Elsevier, Amsterdam (1986) 349

130. Differential geometry of a model membrane consisting of a lipid bilayer with a regular array of protein units (with S. T. Hyde and S. Andersson). Z Kristallographie 174 (1986) 237

131. Periodic minimal surface structures of cubic phases formed by lipids and surfactants. J Colloid Interface Sci 113 (1986) 299

132. The structure of gluten gels (with A.-M. Hermansson). Food Microstructure 5 (1986) 233

133. On the structure of native starch- An analogue to the quartz structure (with A.-C. Eliasson, S. Andersson, S. T. Hyde, R. Nesper and H.-G. von Schnering). Starch 39 (1987) 147

134. Cubic lipid-protein-water phases (with W. Buchheim). J Colloid Interface Sci 117 (1987) 207

135. A study of fat oxidation in microemulsion systems (with L. Moberger, W. Buchheim and H. Timmen). J Dispersion Sci Techn 8 (1987) 207

136. Cross-linking of wheat storage protein monolayers by compression / expansion cycles at the air / water interface (with G. Lundh and A.-C. Eliasson). J Cereal Science 7 (1988) 1–9

137. Physical properties – Structural and physical characteristics of lipids. In: "Lipid Handbook" (F. D. Gunstone, J. L Harwood & F. B. Padley, eds.). Chapman and Hall, London 1986, pp 321–384

138. Anesthetic effect and a lipid bilayer transition involving periodic curvature. Langmuir 4 (1988) 215–217

139. Functionality of wheat lipids in relation to gluten gel formation. p. 62 in "Chemistry and Physics of Baking". Eds. J. M. V. Blanshard, P. J. Frazer and T. Gaillard. Royal Soc Chem 56 (1986)

140. Minimal surfaces and structures: From inorganic and metal crystals to cell membranes and biopolymers (with S. Andersson, S. T. Hyde and S. Lidin). Chem Rev 88 (1988) 221–242

141. Two HII types of phases in the same monoglyceride-water system. J Colloid Interface Sci 122 (1988) 298

142. Lipid phase transitions in membranes involving intrinsic periodic curvature. Chem Phys Lipids 49 (1988) 65–67

143. Insulin adsorption on platinium (with V. Razumas, J. Kulys, T. Arnebrant and T. Nylander). Elektrokhimiya 24 (1988) 1518 (This journal is also translated into English)

144. Lipid-water phases for controlled-release drug delivery (with S. Engström and B. Lindman). Proc Intern Symp Controlled-Release, Controlled Release Soc Inc (1988) 105

145. Molecular aggregation in lipid-water dispersions phases. J Disp Sci Techn 10 (1989) 351

146. Cubic lipid-water phases: Structures and biomembrane aspects. J Phys Chem 93 (1989) 7304–7314

147. Molecular aggregation in lipid-water dispersed phases. Dispersion Sci Techn 10 (1989) 35

148. Evaluation of heparin-coated surfaces in vitro and in vivo (with S. E. Bergentz, H. Håkansson and B. Lindblad). Chimica Oggi 10 (1991) 14

149. A study of membrane lipids from dehydrated-acclimated Brassica Napus root cells: Formation of a cubic phase under physiological conditions (with P. Nordberg and C. Liljenberg). Biochem Cell Biol 68 (1990) 102

150. A study of wheat storage protein monolayers by Faraday wave damping (with D. M. Henderson and Y. K. Rao). Langmuir 7 (1991) 2731

151. Oat lipids – Interaction with water and characterization of aqueous phases (with G. Jayasingha, Y. Miezis and B. Sivik). J Dispersion Sci Techn 12 (1991) 443

152. Structure of the starch granule – a curved crystal. Acta Chem Scand 45 (1991) 840

153. Emulsions of reversed micellar phases and aqueous dispersions of cubic phases of lipids: Some food aspects. In "Microemulsions and Emulsions in Foods". Ed. M. El-Nokaly, ACS Series 448 ACS (1991)

154. Crystallization at interfaces in food emulsions – a general phenomenon (with N. Krog). Fat Sci Techn 94 (1992) 55

155. Emulsions in the food industry. In: Emulsions – a fundamental and practical approach, Ed. J. Sjöblom, Kluwer Academic Publ., Holland NATO ASI Ser. C 363 (1992)

156. On the structure of the liquid state in triglycerides. J Amer Oil Chemists' Soc 69 (1992) 835

157. Formulation of a drug delivery system based an a mixture of monoglycerides and triglycerides for use in the treatment of paradontal disease (with T. Norling, P. Lading, S. Engström, N. Krog and S. S. Nissen). J Periodontal Research, 19 (1992) 687

158. Effects on phase transitions in tripalmitin due to the precence of dipalmitin, sorbitan-monopalmitate and sorbitan-tripalmitate (with H.M.A Mohamed). Fat Sci Techn 94 (1992) 338

159. Dynamic rheological studies on an interaction between lipid and various native and hydroxypropyl potato starches (with H. K. Kim and A.-C. Eliasson). Carbohydrate Polymers 19 (1992) 211

160. Organized interfaces in emulsions. In: Organized solutions eds. S. Friberg and B. Lindman, Surfactant Science Series 44 (1992) 249, Marcel Dekker Inc

161. Effects of phosphatidylcholine and phosphatidylinositol on acetic-acid induced colitis in rats, (with R. Fabia, A. Arrajab, R. Willen, R. Andersson, B. Ahren and S. Bengmark). Digestion 53 (1992) 35

162. Gastroprotective capability of exogenous phosphatidylcholine in experimentally induced chronic gastric cancer in rats (with B. S. Dujic, J. Axelson, J. Arrajab, and S. Bengmark). Scand J Gastroenterol 2 (1993) 89

163. Phospholipase-resistant phophatidylcholine reduces intra-abdominal adhesions induced by bacterial peritonitis (with M. Snoj, A Arrajab, A. Ahren, and S. Bengmark). Res Exp Med 193 (1993) 117

164. Drug delivery system based on a mixture of monoglycerides and triglycerides (with T. Norling, P. Lading, S. Engström, N. Krog and S. Nissen). Proc Controlled Release Bioact Mat 20 (1993) 294

165. Phase behaviour of aqueous systems of enzymatically modified phosphatidylcholine with one hexadecyl and one hexyl or octyl chain (with I. Svensson, P. Adlercreutz; B. Mattiasson and Y. Miesiz). Chem Phys Lipids 66 (1993) 195

166. Green Bananas protection of gastric mucosa against experimentally induced injuries in rats (with B. S. Dunjic, I. Svensson J. Axelson, P. Adlercreutz, A. Arrajab and S. Bengmark). Scand J Gastroenterol 28 (1993) 89

167. Crystallization of fats and oils and an application for production of a margarine without trans-fatty acids. Proc 17th Nordic Lipid Symp (1993) 77

168. Starch complexing by enzymatically prepared 2-monoglycerides compared to effects by 1-isomers (with A. Millquist, P. Adlercreutz, B. Mattiasson and Y. Miesiz). Starch 46 (1994) 347

169. Modification of fats by lipase interesteri-fication. Part 2 Effect on crystallization behaviour and functional properties (with H. M. A. Mohamed). Fat Sci Techn 96 (1994) 56

170. Electrochemical biosensors for glucose, lactate, urea, and creatine based on enzymes entrapped in a cubic liquid crystalline phase (with V. Razumas, J. Kanapieniene, T. Nylander and S. Engström). Anal Chem Acta 289 (1994) 155

171. On phospholipids and hydrophobicity of the gastric wall. J Disp Sci Techn 15 (1994) 353

172. Tayloring lipid functionality. Trends in Food Sci Techn: 5 (1994) 311

173. Physical properties: structural and physical characteristics (with A. J. Quinn). In: Lipid Handbook (1994) 401

174. The effect of Salinum on the symtoms of dry mouth (with G. Johansson, G. Andersson, R. Attström and P. O. Glantz). Gerodontology 11 (1994) 46

175. Gut mucosa reconditioning with species-specific lactobacilli, surfactants, pseudo-mucus, and fibers – an invited review (with S. Bengmark and G, Molin). Biotechnol Therap 5 (1994) 171

176. Preparation and Characterization of a zero-trans margarine (with H. M. A. Mohamed, M. H. Iskandar and B. Sivik). Fat Sci Techn 97 (1995) 336

177. Comparison of the effect of the linseed extract Salinum and a methyl cellulose preparation on the symptoms of dry mouth (with G. Andersson, G. Johansson, R. Attström, S. Edwardsson and P. O. Glantz). Gerodontology 12 (1995) 13

178. Structure of the cubosome – a closed lipid bilayer aggregate (with S. Andersson, M. Jacob and S. Lidin). Z Kristallogr 210 (1995) 315

179. Effects of distearoylphosphatidylglycerol and lysozyme on the structure of the monoolein-water cubic phase: X-ray diffraction and Raman scattering studies (with V. Razumas, Z. Talaikyte, J. Barauskas, Y. Miezis and T. Nylander). Chem Phys Lipids 84 (1996) 123

180. Lipid bilayer standing waves in cell membranes (with M. Jacob and S. Andersson). Z Kristallogr 211 (1996) 875

181. Cubic lipid-water phase dispersed into submicron particles (with J. Gustafsson, H. Ljusberg-Wahren and M. Almgren). Langmuir 12 (1996) 4611

182. Incorporation of proteins in sphingomyelin-water gel phases (with H. Minami, T. Nylander and A. Carlsson). Chem Phys Lipids 79 (1996) 65

183. A comparative study of gelatinization of cassava and potato starch in an aqueous lipid phase (L2) compared to water (with J. da Cruz, J. Silverio and A.-C. Eliasson). Food Hydrocoll 10 (1996) 317

184. Dispersion of the cubic liquid crystalline phase - Structure, preparation and functional aspects (with H. Ljusberg-Wahren and L. Nyberg). Chimica Oggi (1996) 40

185. A cubic monoolein-cytochrome c-water phase: X-ray diffraction, FT-IR, differential scanning calorimetry and electrochemical studies (with V. Razumas, Y. Miezis and T. Nylander). J Phys Chem 100 (1996) 11766

186. Lipid bilayer standing wave conformations in aqueous cubic phases (with M. Jacob and S. Andersson). Z Kristallogr 212 (1997) 5

187. Neglected aspects of food flavor perception (with M. Larsson). Coll Surfaces A123 (1997) 651

188. On periodic curvature and standing wave motions in cell membranes. Chem Phys Lipids 88 (1997) 15

189. Submicron particles of reversed lipid phases in water stabilized by a nonionic amphiphilic polymer (with J. Gustafsson, H. Ljusberg-Wahren and M. Almgren). Langmuir, 13 (1997) 6964

190. On standing wave oscillations in axon membranes and the action potential. Coll Surfaces A, 129 (1997) 267

191. Structural effects, mobility and redox potential of vitamin K1 hosted in a monoolein-water liquid-crystalline phase (with F. Caboi, T. Nylander, V. Razumas, Z. Talaikyte and M. Monduzzi). Langmuir, 13 (1997) 5476

192. Microemulsions in foods (with S. Engström). In Microemulsion Handbook, Marcel Dekker, in press

193. Polymers that reduce intraperitoneal adhesion formation (with K. Falk, L. Holmdahl, M. Halvarsson, K. Larsson, B. Lindman and S. Bengmark). Br. J. Surg., in press

Scientific Books:

Food Emulsions. S. Friberg and K. Larsson (eds.), Marcel Dekker Inc., 1990 (new updated edition 1997)

Cereals in breadmaking – A Molecular Colloidal Approach. A-C. Eliasson and K. Larsson. Marcel Dekker Inc., New York, 1993

Lipids. Molecular Organization, Physical Functions, and Technical Applications. The Oily Press, Dundee, 1994

Language of Shape. S. T. Hyde, S. Andersson, K. Larsson, T. Landh, S . Lidin, Z. Blum, and B. W. Ninham. Elsevier. Amsterdam (1997)

Inventions which have lead to industrial products covered by patents world-wide.

1. Ointment base and dermatologic preparations. English patent 1174672 (1967) and follow-up patents.
   PETROGARD, an "invisible glove" for skin protection. Produced by Beiersdorf AG, Hamburg
   MICANOL, against psoriasis. Produced by Bioglan AB, Malmö
   MICACID, against impetigo. Produced by Bioglan AB, Malmö
   LHP (lipid-stabilized hydrogen peroxide), Antimicrobial formulation. Produced by Bioglan AB, Malmö

2. Cubic lipid-water phases for drug delivery. US patent 5151272 (1992), with S. Engström and B. Lindman.
   ELYZOL dental gel, against periodontal diseases such as paradontitis. Produced by Dumex AS, Copenhagen

3. Saliva substitute. German patent C44374/1 (1993). With R. Attström and P.-O. Glantz.
   SALINUM. Produced by Miwana AB, Gällivare, Sweden

4. Functional food products based on an oat flour formulation fermented by lactic acid bacteria able to colonize the gastrointestinal system. The initial invention together with S. Bengmark and S. Molin is now protected by a series of patents.
   PROVIVA, probiotic fruit drinks. In Scandinavia produced by Skånemejerier AB

Prog Colloid Polym Sci (1998) XI
© Steinkopff Verlag 1998

CONTENTS

Progr Colloid Polym Sci (1998) 108:1–3
© Steinkopff Verlag 1998

B.W. Ninham
B. Lindman

Conceptual locks and cubic phases

In "Eight Little Piggies", one of Stephen J. Gould's splendid collections of essays on the puzzle of evolution, he remarks: "I have long maintained that conceptual locks are a far more important barrier to progress in science than factual lacks". The statement is opposite to the work of Kåre Larsson that has to do with the importance of cubic phases in self-assembly.

To put what we have in mind into perspective, we remark that as long ago as 1836, in two reports to the British Association for Advancement of Science (in which he coined the term Mathematical Physics for what we now call Colloid and Surface Chemistry), the Rev. Challis of Trinity College laid out a program that has concerned us since. D'Arcy Thomson in his famous book on "Growth and Form" tells us that the early founders of the Cell theory of Biology and of Physiology then beginning made an urgent plea and recognized that progress in these sciences had to depend on understanding of the forces and problems of self-organization, which are the preoccupation of the physical chemist. Yet the fact remains that despite enormous progress in molecular biology, which uses the tools of physical chemistry and physics, the physical sciences have contributed nothing conceptually to modern biology during the last five decades. This fact poses a real dilemma and a challenge, because they ought to have contributed.

The reasons are only now becoming clear. To see how this situation came about, let us agree on the main goal: to understand function and structure. We would like to know how it is that molecular forces conspire with the geometry of molecules and the conformations available to macromolecules through the laws of statistical mechanics to give rise to the self-assembled equilibrium or dynamic steady states of matter that form cells and dictate biochemical reactivity. So posed, we can identify several places where such a program will have bogged down.

1. Theories of forces between surfaces did not include all important specific ion (Hofmeister) effects. The classical theories of colloid science are simply incorrect in the biological milieu. We do not know why and how to improve the situation. Words like hydrophobic, hydrophilic, and hydration are as ill-defined as the phlogiston theory of heat or that of the aether of last century. That is certainly a problem that we have reviewed elsewhere.

2. Direct measurements of molecular forces between surfaces appeared to have confirmed the classical theory of lyophobic colloid stability and its extensions due to Verwey and Overbeek and Deryaguin and Landau. But as time has gone on practically *all* measurements interpreted in terms of these ideas invoke fitting parameters like effective charge of interacting surfaces. These vary from surface to surface and electrolyte to electrolyte in a bewildering manner. Even allowing such fitting parameters, theory still failed in many cases, and new forces have been called in. These are variously called hydration, secondary hydration, and that bugbear of the 1990s long range "hydrophobic" attraction.

These problems have arisen in part because of ignorance of the Gibb's adsorption isotherm, the two dimensional analogue of the second law of thermodynamics. This is unforgiveable and its resolution has to await the passing of the present generation.

They have also arisen because the ideal smooth surfaces of theory are usually not those of Nature. (Silica and aluminium hydroxide and protein surfaces and just about everything else are not mica). This is unfortunate, but makes life more interesting.

B.W. Ninham, B. Lindman · (✉)
University of Lund
Physical Chemistry 1
P.O. Box 124
S-22100 Lund
Sweden

3. Theories of colloid stability of solid particles that trace their origin to the ideas of Langmuir and Onsager are on more solid ground, but have little direct relevance to our problem. These theories used entropy to balance hard core repulsions. But when one has to deal with soft condensed matter like the flexible membranes of bilayers or the water–surfactant–oil interface and introduces yet another force due to fluctuations, the Helfrich fluctuation force, we are again moving on shifting sand. The interpretation of the microstructure and bending moduli in terms of indirect and apparently sophisticated methods like SAXS or SANS or light scattering is mostly nonsense. It is nonsense not just because the inverse scattering problem is not unique, but it is so because its interpretation invokes theoretical forces with curve fitting based on an incorrect theory. Just as for the direct force measurements in general.

4. The role of dissolved gas and other solutes in interactions, liquid structure, and in free radical production has been completely ignored. This is important if we are ever understand chemical reactivity, and this is an area virtually untouched. Dissolved gas at atmospheric pressure in water is about 2×10^{-3} molar and about ten times as much in oil. Dissolved gas may well be intimately coupled to the range of the mysterious long range "hydrophobic" interactions. The microstructure of water with dissolved gas and electrolyte, also depends on electrolyte type, and is a subject about which we do not know much about. Work on optical cavitation, sonoluminescence, and related phenomena are beginning to reveal the extraordinary complexity of water. This, the nature of water, is an essential key that remains to be unlocked.

These matters are of some concern. Taken together it would seem that the biologist, who has enough concerns of his own, is not to be bothered about the subtleties of modern extensions of simpler theories and has been right to ignore physical chemistry. We have made a beginning, but after much work and self congratulation and realize at last that we must begin again and have some awareness of what the conceptual locks that constrained us were.

The cubic and bicontinuous states of matter

However, there is yet another conceptual lock. Until recently theories of self assembly of surfactants, lipids, microemulsions, polymers and mixtures thereof were constrained by an intellectual mind set that limited thinking to a particular set of shapes. These are those provided by Euclidean geometry: points, spheres, cylinders, and planes. For example, for surfactants and phospholipids, we tended to think in terms of monomers, micelles, hexagonal phases,

vesicles, lamellar and reverse phases. It turns out that hyperbolic geometries, everywhere bicontinuous, random, or regular with zero (cubic phases) or constant average curvature are the rule in Nature. The same holds for their two dimensional analogues, the mesh phases, which provide a richer framework in which to think of biomembranes and their action than the older Danielli–Davson model. This model relegates the lipids to an inconsequential nonspecific supporting role for the proteins and DNA.

The same holds for polymer and polymer–surfactant mixtures. These hyperbolic geometries also provide a broader framework in which to understand the structure and reactivity of inorganic materials. Such geometries, discovered by Gauss, Lobachevski, Riemann, and Weierstrass in the last century used to be thought of as a mathematical backwater and curiosity. But the structures described by these geometries turn out to be ubiquitous in Nature. For surfactants where local curvature is set by the balance of forces acting at an interface, these self-assembled structures emerge quite naturally as equilibrium phases when the global packing constraints imposed by mass conservation are also required.

For the lipid–polymer–protein mixtures that occur in biological cells we now know through the work of Larsson and his student Landh that there are structures, hitherto unrecognized, called "cubosomes" everywhere. These direct cell traffic and are involved in cell fusion. Their genesis is different to the surfactant mesophases. The dimensions of the connected channels that occur within them are macroscopic on the order of a thousand Ångstroms rather than the typical dimensions of about 20 Å connected with surfactant self-assembled structures. (Such structures emerge as a kind of three dimensional analogue of the Gibbs–Marangoni effect that occurs with double diffusion gradients at an interface. They form as steady state structures due to the need to generate energy flows outwards and reagent flows in).

Awareness of the existence and the consequences of such states especially the cubic phases of self assembly adopted by phospholipids and lipid–protein mixtures, is in large measure due to the pioneering insistence of Larsson that these structures were indeed bicontinuous cubic phases and the consequences thereof important. Others have contributed, especially Andersson, Hyde, Fontell and Luzzatti who found them first in phospholipids.

In two dimensions the analogue of the 3-D cubic phases are the newly discovered and controversial mesh phases. Phospholipids self-assemble, it seems, into a whole rich diversity of bilayer membrane phases that contain catenoidal holes designed to accommodate the proteins. Just like the bulk cubic phases they can transform from one form to another with extravagant ease. Such structures and their implications for problems like conduction

Progr Colloid Polym Sci (1998) 108:1–3
© Steinkopff Verlag 1998

of the nervous impulse were postulated by Larsson a long time ago. The proteins and lipids are coupled together, and the lipids have a much more vital role than simply to serve as a passive sea in which the proteins do their work. The "vesicles" that transport calcium and acetylcholine across the synaptic junction are cubic phases. Long surfactants at the alveola surface are in equilibrium with a cubic phase which is intimately and essentially coupled to the business of oxygen and carbon dioxide transport without which we cannot live. The omega-3 lipids so abundant in the brain may well be there because they may form bicontinuous cubic phases at the nerve membrane surface which act as the necessary reservoirs for calcium.

At the present time we have only just begun to dimly perceive that something is afoot. There is movement in the world of physical chemistry, and its relation and relevance to biology is about to change. It will do so in large measure because of the opening of the conceptual lock associated with our restriction in imagination to Euclidean geometries. This may well be more important than other limitations of present theories in physical chemistry. There is emerging a new language of shape. This represents a paradigm shift of great moment that deserves awareness and proper recognition.

It is for these reasons that we honor Kåre Larsson and his contributions in this commemorative volume.

Progr Colloid Polym Sci (1998) 108:4–8
© Steinkopff Verlag 1998

P. Walstra

Secondary nucleation in triglyceride crystallization

P. Walstra
Department of Food Science
Wageningen Agricultural University
P.O. Box 8219
6700 EV Wageningen
The Netherlands

Abstract From published results on nucleation and crystallization of emulsified fats, on crystal size observed in the emulsion droplets, and on permeabilities as a function of the fraction of solid in the same fats in bulk, it is concluded that copious secondary nucleation can occur, explaining the high number and small size of the crystals often observed. The effect is very large for milk fat (a mixture of many triglycerides), somewhat less for a typical margarine fat, much less for a fairly simple triglyceride blend, and absent for a paraffin blend.

A hypothetical explanation is given, which implies that clusters of partially oriented molecules diffuse away from a crystal face that grows via kinetic roughening; some of these clusters may give rise to nuclei. Kinetic roughening needs a high supersaturation. A further condition would be that crystal growth is small enough for the diffusing clusters to have a reasonable life time. This hypothesis can qualitatively explain the results discussed, but would need further testing.

Key words Triglyceride crystallization – crystal size – secondary nucleation

Introduction

In several applications of "plastic fats", i.e. mixtures of triglycerides that are partly crystalline and partly liquid, the size of the crystals is an important variable. It greatly affects (i) mechanical (rheological) properties, (ii) "oiling off" and (iii) for emulsified fats, stability against partial coalescence. Something similar holds for paraffin mixtures and comparable substances. There is little understanding of the factors governing fat crystal size.

The general idea is that crystal size is governed by the rates of crystal nucleation and crystal growth, which both increase with supersaturation or supercooling, but in a very different manner. Three types of crystal nucleation are generally distinguished [1]:

Homogeneous. This happens in a pure liquid. Generally, considerable supercooling is needed for homogeneous nucleation to occur at an appreciable rate.

Heterogeneous. This happens at a foreign surface (F), be it of a particle or of the vessel wall. It greatly depends on (i) the three interfacial free energies involved: L–C, L–F and C–F (L = liquid and C = crystal phase) and (ii) the shape of the foreign surface. Very small foreign particles may suffice to induce heterogeneous nucleation, and such particles are called "catalytic impurities". The number of impurities that can be catalytic generally increases with supercooling. In most practical situations, where supercooling is not very great, nucleation is predominantly heterogeneous.

Secondary. Here nucleation occurs near (not at) the surface of an already existing crystal of the solute. It is

Progr Colloid Polym Sci (1998) 108:4–8
© Steinkopff Verlag 1998

a somewhat mysterious phenomenon. It has especially been studied in (situations mimicking) industrial crystallizers, involving considerable agitation. It can, however, also happen under quiescent conditions.

Homogeneous and heterogeneous nucleation are thus variants of primary nucleation, which occurs in the absence of the crystalline phase. Nucleation rate theory for crystallization is at present in a stage of some onfusion, but classification into the three types just given still holds.

Various studies on triglyceride crystallization, for the greater part involving milk fat, have led the author to the conclusion that copious secondary nucleation can occur, also in the absence of agitation. In this manner, very small crystals could be formed. There would be considerable variation in secondary nucleation rate among various fats. The evidence discussed here is, however, of a circumstantial nature. This article does not include any new data, but it does give some new considerations.

Experimental evidence

Studies of nucleation and its rate are generally performed in emulsified material. This has been done with reasonably pure triglycerides [2], simple mixtures of triglycerides [3] and milk fat [4], which is a mixture of at least 10^5 different triglycerides. The results of these studies were consistent and point to homogeneous nucleation occurring at a supercooling of about 20 K below the final melting point of α-crystals in the mixture. Hence, under most conditions nucleation would be heterogeneous. For milk fat (melting range about 230–310 K, final melting of α-crystals about 295 K [5]), extensive studies were done, varying emulsion droplet size, temperature and methods of determining the crystalline proportion. Consistent results for the number concentration of catalytic impurities N_{cat} were obtained (surface nucleation at the droplet boundary could be excluded). These results are summarized in Fig. 1. It is seen that $\log N_{cat}$ decreases linearly with increasing temperature, which is commonly observed for heterogeneous nucleation.

On the other hand, partly crystalline droplets of various fats were viewed under the light microscope, using polarized light with crossed nicols [6–8]. The droplets were typically some μm in diameter and the temperature was mostly 20 °C; here, only those studies where no precooling to a lower temperature was applied will be considered. Several types of droplets could be observed, and those relevant for the present case can be classified as follows:

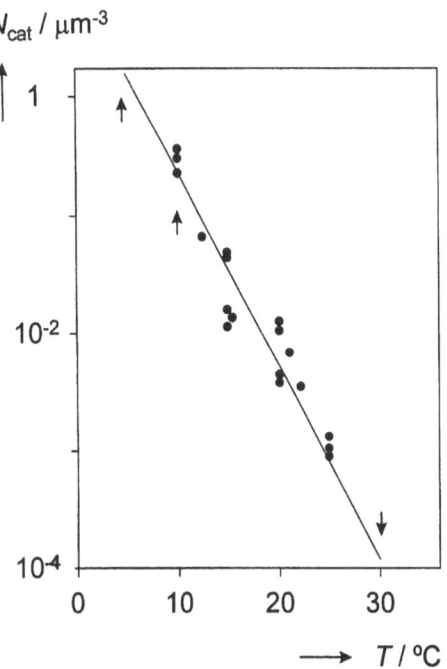

Fig. 1 Number concentration of catalytic impurities N_{cat} for heterogeneous nucleation in milk fat at various temperatures. Arrows indicate minimum or maximum values. From results in [4]

Type O. No crystals are observed. However, in most droplets, part of the fat was solid, since all methods for determining the solid fraction, and also X-ray diffraction studies, showed so. Moreover, small crystals could vaguely be seen by phase contrast microscopy. It must be concluded that the crystals are too small or too thin to be observed by polarized light microscopy.

Type N. Some needle-like crystals appear to be present, but they are more likely long platelets. Typically, 20 crystals can be observed and they have random orientation throughout the droplet. The droplet most likely contains "invisible", i.e. smaller, crystals also.

Type C. Here, the droplet contains one, or sometimes two crystals of elongated shape, length comparable to droplet diameter.

We will first consider milk fat, which almost exclusively shows Type O droplets. Figure 1 shows that at the relevant conditions, N_{cat} would be of the order of one per 100 μm^3, which is also about the size of the droplets examined under the microscope. Hence, there should be zero, one, or occasionally two, crystals in a droplet. However, a typical droplet would contain about 25% solid fat. Electron micrographs show, at comparable conditions, crystals that are typically $0.7 \times 0.15 \times 0.05$ μm in size [9], i.e. making up 0.005 μm^3. For a drop of 100 μm^3, this would result in a number of $100 \times 0.25/0.005 = 5000$ crystals. In other words, there is enormous discrepancy between the number of crystals present and the number that heterogeneous nucleation could give; homogeneous nucleation is out of the question (see Fig. 3, further on).

For other fats, comparable detailed studies are not available, but we have microscopic observations of the droplets and also some ideas about crystal size. This can be derived from permeability studies on a plastic fat [10]. Darcy's law states

$$v \equiv \frac{Q}{A} = \frac{B}{\eta} \nabla p, \tag{1}$$

where Q = volumetric flow rate of liquid (oil in the present case) of viscosity η through a cross-sectional area A as caused by a pressure gradient ∇p. This allows determination of the permeability coefficient B (m^2), which is a measure of the pore size, hence the crystal size, in the fat. Assuming the fat crystal network to be built of fractal aggregates of spherical particles of radius a, the relation would be [11]:

$$B = (a^2/k)\varphi^{2/(D-3)}, \tag{2}$$

where φ is the volume fraction of particles, D is the fractal dimensionality and k is a proportionality constant of order 100 [11]. By determining B for various φ, Eq. (2) can be checked as a scaling law and D can be derived; it is found to equal about 2 [10, 12].

Some results for various fats are given in Table 1; it concerns bulk fat, though in fairly small quantities that had been rapidly cooled. The table also gives results on the droplet types (crystals observed) of the same fats in emulsified form. Fat blend 1 was a typical margarine fat, blend 2, a mixture of fully hardened palm oil and sun flower oil; crystals in blend 2 were exclusively in the β' form.

The calculated crystal sizes are very approximate. The crystals are not spheres, and the meaning of a is thus unclear. The value of a is obviously smaller than the radius of a sphere of equal volume. Moreover, the values of D and k are not precisely known. Nevertheless, the differences among the fats are so large that it can be concluded that the crystals must have differed widely in size. This is in agreement with the microscopic observations. Note that the paraffin blend gave indeed droplets with one or two crystals, as would be predicted from heterogeneous nucleation theory. Also in the case of the fairly simple blend 2, the crystals must have been fairly large, but the droplets still contained of the order of 20 crystals.

The conclusion appears to be that the numerous small crystals formed must have originated from secondary nucleation, as the author had presumed before [4]. However, the conditions in the experiments discussed may have been rather special, in the sense that the crystallization in droplets was isothermal, starting at very high supersaturation. It is generally known that in a bulk fat that is cooled to a few degrees below the final melting point of the blend, and that is further cooled at a slow rate after the first crystals have appeared, rather big crystals are formed, possibly even visible to the naked eye. This cannot happen in small emulsion droplets, where considerable supercooling is needed for nucleation to occur in the great majority of droplets. Larger crystals can, however, be formed by the following procedure. The emulsion is cooled to, say, 5 °C and kept until considerable fat has crystallized. It is then warmed to a temperature (say, 30 °C) that is 5–10 K below the final melting point, whereby most of the fat is melted; presumably only a few crystals remain in each droplet. Now the emulsion is cooled again to 5 °C. If this is done slowly, in about 30 min, the droplets show fairly large crystals, up to several μm long. If the second cooling is fast, in about 5 min, the crystals in the droplets are very small, again [8, 13].

Presumably, existing crystals just grow if the supersaturation is kept small (slow cooling), whereas a high supersaturation (fast cooling) would lead to secondary nucleation as soon as the first crystal has formed. There is, however, considerable variation among fats. Of the fats studied, the crystal growth rate is very slow for milk fat, faster for a typical margarine fast (blend 1), still faster for a simple mixture like blend 2, and fastest for paraffin.

A tentative explanation

It has been postulated by Larsson [14, 15] that in liquid triglycerides close to the melting point fluctuating (i.e. short-lived) ordered clusters of molecules occur. In such clusters, the molecules would be oriented more or less parallel, with an average coordination number of 4, as compared to 6 in an α-crystal. The clusters would be about 10 nm in size, comprising several times 10 molecules. It has been concluded from studies on homogeneous nucleation, that a critical embryo, i.e. a nucleus, would comprise of the order of 10 molecules [3]. This leads to the hypothesis that clusters of more or less oriented molecules may diffuse away from a growing crystal, and subsequently have

Table 1 Results on crystal size in various fats. Permeability (B) for a fraction solid of 0.3. a is the calculated particle radius. The morphology type of the drops is explained in the text. From results in [6–8, 10, 12]; see also [5]

Type of fat	B (m^2)	a (nm)	Drop type
Milk fat[a]	4×10^{-18}	6	O
Fat blend 1[a]	10^{-16}	50	O and N
Fat blend 2	10^{-15}	95	N
Paraffin blend	4×10^{-14}	600	C

[a] B extrapolated from lower volume fractions solid.

Progr Colloid Polym Sci (1998) 108:4–8
© Steinkopff Verlag 1998

a chance of forming a new nucleus; this would then be true secondary nucleation. Two conditions should be fulfilled: (i) that the crystal surface is rough, and (ii) that the life time of the mentioned clusters is long enough.

According to Bennema [16] three crystal growth regimes can be distinguished:

A. Thermodynamic roughening,
B. Two-dimensional nucleation,
C. Kinetic roughening.

Regime A need not concern us here. In regime B the activation free energy ΔG^* for formation of a two-dimensional nucleus on a crystal face is considered. Apart from geometrical constants, it can be written as

$$\Delta G^* = \gamma_x \gamma_y / \Delta \mu, \tag{3}$$

where γ_x and γ_y are the values of the edge free energy per cell (generally comprising one or two molecules) of a two-dimensional nucleus in the x and y direction, and $\Delta \mu$ is the difference in chemical potential between the material in solution (or in the melt) and in the crystals. ΔG^* thus decreases with increasing supersaturation and when it becomes of the order of kT for a given crystal face, numerous small nuclei are formed on that face; this implies kinetic roughening, i.e. growth mechanism C. Remembering that the supersaturation per molecule $\beta = \Delta \mu / kT$, the condition for kinetic roughening to occur becomes

$$\beta > \sim \gamma_x \gamma_y / (kT)^2 . \tag{4}$$

Unfortunately, we have insufficient data to calculate the edge free energies, which depend on the connected network of the crystal lattice and on the bond energies involved. It has been done for β-crystals of n-alkane triglycerides [17].

Figure 2a gives a schematic drawing of a simple triglyceride crystal, to indicate some crystal faces. It was calculated that both γ's were very high for the $\{001\}$ faces, that one of them is much smaller for the $\{100\}$ faces, and that for the $\{011\}$ and other faces in the [010] direction, both γ's are fairly small. Nevertheless, a fairly high supersaturation would be needed for kinetic roughening to occur.

The relative magnitudes of the various edge free energies are in qualitative agreement with the relative size of the faces, since for a smaller $\gamma_x \gamma_y$, a face will grow faster. For the faces in the [010] direction, kinetic roughening will therefore occur most readily, as has been clearly shown for n-paraffin crystals [18]: the [010] ends of the crystals then become rounded, rather than showing plane faces. It is interesting to note that electron micrographs of milk fat crystals show rounded ends, the $\{100\}$ faces are either flat or somewhat rounded, and the large $\{001\}$ faces are flat, occasionally showing a sharp step [9].

Initial crystallization will certainly not be in the β-polymorph, but most triglyceride crystals will have a

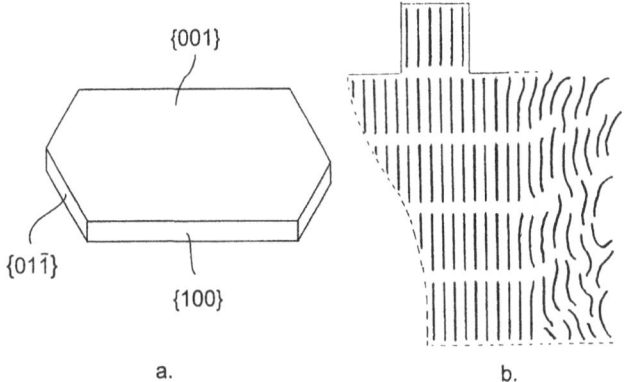

Fig. 2 (a) Schematic diagram of a triglyceride crystal; the Miller indices of some faces are given. (b) Highly schematic diagram of two edges of a crystal lattice of long chain molecules. At the top face (e.g. $\{001\}$) growth is by two-dimensional nucleation, at the right-hand face (e.g. $\{001\}$) kinetic roughening occurs

comparable morphology. Figure 2b shows, in a highly schematic way and for rod-like molecules, how two surfaces of a crystal may grow. Clusters of partially oriented molecules may diffuse away from faces in the [010] direction. The question is, as mentioned, whether they have occasion to do so. Figure 3 gives approximate time scales for nucleation and growth of milk fat crystals as a function of temperature. Curves 1, 2 and 3 are derived from [4]. The growth rate was determined by DSC in very small samples for isothermal crystallization. Curve 4 gives the average time for diffusion t_D over a distance δ of 1 μm, calculated for clusters of about 30 triglyceride molecules, by means of

$$t_D = 3\pi \eta R \langle \delta^2 \rangle / kT , \tag{5}$$

where oil viscosity $\eta \approx 0.07$ Pa s, and hydrodynamic cluster radius $R \approx 5$ nm. It is seen that, for most of the temperature range, the diffusion time is at least an order of magnitude shorter than the half time for growth. This may imply that for milk fat there is sufficient time for secondary nuclei to form.

If the growth would be much faster, triglyceride molecules would be incorporated into the crystal lattice before a cluster could diffuse away. The growth rate is thus an important variable and triglyceride mixtures generally crystallize very sluggishly. This has two main reasons [5]:

1. The fitting of the molecule in the crystal lattice gives an enormous loss of entropy, corresponding to a very large activation free energy. In other words, a molecule may readily be detached from a crystal face before it is fully incorporated, and then has to "try" again.

2. In a multicomponent fat, numerous different molecules are present. This means that the supersaturation for each

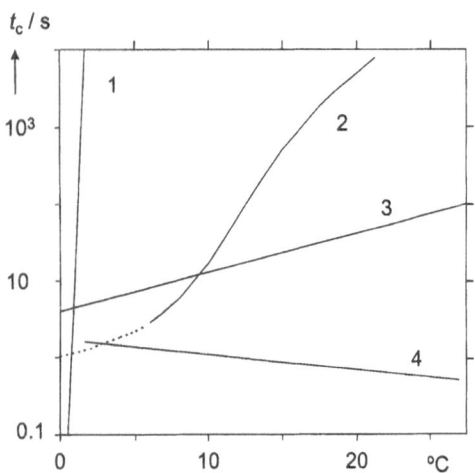

Fig. 3 Some characteristic time scales t_c for crystallization of milk fat as a function of temperature. Curve 1, homogeneous nucleation in 1 μm^3. Curve 2, heterogeneous nucleation in 1 μm^3. Curve 3, half time for crystal growth. Curve 4, diffusion of clusters of about 30 molecules over 1 μm. Curves 1–3 after results in [4], curve 4 calculated

of those is small. Even more important is the very strong competition of very similar molecules for incorporation into the crystal lattice; since these molecules will not completely fit, they will become detached again, to make place for another molecule, that will probably not fit either, etc.

The fewer components a fat contains, the faster its crystals can grow at the same supersaturation; hence, the smaller the probability would be that clusters can diffuse away from a growing crystal; hence, the smaller the rate of secondary nucleation; hence the fewer and the larger the crystals will be. This is in qualitative in agreement with the observed sequence milk fat–fat blend 1–fat blend 2.

Paraffins will crystallize much faster than triglycerides, because the entropy loss on crystallization is far less.

$S = k \ln W$, where S is the entropy and W the number of degrees of freedom. The loss in W on crystallization would roughly equal 3^n, where n is the number of carbon atoms in the molecule. The corresponding increase in free energy on crystallization then is given by $n \ln 3$ times kT. For triglycerides this would be e.g. $51 \ln 3 = 56kT$ (or 134 kJ/ mole), for paraffins e.g. $16 \ln 3 = 18kT$. This would explain why paraffins crystallize much faster than triglycerides and do not show significant secondary nucleation in emulsion droplets.

Concluding remarks

Secondary nucleation would occur if two conditions are fulfilled:

i) Kinetic roughening occurs at least in one of the crystal faces; it needs a high supersaturation and it further depends on the various bond energies between molecules in the connected network of the crystal lattice.

ii) Crystal growth rate is so slow that clusters of partially oriented molecules can diffuse away from a roughened crystal face before these molecules become incorporated in the crystal; crystal growth will be smaller for a molecule with a higher conformational entropy and for a fat containing a broader range of similar molecules.

Circumstantial evidence for this hypothesis has been given, but it remains a hypothesis. It can, however, be tested. Considering the importance of understanding the factors that govern fat crystal size, it is hoped that such a study will be performed.

Acknowledgment The author is indebted to Professor P. Bennema for useful discussions.

References

1. Garside J (1987) In: Blanshard JMV, Lilford P (eds) Food Structure and Behaviour. Academic Press, London, pp 35–49
2. Phipps LW (1964) Trans Faraday Soc 60:1873–1883
3. Skoda W, van den Tempel M (1963) J Colloid Sci 18:568–584
4. Walstra P, van Beresteyn ECH (1975) Neth Milk Dairy J 29:35–65
5. Walstra P, van Vliet T, Kloek W (1994) In Fox PF (ed) Advanced Dairy Chemistry, Vol 2: Lipids. Chapman & Hall, London, pp 179–211
6. Walstra P (1967) Neth Milk Dairy J 21:166–191
7. van Boekel MAJS (1980) Agr Res Reports Wageningen 901
8. Boode K (1992) Partial coalescence in oil-in-water-emulsions. PhD thesis, Wageningen Agricultural University
9. Mulder H, Walstra P (1974) The Milk Fat Globule: Emulsion Science as Applied to Milk Products and Comparable Foods. Pudoc, Wageningen and CAB, Farnham Royal
10. Boode K, Walstra P, de Groot-Mostert AEA (1993) Colloids Surfaces A 81: 139–151
11. Bremer LGB, van Vliet T, Walstra P (1989) J Chem Soc Faraday Trans 1, 85:3359–3372
12. de Groot-Mostert AEA, Boode K, van Vliet T (1998) to be published
13. Boode K, Bisperink C, Walstra P (1991) Colloids Surfaces 61:55–74
14. Larsson K (1972) Fette Seifen Anstrichmittel 74:136–142
15. Larsson K (1986) In Gunstone FD, Harwood JL, Padley FB (eds) The lipid handbook. Chapman & Hall, London, pp 321–384
16. Bennema P (1993) In Hurle DTJ (ed) Handbook of crystal growth, Vol 1. Elsevier, London, pp 477–581
17. Bennema P, Vogels LJP, de Jong S (1992) J Crystal Growth 123:141–162
18. Xiang-Yang Liu (1993) PhD thesis, Nijmegen Catholic University

Progr Colloid Polym Sci (1998) 108:9–16
© Steinkopff Verlag 1998

S.E. Friberg
Z. Zhang
R. Patel
G. Campbell
P.A. Aikens

Kinetics of formation of structures in a three-phase system water/lamellar liquid crystal/water-in-oil microemulsion after shear

Dr. S.E. Friberg (✉) · Z. Zhang · R. Patel
Department of Chemistry
Clarkson University
Potsdam, New York 13699-5810
USA

G. Campbell
Department of Chemical Engineering
Clarkson University
Potsdam, New York 13699-5705
USA

P.A. Aikens
ICI Surfactants
Concord Plaza, Bedford Bldg.
3411 Silverside Road
Wilmington, DE 19850-5391
USA

Abstract A water-in-oil micro-emulsion with low water content was combined with water in a stop-flow equipment to form a system that at equilibrium would be a (water + lamellar liquid crystal)-in-(water-in-oil) microemulsion ((W + LLC)/(W/O) μem) emulsion and the growth of droplets and liquid crystal particles was followed by measuring the intensity of scattered light; both of total scattered light and that passing through crossed polarizers.

The total scattered light intensity was reduced during 5 s with a subsequent slow increase, while the bi-refringent part showed an initial growth during 1.2 s followed by a slow reduction.

The results are interpreted as the primary formation of lamellar liquid crystal particles at the water/micro-emulsion interface. These are dispersed in the oil phase and accounting for the initial growth of the intensity of scattered light through crossed polarizers. Subsequent dissolution of the liquid crystal to form W/O microemulsion droplets accounts for the reduction of this part of the intensity of scattered light. Separate experiments under non-equilibrium conditions demonstrated the growth of a liquid crystal to be significantly faster than the formation of W/O microemulsion droplets and that the transition of a lamellar liquid crystal to W/O microemulsion droplets was monitored by the droplet diffusion rate away from the interface.

Key words Emulsions – detergents – liquid crystals – fragrances – solubilization

Introduction

Amphiphilic association structures are essential for understanding the properties of macro-dispersed systems; and Professor Larsson's early contributions in the general area of molecular organization of lipids [1–4], has provided a solid foundation, not only for his later pioneering research into complex nano-dispersed system [5, 6], but also for the science of multiphase dispersed systems in general.

The influence of association structures on macrodispersed systems is indirect but obvious for micelles, which limit the amount amphiphile adsorbed to a macroscopic interface [5, 7] or for the hydrophobe/hydrophile combinations forming microemulsion droplets [8], but is also direct as for the stabilizing action per se of micelles [9, 10], liquid crystals [11, 12] and liposomes [13]. The literature about equilibrium association structures in the area is by now a rich source, from which one can obtain sophisticated information, but, unfortunately, is deficient in essential parts: the kinetics of formation of the decisive structures of some of the intermediate structures during the formation of a macrodispersion.

10

S.E. Friberg et al.
Kinetics of complex emulsion formation

In the area of dynamics of amphiphilic association structures that of micelles is extremely well described by Zana [14] following the pioneering contributions from Aniansson [15] and Kahlweit [16]. Recent review articles as part of the assembly of articles edited by Gast and Robinson [17] provide excellent critical reviews of the dynamics in micellar and microemulsion systems [18, 19].

The dynamics of vesicles at the equilibrium state has also been treated extensively; especially for systems of biological interest [20–23]. The equilibrium structures during the transformation from micelles to vesicles has also been studied [24–29] and its time dependence has recently been investigated by Egelhaaf and Schurtenberger [30, 31]. These investigations are concerned with the extremely important transition from micelles to vesicles, a phenomenon, the slow rate of which was earlier revealed by Kaler and his collaborators [32].

In the present investigation we are concerned with a more complex, but also more essential phenomenon for the formation of macrodispersed systems, more specifically, emulsification; the simultaneous formation of liquid crystal particles (or liposomes) and inverse micelles. The formation of a lamellar liquid crystal from inverse micelles has been investigated by Tondre et al. [33, 34] showing fast kinetics. The transport of hydrocarbons from oil droplets to inverse micelles has recently been studied by light scattering [35].

We have recently been able to obtain information about the kinetics of vesicle formation from a monomerically dispersed commercial nonionic surfactant [36, 37]. The surfactant was solubilized in a water/hydrotrope system [38], in which the association structures were disintegrated to a molecular dispersion during the intense shear in a stop-flow equipment [39], the shear in which is well in excess of $10^4 \, s^{-1}$, at which W/O microemulsion droplets are disintegrated [40].

In the experiment, the build-up of vesicles was followed using a light scattering device combined with the stop-flow equipment [41] and the results showed, unexpectedly, that the vesicles were formed by addition of single molecules to vesicular fragments. For surfactant concentrations less than 5% by weight of the total, the plot of the inverse relaxation time versus surfactant concentration showed a negative slope, identifying molecular build-up [15].

We found the combination of the stop-flow equipment with its reduction of amphiphilic association structures to molecular dispersions with a light scattering equipment to determine the re-build-up of colloids and even microdispersions to be attractive for the study of more complex phenomena, such as those encountered in an inverse emulsification process, because the equipment allows concurrent determination of the intensity of total scattered light as well as the intensity of light passed through crossed polarizers [41].

With the present articles, we report the change in intensity of scattered light after initial breakdown in a stop-flow equipment for a system forming three phases at equilibrium: an aqueous solution, a lamellar liquid crystal and a water-in-oil microemulsion after mixing water and dilute W/O microemulsion.

The results are obviously preliminary and a quantitative interpretation has been delayed pending more information. However, the intuitive interpretation provided is not contrary to the fragmented knowledge available, is strongly supported by complementary experiments and will, hopefully, serve as a useful introduction to later quantitative treatments.

Experimental

Materials

Laureth 4 (Brij® 30), ICI Surfactant, Wilmington, DE; and phenethyl alcohol (PEA), Aldrich Chemical Co., Milwaukee, WI; were used as received. Water was deionized and doubly distilled.

Stop-flow measurements

The device designed to make the stop-flow measurements has been extensively described elsewhere [39, 41] and has been modified to detect both the total scattering and that of the anisotropic part. An He–Ne laser, Huges Model 5040, operating at 632.8 nm, was applied as the light source. 90° light scattering passed through two polarizers, with polarization directions perpendicular to each other, and its intensity and that of totally scattered light were determined by separate photomultiplier tubes. The stop-flow measurement was made 12 times and the results reported are an average of the results. The difference between the average curve and the individual results was less than 5% on the voltage scale. 5 ml of a microemulsion, 21% PEA, 67% Laureth 4, 12% water, marked as *Initial* in the phase diagram, were mixed with an equal volume of water in this apparatus, resulting in the point 1 of the phase diagram, Fig. 1.

Microscopic observations during shear

The influence of shear on the anisotropic species formed during mixing was estimated by observation of a sample

Progr Colloid Polym Sci (1998) 108:9–16
© Steinkopff Verlag 1998

Fig. 1 The phase diagram of
the system water, phenethyl
alcohol and Laureth 4.
(Adapted from ref. [42]). LLC:
Lamellar liquid crystal; L₂:
Water-in-oil microemulsion

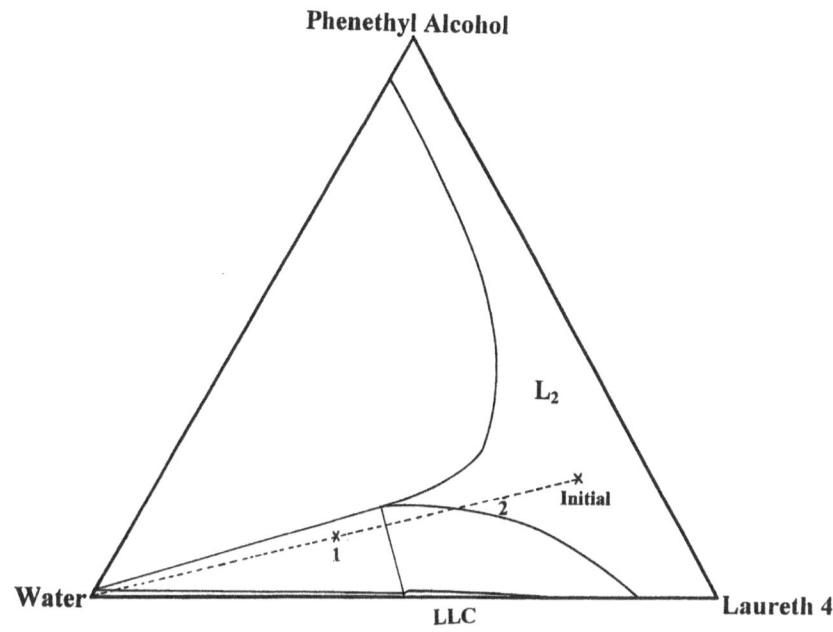

Results

between crossed polarizers in an Olympus BX50 micro-scope and recorded by a Sony SSC-S20 color video camera. A drop of water was deposited on a slide glass, and a drop of microemulsion (*Initial*) was put on top of it. Another slide was placed on top of the sample, and shear was exerted by the relative movement of the two slides.

Static experiments

In the static experiments, the W/O microemulsion marked "Initial" in Fig. 1 was placed on top of an equal amount of water and the changes in amount and structure were followed. In another series of experiments to estimate the rate determining step of the formation of inverse micelles from a lamellar liquid crystal, the latter with maximum water content was placed in contact with excess W/O microemulsion of the composition 7.5% water, 42.5% phenethyl alcohol, and 50% Laureth 4. In the first of these experiments, the microemulsion phase was at rest and the growth of a more concentrated microemulsion followed by observation of the movement of the distinct change in refractive index through the microemulsion phase. In the second experiment, the influence of the diffusion process in the microemulsion on the dissolution rate of the liquid crystal was extirpated by gently stirring the microemulsion to create an irregular movement in the liquid without perturbing the liquid crystal/microemulsion interface and the reduction in the amount of liquid crystalline phase was followed.

The phase diagram, Fig. 1, has been described earlier [42]; the present investigation is concerned only with the combination of three phases in it: W/O microemulsion (L₂), the lamellar liquid crystal (LLC) and the aqueous solution of the phenethyl alcohol.

The intensities of scattered light after the shear in the stop-flow equipment are given in Fig. 2. The intensity of total scattered light is described by two exponential functions; a fast reduction of intensity with a decay time of 1.4 s and a growth with a decay time of 22 s. The intensity of light through crossed polarizers shows a fast growth function, decay time 0.25 s and a reduction of size with a decay time of 2.6 s. Microphotographs of the dispersion after treatment in the stop-flow equipment showed the presence of liquid crystalline particles, Fig. 3.

The static experiment showed the formation of a liquid crystalline dispersion in the oil phase with no water macro-droplets observed. The change of water amount with time was used to calculate the upper limit of the amount of liquid crystal formed. The amount of the liquid crystal dispersion per se was obtained from the height in the test vessel. Figure 4 shows the amount of liquid crystal and the weight of the total dispersion versus time. The amount of liquid crystal in the dispersion remained at an approximate weight fraction of 0.2 for the entire experiment.

The photographs from the shear experiments are shown in Fig. 5. During the shear, Fig. 5A, elongated

12

S.E. Friberg et al.
Kinetics of complex emulsion formation

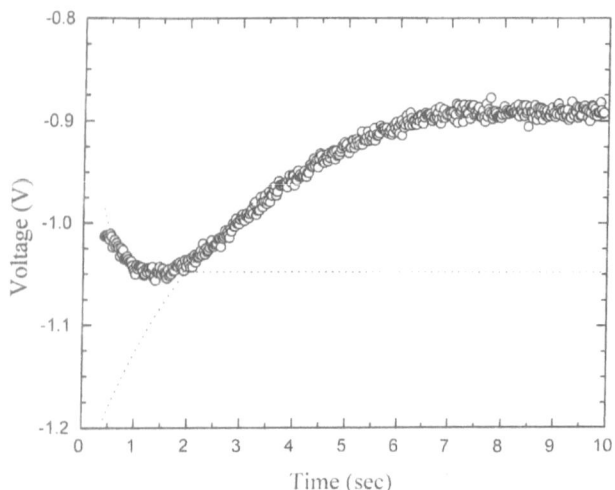

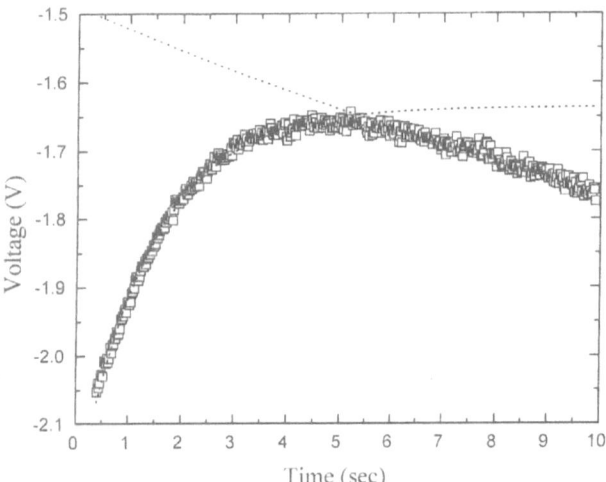

Fig. 2 The intensity of total scattered light (bottom) and that through crossed polarizers (top). The curves were adapted to the following equations: Top: $V = -1.05 + 0.31e^{-t/0.25}$ and $V = -0.88 - 0.36e^{-t/2.56}$; Bottom: $V = -1.64 - 0.58e^{-t/1.37}$ and $V = -2.22 + 0.71e^{-t/21.8}$

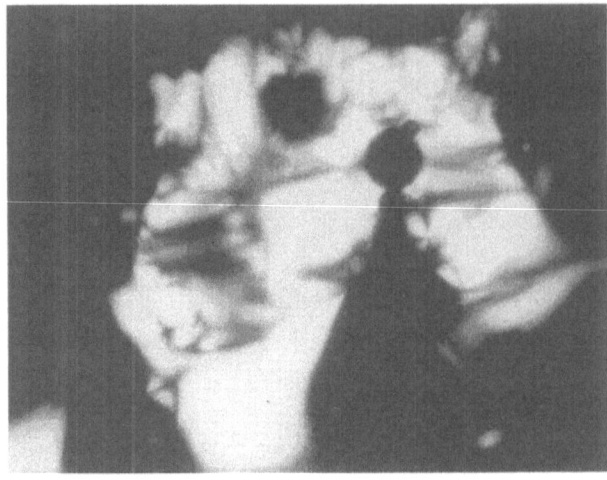

Fig. 3 Microphotographs of the dispersion after shear in the stop-flow equipment showing the presence of large liquid crystal particles

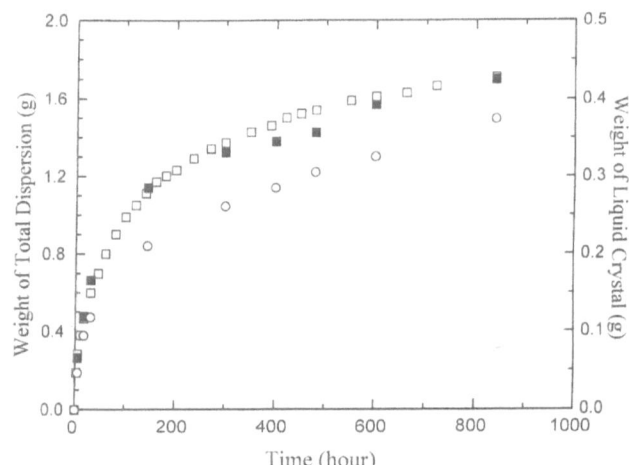

Fig. 4 The growth of the liquid crystal (○) and that of the total dispersion (■, □) plotted against time, in the static experiment with the microemulsion of composition "Initial", Fig. 1, on top of equal amount of water. The data represented by ○ and ■ were calculated from the reduction in size of the water and microemulsion layers, and those by □ from the height of the total dispersion

patterns of the birefringent liquid crystal were observed. These patterns rapidly formed discrete particles after terminating the shear, Fig. 5B, and these particles were reduced in size with time, Figs. 5C and D.

In the experiment to determine the dissolution rate of the liquid crystal the extent of the more concentrated microemulsion layer in the unstirred microemulsion is shown in Fig. 6 plotted as the square root of time. The values form a straight line with an R value of 0.998. The weight of dissolved liquid crystal in the experiment with stirring is given in Fig. 7. It is characterized by three stages: a fast initial linear dissolution during 7.5 h followed by a slower rate to 50 h in turn followed by a slower dissolution to 100 h.

Discussion

The results provide introductory information about the phenomena taking place, when an organic polar solvent containing weakly associated inverse micelles of a nonionic surfactant and small amounts of water is intensely mixed with water in a stop-flow equipment to form a three-phase emulsion. The fact that the original association structure, the inverse micelles, are disintegrated [39] during the stop-flow process, is worth emphasizing.

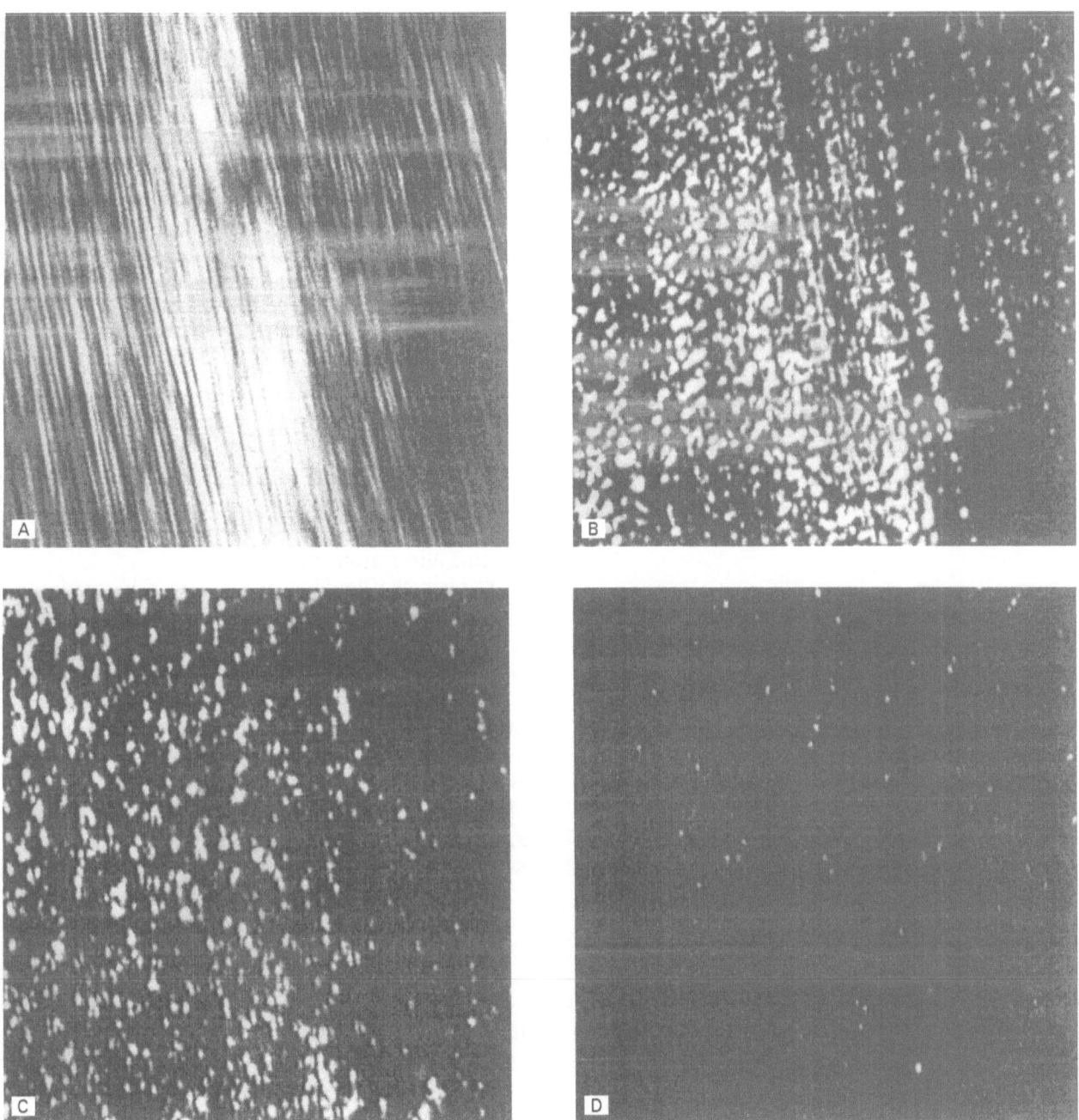

Fig. 5 Microphotographs of sample "Initial" Fig. 1 combined with water during shear (A) and after rest (B) 3 s, (C) 10 s and (D) 60 s

The phase diagram, Fig. 1, provides information about contribution of different phases. At equilibrium, the W/O microemulsion comprises 64% by weight of the total, while water and the lamellar liquid crystal each contribute 18%. These numbers strongly indicate final emulsion of the form (water + lamellar liquid crystal)-in-(water-in-oil) microemulsion {(W + LLC)/(W/O)μem}. At equilibrium, the continuous oil phase, the W/O microemulsion, will,

hence, contain both aqueous macroemulsion droplets and microemulsion droplets, in addition to dispersed particles of the lamellar liquid crystal. The presence of the later are demonstrated in Fig. 3.

Other possibilities are a double emulsion of the form LLC/W/(W/O) μem or W/LLC/(W/O) μem. The first structure is less probable, because the presence of the liquid crystal inside the water droplets would cause an

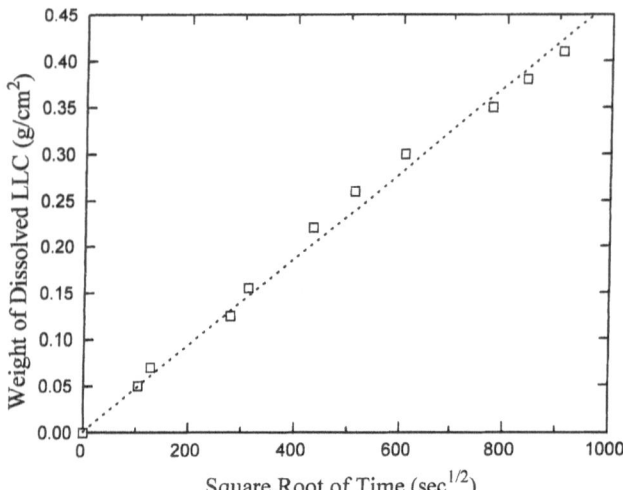

Fig. 6 The dissolution of the lamellar liquid crystal into a micro-emulsion (42.5% PEA, 50% Laureth 4, and 7.5% water) on top of the liquid crystal. The dotted line: $m_{LC}/A = 0.00046 t^{1/2}$

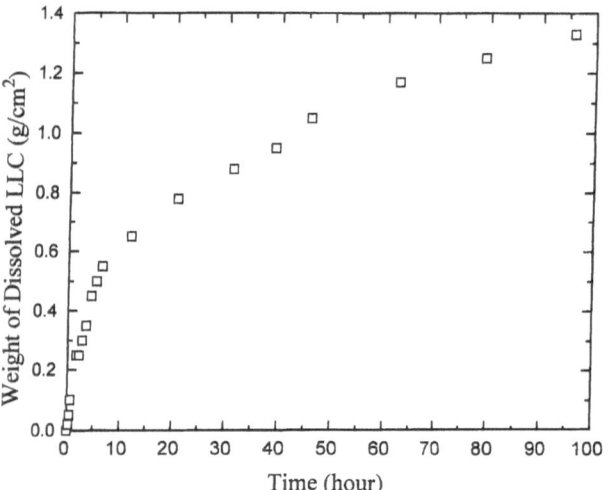

Fig. 7 The dissolution of the lamellar liquid crystal, Fig. 6, into a stirred microemulsion

increase of the size of the oil–water interface. The interfacial free energy $\gamma_{o/w}$ is significantly greater than between the liquid crystal and the liquid phases [43] and this type of dispersion is, hence, highly unlikely. This conclusion is also strongly supported by the fact that the liquid crystal was dispersed in the oil phase in the separate static experiment and that no water droplets could be observed. The second alternative, the formation of water droplets surrounded by liquid crystalline layers certainly satisfies the condition of minimizing the interfacial free energy, but, as has been shown by Westesen [13], intense emulsification removes such layers; forming vesicles instead.

With the available information, the following interpretation of the light scattering results, Fig. 2, is reasonable. During the mixing in the stop-flow equipment the weakly associated surfactant aggregates [44] are dispersed due to the intense shear. Furthermore, the water is finely divided into macroscopic droplets located in the oil phase, because this phase containing the surfactant will rapidly become the continuous phase during the mixing.

The water droplets are, hence, dispersed in the surfactant–oil phase and a lamellar liquid crystal will primarily form at the water–oil interface as demonstrated in the present static experiments with W/O microemulsion systems and also earlier [33, 34, 45]. The lamellar liquid crystal subsequently disperses into the oil phase; not into the water droplets, as demonstrated by the static experiment. The formation of the lamellar phase causes a reduction of the water droplet size, because the amount of liquid crystal at equilibrium is equal to the amount of water. The water removed to form the lamellar liquid crystal is, hence, significant and the reduction in total scattering intensity is expected and reasonable.

The lamellar liquid crystal is now dispersed in an oil phase with less than maximum water solubilized. Hence, the liquid crystal will partially change to inverse micelles with a change in total composition along the line *Initial*-2 in Fig. 1 and the intensity of scattered light passing through the crossed polarizers is reduced. This interpretation is supported by the pattern in the microphotographs after shear showing the primary formation of liquid crystals, which are slowly reduced in size. The final slow growth of the total scattering is referred to coalescence of the water droplets.

The essential thesis is, hence, that the formation of the lamellar liquid crystal particles is a faster process than the one by which the W/O microemulsion droplets are formed, i.e. the dissolution of the liquid crystal is a slow process compared to its formation. This conclusion is supported by the results of the static experiments dissolving the liquid crystal into the microemulsion. Of these the kinetics in the first experiment is governed by the diffusion of W/O microemulsion droplet from the liquid-crystal–microemulsion interface as evidenced by the difference in dissolution rate in Figs. 6 and 7. This interpretation is supported by an estimation of the diffusion coefficient from the value in Fig. 6. One has [46] the flux J.

$$J_0 = DC_0/\sqrt{\pi Dt} \tag{1}$$

with D the diffusion coefficient, C_0 the concentration in one semi-infinite half-sphere and t time, from which

$$D = B^2\pi/4C_0^2 \tag{2}$$

Progr Colloid Polym Sci (1998) 108:9–16
© Steinkopff Verlag 1998

in which B is the slope obtained from the linear fit in Fig. 6.

The results in Fig. 6 give $D = 1.7 \times 10^{-11} \, m^2 s^{-1}$ in good agreement with NMR determinations [47].

The dissolution rate of the liquid crystal is obviously an important factor and a preliminary analysis is justified. The weight of initially dissolved liquid crystal from Fig. 7 amounts to $3.10 \times 10^{-5} \, g/cm^2 s$ or, 76 double layers per second assuming a lamellar double layer thickness of the order of 40 Å. This value is extremely small and some structural entity obviously retards the dissolution. An estimation of the absolute minimum rate may be made using the molecular exchange rate between the lamellar layers and the environment. These values are not available for liquid crystals, but the determinations by Zana [48] of the exit rates from vesicles may provide some guidance as to reasonable rates. These values vary strongly, but for the present state a first order reaction constant of $1.0 \, s^{-1}$ is considered at an appropriate magnitude. Applying the reaction constant to single layers in a lamellar liquid crystal certainly is an approach open to criticism, but will at least give some indication of the magnitude; the reaction rate gives a dissolution of approximately two layers per second.

This values is, as mentioned, the absolute minimum and the experimental values are one to almost two magnitudes greater and, obviously, an additional mechanism to the molecular exit rate is operating. Intuitively it seems reasonable to assume that the initial exit of individual molecules causing disordering in the lamellar layer, should facilitate the necessary reorganization of surfactant and water molecules to inverse micelles and, hence, rapid dissolution from the outermost lamellar layer.

We are well aware that in these experiments, especially in the stirred one, the fact that the liquid crystal is not in equilibrium with the liquid will result in a transport also from the liquid to the liquid crystal especially of alcohol. However, the rapid rise in chemical potential of the phenethyl alcohol within the lamellar liquid crystal [42] implies that the amount of alcohol entering the lamellar structure remains small and its influence on the total amount may be neglected; at least under initial conditions.

Conclusion

The intense mixing in a stop-flow equipment of water with an oil solution of a nonionic surfactant and phenethyl alcohol with small amounts of solubilized water give a primary formation of a lamellar liquid crystal followed by its transformation to inverse micelles.

Acknowledgement This research was financed in part by ICI Surfactants, Wilmington, DE.

References

1. Larsson K (1964) Arkiv Kemi 23:29–36
2. Larsson K (1967) Nature 213:383–384
3. Larsson K (1989) J Phys Chem 93:7304–7314
4. Larsson K (1997) In: Friberg SE, Larsson K (eds) Food Emulsions. Marcel Dekker, New York, pp 111–139
5. Buchheim W, Larsson K (1987) J Colloid Interface Sci 117:582–583
6. Guslafsson J, Ljusberg-Wahren H, Angren M, Larsson K (1996) Langmuir 12:4611–4613
7. Sjöblom J (ed) (1996) Emulsion and Emulsion Stability. Marcel Dekker, New York
8. Fletcher PD (1996) Curr Opin Colloid Interface Sci 1:101–106
9. Manev E, Sazdanova SV, Wasan DT (1982) J Disp Sci Technol 3:435–463
10. Nikolov AD, Wasan DT (1992) Langmuir 8:2985–2994
11. Friberg SE, Jansson PO, Cederberg E (1976) J Colloid Interface Sci 55:614–623
12. Jansson PO, Friberg SE (1976) Mol Cryst Liq Cryst 34:75–79
13. Westesen K, Wehler T (1992) J Pharm Sci 81:777–786
14. Zana R (1993) Polym Mat Sci Eng 69:124–125
15. Anainsson EAG, Wall SN, Almgren M, Hoffman H, Kielman I, Ulbricht W, Zana R, Lang J, Tondre C (1976) J Phys Chem 80:905–921
16. Kahlweit M (1982) J Colloid Interface Sci 90:92–99
17. Gast AP, Robinson BH (1996) Current Opin Colloid Interface Sci 1:771–772
18. Schurtenberger P (1996) Current Opin Colloid Interface Sci 1:773–778
19. Engbert JBFN, Kevelam J (1996) Current Colloid Interface Sci 1:779–789
20. Groll R, Bottcher A, Jager J, Holzworth JF (1996) Biophys Chem 58:53–65
21. Jorgensen K, Klinger A, Braiman M, Biltonen RL (1996) J Phys Chem 100:2766–2769
22. Kamp F, Zakim D, Zhang FL, Nay N, Hamilton JA (1995) Biochemistry 34:11 928–11 937
23. Zvelmdovsky AV, Vanderlinden E, Bedeaux D (1995) Physica A 218:319–334
24. Cantu L, Corti M, DelFavero E, Maurer N (1995) Progr Colloid Polym Sci 9:197–200
25. Edwards K, Silvander M, Karlson G (1995) Langmuir 11:2429–2434
26. Fatal DR, Andelman D, Ben-Shaul A (1995) Langmuir 11:1154–1161
27. Huang JB, Zhao GX (1995) Colloid Polym Sci 273:156–164
28. Long MA, Kaler EW, Lee SP (1994) Biophys J 67:1733–1740
29. Pedersen JS, Egelhaaf SU, Schurtenberger P (1995) J Phys Chem 99:1299–1305
30. Egelhaaf SU, Schurtenberger P (1997) Progr Colloid Surface Sci 104:152–156
31. Egelhaaf SU, Schurtenberger P (1997) Physica B 234:276–278
32. Yatcilla MT, Herrington KL, Brasher LL, Kaler EW, Chiruvolu S, Zasadzinski JA (1996) J Phys Chem 100:5874–5879
33. Tondre C, Burger-Guerrisi C (1987) J Phys Chem 91:4055–4059
34. Burger-Guerrisi C, Tondre C, Canet D (1988) J Phys Chem 92:4974–4981

35. Weiss J, Compland JN, Braithwaite D, McClements DJ (1997) Colloids Surf 121:53–60

36. Campbell S, Yang H, Patel R, Friberg SE, Aikens PA (1997) Colloids Polymer Sci 275:303–306

37. Friberg SE, Campbell S, Fei L, Yang H, Patel R, Aikens PA (1997) Colloid Surf 129–130:167–173

38. Friberg SE, Yang H, Fei L, Sadasivan S, Rasmusen DH, Aikens PA (1998) J Disp Sci Technol 19:19

39. Patel RC (1976) Chem Instrum 7:83–89

40. Warr GG (1995) Colloids Surf (1995) 103:273–279

41. Hsu WP, Patel RC, Matijević E (1987) Appl Spectrosc 41:402–407

42. Friberg SE, Huang T, Fei L, Vona Jr SA, Aikens PA (1996) Progr Colloid Polymer Sci 101:18–22

43. Ghosh O, Miller CA (1987) J Colloid Interface Sci 116:593–597

44. Christenson H, Friberg SE (1980) J Colloid Interface Sci 75:276–285

45. Friberg SE, Podzimek M, Neogi P (1986) J Disp Sci Technol 7:57–79

46. Crank J (1970) The Mathematics of Diffusion. Clarendon Press, Oxord, pp 30–32

47. Lindman B, Söderman O, Wennerström H (1987) In Zana R (ed) Surfactant Solutions, New Methods of Investigation. Marcel Dekker, New York, pp 295–357

48. Zana R (1986) In Mittal KL, Bothorel P (eds) Surfactants in Solution, Vol 4, Plenum Press, New York, pp 115–117

Progr Colloid Polym Sci (1998) 108:17–20
© Steinkopff Verlag 1998

D. Chapman

New biomaterials based upon biomembrane mimicry

Prof. D. Chapman (✉)
10 One Tree lane
Beaconsfield
Bucks HP9 2BU
United Kingdom

Abstract This talk illustrates how studies of biomembranes have led to a technique for producing new haemocompatible materials.

Key words Biomembranes – biocompatibility – phosphorylcholine

Introduction

New biomaterials

Many materials used for fabricating medical devices produce adverse reactions when in contact with body tissues and fluids and especially blood. Such materials include metals and various polymers such as polyethylene, polyvinylchloride, polyurethane, cellulose and its derivatives and materials such as silicone. Many of these materials have good mechanical properties but have unsatisfactory biocompatible characteristics. Indeed many of the materials commonly used in medical devices were found by empirical means without considering their interaction with the biological fluids. In particular, the contact of materials with blood causes protein deposition as well as platelet adhesion and activation.

To overcome these effects, anticoagulants such as heparin are sometimes used. This can lead to various complications including uncontrolled bleeding and can lead to a lowering of blood quality.

The search for satisfactory materials which can be placed in blood without causing blood to clot has led to many investigations but with little success. Negatively charged surfaces, positively charged surfaces and hydrophobic surfaces have all been tried. Andrade and Hlady [1] commented "Virtually every physical and chemical characteristic of materials has been suggested as being important in blood coagulation and thrombosis". In addi-

tion to problems with materials contacting blood, there are many other situations where adverse reactions occur between polymers and body fluids. One example of this is with Hygrogel contact lenses. These are known to suffer from the problem of protein deposition. The issue of biocompatibility (i.e. the property of interfacing with a biological system without modifying or adversely affecting its normal function) is being increasingly seen to be an important one which requires resolution. Associated with this requirement of biocompatibility is the related problem of obtaining surfaces which are protein resistant and which prevent cell attachment, e.g. the requirement to produce anti-fouling surfaces.

This feature, that proteins adsorb on to a variety of polymer and metal surfaces appears to be a key event in stimulating the blood coagulation process. Thus, when body fluids are in contact with the usual polymeric synthetic materials, protein adsorption takes place rapidly (in milliseconds). This means that a polymer when placed in blood or plasma has within a few minutes a coating of protein on its surface. The predominant adsorbed protein from blood is fibrinogen. Furthermore, fibrinogen is known to bind the platelets as well as to factors which can cause platelet activation. As well as this, fibrinogen is converted by thrombin to fibrin producing an insoluble polymer. In this way clots and microemboli arise when a material is placed in blood (see Table 1).

We have introduced a new approach to this problem which has been shown to produce a satisfactory haemocompatible material. This is based upon a mimicry of

Table 1 Events occurring on biomaterials leading to abnormalities in blood function

Protein deposition	Fibrinogen
	Immunoglobulin
	Albumin
	C-reactive protein
Platelet interactions	Adhesion
	Activation
Activation phenomena	Clotting cascade – in particular Factor XII
	Complement cascade – C3 to C-3a

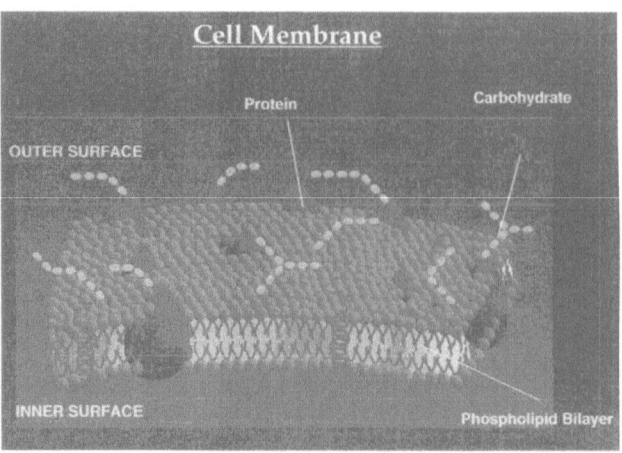

Fig. 1.

the outer lipid surface of the biomembranes of erytrocytes and platelet cells. The biomembranes of these cells are built upon a lipid matrix (usually a bilayer) into which the membrane proteins (integral proteins) are inserted (see Fig. 1).

It is known that the lipid bilayer matrix of red blood cells and platelet cells has lipid class asymmetry, i.e. the negatively charged phospholipids are predominantly located in the inner leaflet of the plasma membrane, while the zwitterionic phopholipids are found predominantly in the outer lipids (see Fig. 2).

It was suggested some time ago that this lipid asymmetry may serve a biological purpose for the maintenance of a delicate balance between haemostasis and thrombosis [2]. It was consideration of this lipid class asymmetry of red blood cells and platelet cells that led us to the idea of mimicking the outer lipid surface of red blood cells and platelet cells so as to produce a new type of haemocompatible biomaterial [3]. A coating of any one of these lipids, either phosphatidylcholine or sphingomyelin, provides a 90% mimicry of the outer lipid surfaces of erythrocytes.

We next began to realize that the total phospholipid structure of phosphatidylcholine lipids, including the fatty acids residues was not essential to producing these excellent haemocompatibility characteristics, rather it was the phosphorylcholine headgroup which is important. We therefore started the study of derivatives of the phosphorylcholine headgroup to see how it could be attached to plastics and metal surfaces. These experiments also showed that improvements in haemocompatibility occur. For example, in collaboration with Biocompatibles Ltd., we synthesized physisorbable phosphorylcholine containing polymers for coating hydrophobic surfaces such as PVC, polyethylene, polypropylene, etc. (see [4]). These polymers, based on methacrylate chemistry, have high molecular weights and therefore form very stable coatings owing to their multipoint attachment. In addition, a variety of surface modifications have been used, such as plasma discharge and chemical procedures, so as to be able to attach or graft the PC polar groups to these

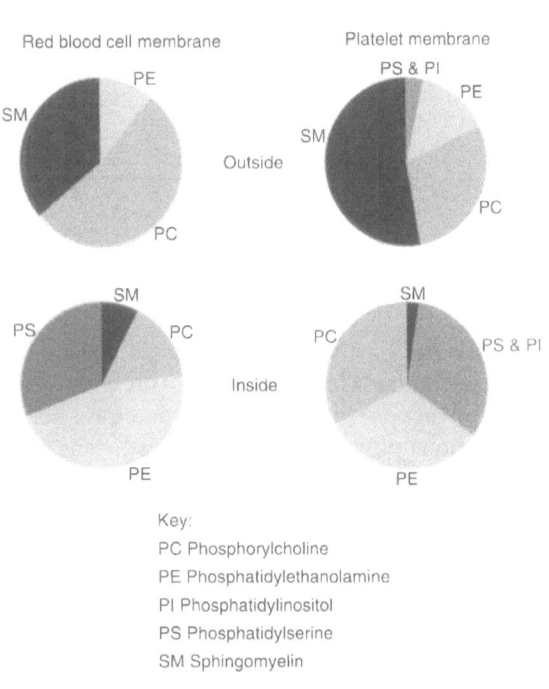

Key:
PC Phosphorylcholine
PE Phosphatidylethanolamine
PI Phosphatidylinositol
PS Phosphatidylserine
SM Sphingomyelin

Fig. 2 Distributional asymmetry of phospholipids in the membrane of platelets and red blood cells

materials. Materials such as celluloses and stainless steel have been successfully coated [4].

Functionally active phosphorylcholine derivatives have been synthesized (see Fig. 3), either as individual functionally active headgroups [5] or as methacrylate based polymers, and these molecules have been chemically attached to a variety of hydrophilic surfaces. Various surface modifications have been made to introduce partner functional groups where these groups are not already

Progr Colloid Polym Sci (1998) 108:17–20
© Steinkopff Verlag 1998

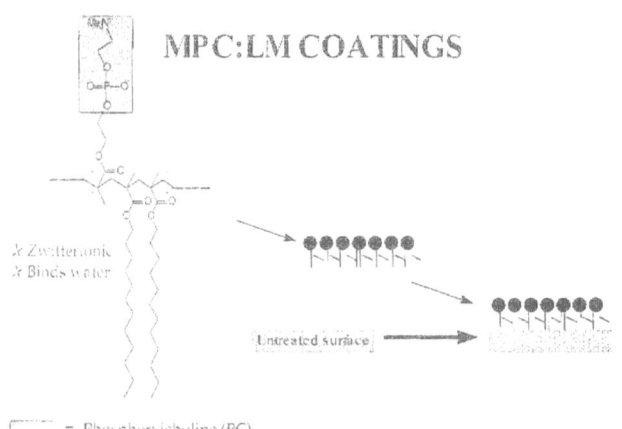

Fig. 3.

Table 2 Biocompatibility of PC-coated surfaces in vitro

Substrate	Reduction in adsorption (%)		Platelet activation
	Fibrinogen	Platelets	
PVC	90	99	No
Polypropylene	83	86	No
Polyethylene	86	99	No
Polystyrene	90	80	No
Polyurethanes	69	97	No
Polyimides	91	96	No
Steel	84	97	No
PTFE	90	90	No
Polyamide	84	90	No

present, e.g. introducing acid groups into stainless steel for the subsequent reaction of aminophosphorylcholine derivatives.

New haemocompatible biomaterials have also been made either by introducing phosphorylcholine derivatives as plasticisers into polymers such as PVC or polyurethane [6], or by copolymerizing phosphorylcholine monomers into the polymer backbone of polyurethanes and polyesters [7, 8].

These PC treated surfaces have been treated for fibrinogen adsorption, platelet adsorption, platelet activation and complement activation (see [4]). Fukushima et al. [9] have also pursued this concept of development of phosphorylcholine polymers produced from vinyl-containing monomer derivatives and have produced new materials and surface coatings which also show good compatibility with biological fluids. Using in vitro platelet adhesion and activation assays they also demonstrated that polymeric phosphorylcholine-based materials are non-thrombagenic (see Table 2).

Coagulation patterns in left heart bypass operations with bovine experiments (the left heart bypass is used for mechanical circulatory support). Experiments have been carried out by Segesser et al. [10] in Zurich. Heparin coatings and phospholipid coatings were compared with systemic heparinization was absent. The coagulation studies included platelet counts, activated coagulation time, thrombin time, fibrinogen (factor 1), antithrombin III and fibrinopeptide A (FPA) measurements. With both the heparin and the lipid coating, no macroscopic red clots were observed. Segesser concluded that the antithrombotic properties of the phospholipid coating is similar and even superior to that of the bonded heparin. A particular feature noted by Segesser is that the fibrinopeptide A levels remain constant, i.e. the lipid coatings do not trigger the coagulation cascade. This is in contrast to the heparin bonded surfaces which show an increase with time of this peptide level. Segesser pointed out that there are situations where even bonded heparin is not recommended as occurs with patients with heparin induced thrombocytopenia.

Experiments have also been made by Hunter and Angelini [11] with chest drainage tubes following open heart surgery. Occlusion of these tubes by thrombus is fairly common after such an operation. The experiments were carried out with 102 patients and the comparison was made with non-coated polyvinylchloride (PVC) chest drainage tubes. In these experiments, patients with PC coated tubes had, after removal, less post-operative supraventricular arrhythmias and a shorter hospital stay (8.4 vs 9.7 days).

From the many various in vitro and in vivo experiments that have been performed, it is clear that these new biomaterial surfaces containing well packed PC polar groups provide a new satisfactory haemocompatible biosurface. As we have seen, these results are in marked contrast to previous attempts to produce such haemocompatible surfaces using either positively or negatively charged surfaces or hydrophobic surfaces. The phosphorylcholine polar group on the other hand has zwitterion structure, i.e. it contains within the same polar group both a positive and also a negative charge. It is in fact isoelectric over a wide pH range (3–10). It also binds large amounts, ~ 12 molecules, of water as indicated by a range of experiments including our own early studies using differentials scanning calorimetry and also deuterium nmr spectroscopy.

Close packed arrays of these phosphorylcholine PC groups are protein resistant. This has been shown in experiments with fibrinogen and also with lysozyme. Some recent experiments with albumin were carried out in our laboratory using the surface plasmon resonance technique. This shows considerable reduction in protein adsorption which occurs after treating the metal surface so that it is covered with an array of PC-polar groups. We conclude that it is this protein resistant characteristic of the close packed arrays of the phosphorylcholine groups which is the key for its important and successful haemocompatible characteristics. A further important feature of these new haemocompatible materials is that because proteins do not bind to their surfaces, cell attachment is inhibited thereby reducing infection by bacteria.

The production of these new biomaterials provide polymers and coating which have considerable potential for a range of medical applications. These include blood contacting devices, catheter, in-dwelling biosensors, stents and filtration membranes. The important characteristic of protein resistance has led Biocompatibles Ltd. at Farnham to produce new types of hydrogel polymers which are synthesized so as to contain arrays of these PC groups. These are then used to make contact lenses. The PC contact lenses have new and improved characteristics. They show no protein deposition, do not readily "dry-out"

and also have excellent oxygen permeability characteristics.

Conclusion

1. A fundamental understanding of biomembrane protein structure remains an important subject for the application of physical techniques. It is by combining both static and dynamic information, obtained from these different techniques, that a better understanding of signal-transduction and transport processes will be possible.

2. Knowledge gained from basic studies of biomembranes can be extended to biotechnological applications. Biomembrane mimicry has already been shown to provide useful technological applications in a range of areas including medicine. Application of this approach in related non-health care area including the topic of anti-fouling may become possible.

References

1. Andrade JD, Hlady V (1986) Adv Polym Sci 79:1 (Chapman D (1979) European Patent 32622)
2. Zwaal RFA, Beavers EM (▮▮▮▮) In: Roodyn DB (ed.) Haemostasis Subcellular Biochemistry, Vol 9. Plenum Press, New York, p 299
3. Hayward, Chapman D (1984)
4. Chapman D, Charles SA (1992) Chem Britain 28:253
5. Chapman D, Durrani A (1984) European Patent, 157469
6. Valencia GP (1985) European Patent, 247114
7. Chapman D, Valencia GP (1984) European Patent, 199790
8. Durrani AA (1986) European Patent 275293
9. Fukushima S, Kadoma Y, Nakabayashi N (1983) Kobunshi Ronbunshu 40:785
10. Segesser LV, Olan M, Leskosek B, Turina M (1992) ASAIO Abstract, 38th Annual Meeting, p 25
11. Hunter S, Angelini GD (1993) Ann Thoracic Surgery 56:1339

Progr Colloid Polym Sci (1998) 108:21–33
© Steinkopff Verlag 1998

S. Bengmark

Polymers for bioadhesion, absorption control and tissue separation

Introduction

All tissues which come into contact with the exterior and "hostile" environment need to be protected. Such a protection is provided in some organs through a mucus gel, often in combination with a thin layer of special lipids, called lipid surfactants or biosurfactants. The covering layers do not only constitute a protective barrier, they provide unique lubrication and play a crucial role in the selection of substances for binding, interaction and uptake by the cells.

Mucus in itself is a weak viscoelastic gel. Its basic component is mucin, a glycoprotein, with an estimated mol. wt. of $2–14 \times 10^6$ Da [1]. The glycoprotein molecules form a gel matrix through their molecules being associated by non-covalent interaction. Mucus covers the epithelium of the digestive, respiratory and genito-urinary tracts. A mucus-like layer is also present on the synovia-covered surfaces such as joints, tendons and tendon sheets. However, mucus is not available on mesothelium-covered surfaces such as the peritoneal, pericardial and pleural membranes, or on endothelium-covered surfaces such as the intima – inside of blood vessels. Here the covering and protecting layer consist only in a mixture of hyaluronic acid and lipid surfactants (mesothelium) or no layer at all (endothelium). This section deals with possibilities and attempts made to enforce surface protection through

S. Bengmark (✉)
Suite A 361
Ideon Research Center
Beta House
S-22370 Lund
Sweden
E-mail: Stig.Bengmark@kir.lu.se

exterior supply of polymeric substances with ability to adhere to biological surfaces. As the protecting polymers are tissue adhesive they can also be used for drug delivery.

Adhesion and absorption control

Bioadhesion

Synthetic and biological macromolecules under certain conditions adhere to biological tissues [2], a phenomenon usually referred to as bioadhesion. It is called mucoadhesion, when applied to mucosal epithelium, where the mucosa and its mucus layer primarily is involved [3]. The mucoadhesive capacity has been identified to be limited to *hydrophilic macromolecules*, which possess numerous hydrogen bond-forming groups. It is known that the rheological properties can be manipulated in an effort to further strengthen the gel, and more pronounced mucoadhesion does usually not occur without considerable change in rheological properties of the mucus layer. Macromolecules are known to be capable of aggregating to each other in solution, forming intermolecular complexes and resulting in phase separation phenomena; precipitation, coacervation, emulsification, crystallization and gelation [4]. It has been suggested that intermolecular forces are formed as a result of secondary binding forces between the molecules, based on electrostatic interactions (Coulombic forces), hydrogen bonds, van der Waal's forces, hydrophobic interactions and charge transfer complexes [5].

Allen et al. [6] demonstrated in 1986 that the viscosity of gastric mucus glycoprotein could be increased by addition of a poly(acrylic) acid, Carbapol 934P, a function heavily depending on pH, with optimum at pH 6.0 [7]. Montazavi and Smart [8] studied and compared the gel-strengthening effects of several poly(acrylic) acids with molecular wts varying between 5 and 3000 kDa. Anionic polymers were found to strengthen the mucus gel more than neutral or cationic polymers, and an optimal gel-strengthening effect was observed, when Carbopol with a molecular wt. of 750 kDa (Carbabol 810) was tried.

Absorption enhancing

The oral route is clearly the most natural and also convenient way to administer drugs and vaccines. However, universal use of the gastrointestinal route is to a large extent hampered by the fact that many the administered substances are either unstable and metabolized before they reach the target organ, or/and they are poorly absorbed by the gastrointestinal mucosa. To overcome this, many attempts have been made to develop systems to enhance absorption of complex drugs, especially those of peptide or lipid nature. These efforts include use of peptidase inhibitors [9], surface active agents [10], modification of the hydrophobic prodrugs [11] and coupling to the drug of a ligand (hybridization) [12], in order to promote active transport.

Passive absorption of drugs occurs mainly via two different routes, the transcellular and the paracellular. Only small hydrophobic compounds capable of passing through the lipophilic cell membrane can utilize the transcellular pathway, leaving the paracellular route as the only possibility for hydrophilic compounds [13]. The capacity of the paracellular route, however, is rather limited, which explains why the absorption of orally administered hydrophilic drugs, such as peptides, proteins and antibiotics is severely restricted. As pointed out in a recent paper by Schipper et al. [14], numerous attempts to enhance the absorption of hydrophilic drugs have been made. For review see also [15]. Unfortunately, most of the attempts have resulted in mucosal damage, induced by the enhancer per se [16]. Only a few enhancing compounds have demonstrated the ability to improve absorption without subsequent harmful side effects, and this even when applied to rather robust tissues such as the skin.

Transdermal drug delivery

It is a most attractive suggestion to use the transdermal route, the intact skin, to deliver drugs, also for expected systemic effects. One of the advantages of transdermal delivery systems (TDSs), especially when compared to the oral administration, is that it constitutes a method to by pass the liver, and to avoid most of the active drug from being absorbed and/or broken down before reaching its target organ. The liver is, indeed, an effective filter, which effectively absorbs most of orally administered drugs, hereby limiting the efficacy of oral drug delivery. Many attempts have been made, especially in the last ten years, to develop effective TDSs. As is commonly seen in the sensitive GI tract, inflammatory and sensitization reactions do often occur also in the skin [17]. Many carriers of the active drugs have been tried over the years, but skin reactions, as often directed towards to the carrier as towards the drug, have limited their clinical usefulness [18, 19]. One possible approach to minimize carrier/device-associated skin reaction to TDSs has recently appeared through the availability of highly biocompatible and non-antigenic biopolymers. One such polymer is chitosan, a natural polycationic polymer. Chitosan and its derivates are being currently tried extensively in different formulations for sustained release of drugs [20–23] – see further below. Thacharodi and Rao have in a recent series of works [24–26] demonstrated that the absorption of the active drug in TDSs, in addition to the size of the applied area (skin patch), mainly depends on the degree of cross-linking. With a high degree of cross-linking a minimum degree of release is obtained.

Oral drug delivery

Polymeric microparticles that escape absorption in the upper GI tract have a great ability to be taken up by the GALT (Gut Associated Lymphoid Tissue) system, especially by the so-called Peyer's patches. These groups of cells are in the intestine responsible for the uptake of macromolecular antigens from the intestinal lumen. It is an attractive suggestion to make use of the GALT function to promote absorption of drugs, vaccines and special nutrients. Earlier studies have shown that molecules with diameter smaller than 5–10 μm and with hydrophobic surfaces tend to be taken up the best [27, 28]. Liposomes have a series of characteristics, which render them especially suitable as drug delivery vehicles. It is especially important to know that liposomes encapsulate hydrophilic and hydrophobic substances separately, and that they can be prepared without exposure to the aqueous phase organic solvents – most peptides and proteins lose their activity when exposed to organic solvents. Unfortunately, regular liposomes are not stable to acids, bile salts or enzymes, which seemingly make them unsuitable for oral delivery use. This can, however, be overcome by use of either liposomes based on phospholipase-resistant phospholipids or better polymerized liposomes [29, 30]. The phospholipid molecules in polymerized liposomes are in sharp contrast to regular liposomes, linked together by covalent bonds. Although this principle has been known

Progr Colloid Polym Sci (1998) 108:21–33
© Steinkopff Verlag 1998

for at least 15 y, few attempts seem to have been made to use polymerized liposomes in oral delivery systems. Okada et al. [31] report recently that polymerized liposomes are effective and retain their content in up to 93% and 75%, respectively, when kept during two days in vitro in gastric and intestinal media, which is significantly better than can be obtained with regular liposomes, 60% and 30%, respectively. However, as concluded by the authors, several characteristics of the polymerized liposomes remain to be studied before they can be even tried under clinical conditions. These include toxicity, uptake by the Peyer's patches, and the in vivo cellular processing of polymerized liposomes.

Chitosan and its derivates

Chitosans have in themselves strong mucoadhesive properties. It is suggested that this function is mediated through ionic interaction between positively charged amino groups in chitosan and negatively charged siliac acid residues in mucus or on cell surfaces [32]. Studies have shown that chitosans do enhance absorption of drugs such as insulin and decapeptide over mucosal membranes such as those of the intestine and nose [33, 34]. Rentel et al. [34] suggest in a recent publication that cationic polymers in general should be further investigated with regard to the possibility of improving the mucoadhesive properties in a neutral or slightly alkaline environment.

Chitosans are believed to be non-toxic and are, as a matter of fact, approved as food additive in countries such as Japan. Recent studies have shown that Chitosan and its derivates have the potential to be used clinically as absorption enhancers [33, 34]. It is also suggested that chitosan has a different mechanism of action than other absorption enhancers. Not only cationic chitosan macromolecules, but also other macromolecules such as protamin and polylysin can on the surface of epithelial cells interact with anionic components (siliac acid) of glycoproteins [35]. Schipper et al. [14] compared in an in vitro system of Caco-2 intestinal mucosal cells eight different chitosans with varying degree of acetylation and molecular weight, and demonstrated that chitosans with low degree of acetylation are the most active as absorption enhancers, both at low and high molecular weights. They display, however, all a clear dose-dependent toxicity. Chitosans with a degree of acetylation varying between 35% and 49% seemed to both exhibit low toxicity and enhance well transport. Chitosan with 35% degree of acetylation and molecular weight of 170 kD showed especially advantageous properties; early onset of action, very low toxicity and a flat dose-absorption response relationship. It remains, however, to verify that these conclusions are valid also under clinical conditions.

Liver-controlled delivery

The liver regulates the homeostasis of the blood by adding and extracting substances to or from the blood. In doing so it sequesters various larger molecules, especially waste products. In addition, the liver supplies the blood with the bulk of plasma proteins. Within the liver most of the synthesizing functions do occur within the hepatocytes, which constitute two-thirds or more of the volume and number of cells within the liver. Most of the sequestering functions are dealt with by the so-called reticuloendothelial cells, of which the sinusoidal endothelial cells constitute more than two-thirds, the remaining mainly made up by the so-called Kupffer cells. Liver endothelial cells do take up soluble matters and very small particles, whereas Kupffer cells take up larger particles [36]. The liver has a great capacity for uptake of polymers similar to hyaluronan and chondroitin sulfate. As for example, intravenously injected hyaluronan is rapidly taken up by the endothelial cells of the liver.

Protection of mesothelium covered surfaces

Peritoneum, pericardium and pleurae

The four large body cavities the two pleural cavities, the pericardium and the peritoneum enclose most of the large organs of the body. These cavities allow the internal organs to move freely with a minimum a friction. It is important for respiration that the lungs can expand. Also the abdominal organs, particularly the liver, can move freely with the respiratory movements. These cavities are structured according to the same principles and are all covered with similar mesothelial cells. Although these membranes have unique lubricating abilities they do also exhibit many other important functions, such as secretion and absorption. For review see the handbook of Bengmark 1989 [37]. As most studies to the function of the mesothelial membranes have been carried out on the peritoneum, this review will mainly deal with the peritoneal membrane. Much support, however, is present for an assumption that what is effective in the peritoneum is also effective in the other cavities.

Peritoneum

The word peritoneum was used already by the ancient anatomists to describe the largest serous membrane of the body, a membrane which covers both the inner surface of the abdominal wall and the outer surface of the abdominal organs. Its etymological meaning is "wrapped tightly around". Peritoneum has roughly the same area as the skin, in the adult about 2 m^2, which makes the peritoneum one of the largest surfaces in the body. It consists of a single layer of mesothelial cells, covered by a thin, appr. 5 μm, thick liquid biofilm, the peritoneal fluid. The secretory capacity of the peritoneum has in the past been largely underestimated. The biofilm of all the mesothelium-covered surfaces (peritoneum, pericardium, pleura) is rich in biologically active cells e.g. mast cells, lymphocytes, macrophages, polymorphs and mesothelial cells [38, 39]. In the peritoneum lytic enzymes, oxidative free radicals (to kill invading microorganisms) angiogenesis factors, prostaglandins, cytokines and many other substances are produced. This makes the peritoneum a very active organ. In the absence of a mucus layer, the peritoneum is for its lubrication and protection totally dependent on the surface film, biofilm, which mainly consists in phospholipids, where dipalmitoylphosphatidylcholine, phosphatidylethanolamine and sphingomyelin are the dominant species [40]. The fluid also contains a number of other macromolecules such as albumin, globulins, lipoproteins and cholesterol and also several forms of glucoseaminoglucans, including hyaluronan (hyaluronic acid) [41]. The recent finding that the peritoneal fluid is rich in hyaluronan, especially during inflammation, and that mesothelial cells are an important source of this macromolecule is of particular interest, as the substance is known to be active in water homeostasis, lubrication and repair processes [42]. It is also known that phospholipids in combination with hyaluronan provides an outstanding lubrication, through their rheological properties. Several studies suggest that cytokines such as TGF-beta and PDGF stimulate the synthesis of hyaluronan [43, 44].

The thin mesothelial layer and the protecting biofilm, which covers the mesothelium, is easily destroyed as a result of infection, inflammation or "careless" handling (desiccation, abrasive manipulative tissue contacts with surgical instruments, gloves, sponges, pads, etc.), leading to denudation of the peritoneum and subsequent development of adhesions between the intraabdominal organs. Such adhesions constitute the most common cause of intestinal obstruction, abdominal (pelvic) pain and female sterility. Surgical operations are the most common contributors of adhesions in the peritoneum, pleura and pericardium, adhesions being reported to occur in almost ninety percent of patients after abdominal operations [45] and the adhesion frequency is no better after thoracic operations. However, also approximately 30% of non-operated individuals do [46] show varying degree of adhesions, which most likely is due to a previous, eventually unnoticed, external abdominal trauma and/or clinical or subclinical infection in the cavity. Not only that adhesions lead to extensive suffering, unwanted expenses for the

Society, it makes subsequent operations more difficult and risky, in the worst scenario impossible. As an actual example, it is highly attractive to use the peritoneal cavity for dialysis treatment in patients with kidney insufficiencies, but its universal use has to a large extent been hampered by the risk of developing infections and subsequent adhesion formation. It is also of a growing interest to use the peritoneum as an access organ for administration of specific drugs, especially for the purpose of creating slow release systems. Such systems could improve the administration of hormones like insulin. Also it could improve administration of drugs with the aim to control the growth of cancer or to modulate the immune function. There is today a fast increasing interest to develop systems and equipments for intraperitoneal drug administration, an interest, which can easily develop into a multibillion business, should the problem of peritoneal adhesion prevention find a satisfactory solution. These attempts will not be fully successful before effective tools to control adhesions have been developed.

The role of phospholipids

To control the development of adhesions has been a primary issue for hundreds of years, and hundreds of substances have been tried, of which some are summarized in Table 1, reproduced after Christen and Buchmann [47]. Among the first polymers to be tried were dextrans and polyvinylpyrrolidone (PVP), both biomedical polymers developed appr. 50 yr ago as blood volume expanders. An excellent review of the early experience using polymers has been provided by diZerega [48]. Only when used in highly concentrated and very viscous or gel-like compositions do these early used polymers show some, even if weak, adhesion-preventive effect. A great step forward was taken when Grahame et al. [49] in 1985 demonstrated the presence of surface-active material in the peritoneal effluent from patients maintained on continuous ambulatory peritoneal dialysis (CAPD), suggesting that the surfactants of the biofilm may play a protecting role in ultrafiltration, and in the prevention of peritoneal infection. Soon thereafter DiPaolo et al. [50] demonstrated a decrease in the content of phospholipids in dialysis fluids in parallel with a reduced ultrafiltration and/or development of peritonitis. These early observations lead us, in collaboration with Kåre Larsson, to study experimentally the adhesion preventive effects of different phospolipids using different animal models [51–55]. The results of these studies have been recently summarized [56]. While insoluble phosphatidylcholine (PC) seemed to enhance production of adhesions, soluble PC did prove very effective in preventing adhesions – none or minimal adhesions were found in 77% of

Table 1 Prevention of fibrin formation (Compounds and techniques used to minimize development of peritoneal adhesions. After Christen and Buchmann 1991 [47])

Anticoagulant:
— Na-citrate
— Heparin
— Dicumarol
— Dextran

Fibrin elimination
— Lavage
— Enzymatic (intraperitoneal)
    — Pepsin
    — Trypsin
    — Papain
    — Streptokinase/dornase
    — Actase
    — Urokinase
    — Hyaluronidase
    — Fibrinolysin
    — Protoporphyrin

Surface separation
— Lavage
— Oxygen insufflation
— Stimulation of peristalsis
— Ingestion of iron and frequent passes of a large magnet over the abdominal wall
— Paraffin oil
— Olive oil
— Lanolin
— Amniotic liquid
— Amniotic membranes
— Fish bladder
— Carp peritoneum
— Bovine peritoneum
— Vitreous of calf eye
— Oiled silk
— Silver and gold membranes
— Free omentum graft
— Silicone
— Fibrin sealant
— Concentrated dextrose solution
— Polyvinylpyrrholidone
— Chyme
— Inert polysiloxanes
— Hydroxyethyl starch
— Gelatin
— Dextran

Reduction of fibrosation
— Antihistaminics
— Steroids
— Cytotoxic substances
— Vitamin E
— Ibuprofen
— Colchicine

Surgical technique
— Unpowdered gloves or extensive rinsing of the gloves
— No peritoneal suture under tension unless absolutely necessary
— Cover tended peritoneal sutures with omentum
— Avoid traction or torsion of bowel and mesenterium
— Leave rinsing solution in the abdomen
— Be critical towards omentectomy

the treated animals compared to 28% in the untreated controls. The parietal adhesions as measured in millimeter decreased by 4% to 500%. Of the different phospholipids tried DL-PC (50% phospholipase-resistant) and phosphatidylinositol (PI) seemed to be the most effective in preventing adhesion formation. Supporting results were later obtained from other groups. In a British study, a treatment with phosphatidylcholine during 8 d before and 7 d after induction of adhesions by intraperitoneal irrigation with saline at 40°C for five minutes effectively prevented adhesion development [57]. A German study by Treutner et al. [58, 59] compared the adhesion-prevention effects of a PC, sphingolipid (SL) and galactolipid (GL), and found a pronounced reduction of both primary adhesions induced by abrasion of the peritoneum and of redeveloped, secondary adhesions after adhesiolysis was performed. No differences could be observed in the ability of the three lipids tried to prevent primary adhesions, but GL seemed to have a slightly stronger ability than the other two to prevent secondary adhesions, for review see ref. [60]. Although so far better results have been obtained by phospholipid treatment than with any other substance tried so far, total elimination of the adhesion problem has not been achieved.

The role of mucopolysaccharides

Also other polymers have been recently tried with some success. Urman et al. [61, 62] studied the effects of 0.25% and 0.4% hyaluronic acid (HA) respectively in both primary adhesions, applied before the induction of a laser-induced primary peritoneal lesion, and in secondary adhesions, to prevent reformation of adhesions after adhesiolysis. Both 0.25% and 0.4% were effective to reduce the development of primary adhesions, but only the concentration of 0.4% could reduce the formation of secondary adhesions. In a large multicenter study Burns et al. [63] studied the effects of precoating with hyaluronic acid in a rat cecal abrasion model. The mean cecal adhesions decrease from $1.6 \pm 0.11\%$ to $0.7 \pm 0.09\%$ in parallel with an increased supply of HA from 0% to 0.4% at the same time that the number of animals with cecal adhesions decreased from 89% to 50%. Similar results were also obtained in a dog study, using a method of dessiccation and abrasion of the pericardial and epicardial surfaces [64]. The six animals treated with HA showed no adhesions or only filmy and transparent adhesions, being significantly less severe than in the untreated controls. Experimental studies in rabbits have also shown that HA is effective to prevent adhesions in the sheets of flexor tendons, subjected to surgery [65].

It is fair to conclude that also the results of HA as prophylaxis against adhesion formation is not satisfactory. Of that reason, there is a growing interest to try to further improve the adhesion-preventive effects by trying other bioactive polymers. When 1% sodium carbomethylcellulose (NaCMC, 350 000 mol. wt.) was applied after the induction of the peritoneal injury in a standard laparotomy model some adhesion inhibition was observed both on primary and secondary adhesions [66, 67]. Similar results have been obtained also by other groups [68, 69]. Peck et al. tried more viscous solutions (>1000 centipoise) based on higher molecular weights (>700 000) and appr. twice as concentrated solution, but this did not observe any significant difference in outcome. However, it was observed during these studies that the applied volume is important, and the preventive effect increases with increasing volumes of solution administered.

In the past perfoming randomized trials in humans have been regarded as almost impossible. Such studies would require either a second look operation or direct laparoscopic peritoneal imaging at a certain time interval after the treatment. However, recently, the first prospective, randomized and double-blind study was published, based on laparoscopic peritoneal imaging 8–12 weeks after colectomy for non-malignant disorders. Becker et al. [70] were able to demonstrate a statistically highly significant difference as no adhesion formation was observed in 43/85 patients (51%) when treated with sodium hyaluronidate and carboxymethylcellulose bioresorbable membrane, compared to 5/90 (6%) in the control patients.

Modified mucopolysaccharides

Mucopolysaccharides are in general extremely hydrophilic in nature. It has been suggested that incorporation of photodimerizable groups onto mucopolysaccharides may increase their hydrophobic characteristics [71]. One possibility is to produce hydrogels from hydrosols simply by increasing the degree of crosslinking of the mucopolysaccharides. It has been observed that the contact angle of photocured films increase with increasing degree of substitution. Suitable molecules for photodimerization are, according to this study, cinnamate and thymine groups. It has also been suggested that polysaccharides containing amino sugars, such as chitosan and chitin, or derivates thereof, are especially suitable for production of slowly biodegradable, absorbable polymers, suitable for adhesion prevention [72]. N,O-carbomethyl chitosan (NOCC) is a novel agent with structural similarities to hyaluronan. It is depending on nature and extent of cross-

linking available in a variety of forms. A recent study based on two different animal models has shown that NOCC gives superior adhesion protection compared to hyaluronan [73].

The driving force behind the tendency of surfactants to stick to surfaces – and to form micelles – is the hydrophobic attraction [74]. The surfactant concentration needed for onset of formation of micelles is rather well defined [75]. Hydrophobically modified polymers (HM-polymers) have both polymer and surfactant characteristics. They are often constituted in a hydrophilic backbone to which a low number (usually below 5 mol%) of hydrophobic sidechains have been attached – for review see ref. [76]. Interesting such polymers can be produced by hydrophobical modification of molecules such as hyaluronan, chitosan and cellulose.

Two interesting such modifications of cellulose and hydroxyethylstarch, respectively, have recently been tried in experimental animals, HM-EHEC and LM-200. HM-EHEC is made hydrophobic by grafting a low number of nonylphenol groups to the polymer chain, and LM-200 is a chloride salt of an N,N-dimethyl-N-decocyl derivate of hydroxyethylstarch. Both these polymers demonstrated, in a recent experimental study in collaboration with Kåre Larsson, compared to hyaluronan, a superior ability to prevent experimental peritoneal adhesions [77].

Polymeric barrier films

Non-erodable polymer films appear attractive to use to cover limited areas of tissue surfaces with the aim of preventing local adhesions. But they are unsuitable for general prophylaxis. Examples of polymers, which have been tried to produce such films are crosslinked polydimethylsiloxane (PDMS), and microporous polytetrafluoroethylene (PTFE-Goretex[R]); for review, see refs. [78, 79]. The presently most interesting, and most explored barrier-producing film appears to be based on a combination of hyaluronan and carboxymethylcellulose (CMC). This membrane is chemically derivitized via ionic crosslinking in order to prolong its in vivo persistence to up to 7 d (Seprafilm[R], Genzyme Corp). Experimental and clinical experience with Seprafilm has recently been reviewed [79].

Due to its production price and eventual antigenicity, hyaluronan seems not ideal for total serosal coating and general prevention of adhesions. So far, however, the most attractive solution in order to prevent reformation of secondary adhesions after adhesiolysis seems to be a combination of HA and CMC locally applied. It is, however, not intended to be used as a tool for general prophylaxis,

Progr Colloid Polym Sci (1998) 108:21–33
© Steinkopff Verlag 1998

especially on larger surfaces such as those of the pericard, the pleura and the peritoneum, each with a surface varying between appr. 0.25 and 2 m^2.

Protection of mucosa-covered surfaces

Gastrointestinal tract mucosa

The gastrointestinal mucosa is exposed every day to food with a broad range of content, consistency, pH, osmolarity, temperature, microorganisms and toxins, and occasionally alcohol, tobacco ingredients including nicotine and pharmaceutical drugs. Large variations in shear forces, intraluminal concentration of hydrochloric acid, pepsin, bile and content of enzymes occur in and between different parts of the digestive tract. Occasionally does bile and/or acid cause problems by refluxing into parts of the GI tracts such as the oesophagus, which normally is less protected against these chemicals. The GI mucosa is, in the healthy individual, covered by a protection layer, which varies in thickness and preventing efficacy from one part of the tract to another. Mucus and surfactants are the main ingredients in the protection barrier. It is clear that the integrity of the mucosa is much dependent on a favorable balance between endogenous protective factors and exogenous aggressive factors. Although the existence of the barrier function has been known for more than a hundred years, it is only during the last 10–20 yr that one has started to realize that the quality of the protection is to a large extent dependent on oral supply of the various ingredients. Consequently, attempts have been made in recent years to support the protection through ingestion of gel-forming mucosa adhesive substances and biosurfactants. Mucus glycoprotein (also called mucin) exists as a highly hydrated polymer form consisting in glycoprotein subunits joined by disulfide bridges from their protein cores. Only those capable of gel formation seem to have essential protective functions [80]. Neutral lipids, glycolipids and phospholipids and fatty acids, which are associated with the glycoprotein through hydrophobic forces or covalent bonds, may modify the physicochemical characteristics of the barrier [81–83]. Phospholipids provide a true physical barrier, but it is only the hydrophobicity per se, which seems to be important. The mucosal hydrophobicity of the stomach is high in a healthy state, but has been shown to be significantly reduced in conditions such as gastritis, peptic ulcer disease, and with the occurrence of *Helicobacter pylori* [84, 85]. Furthermore, a lower gastric mucosal phospholipid concentration than normal is demonstrated in patients with gastritis and gastric ulcers [86].

In ulcerative colitis a significantly increased turnover of membrane phospholipids is observed. This has been described as an autocannibalism of membrane lipids. Hereby substrate for production of various membrane-derived mediators is provided, often through the action of phospholipase A$_2$ and/or various cytokines. One such product is Platelet Activating Factor (PAF), which is known to be predominately derived from membrane bound phosphatidylcholine. PAF has been found in high concentrations in mucosal biopsy tissues from patients with ulcerative colitis [87, 88]. Similarly, also large quantities of Prostaglandin E$_2$ (PGE$_2$) [89] and leukotreine B$_4$ (LTB$_4$) [90] are produced from substrates in the lipid membranes. Although the different cytokines and other mediators are produced by the action of different inflammatory cells, the gut epithelial cells, which is the main focus for the destructive process appear also to be the main contributor of phospholipids to the production of the different lipid-based mediators [91]. With this in mind, it seems logical to try to increase the protection by exogenous supply of substrate/membrane lipids.

The first layer of protection is to be found extra-mucosally and consists in mucus and phospholipids. The outer layer consists almost only in surfactant-like lipid molecules, which should render the mucosa "nonwettable" and provide special protection to the mucosal epithelium against chemical and mechanical injuries [92]. This most outer layer of phospholipid represents the first goal for injurious agents and its damage is an important factor for initiation of inflammation and ulceration. Mucus is important as a physical barrier, but also as the matrix/carrier of immune cells, immunoglobulins, protective flora and important enzymes. Mucin in itself has been shown to modulate the potency of enteric toxins [93]. Awareness of the crucial role of mucosal protection by surfactants and mucus led us in collaboration with Kåre Larsson to attempt to build new protection layers through external supply of different biosurfactants, mucus substitutes and also preventive flora, with the aim to prevent inflammation and ulceration, and to improve the barrier function and prevent leakage into the body of bacteria and toxins – a process usually referred to as translocation.

The gastric mucosa

In collaboration with Kåre Larsson we [94] induced acute mucosal erosions in the gastric corpus of starved rats by intragastric administration of 2 ml absolute ethanol (EtOH) or by subcutaneous injection of indomethacin (Ind) in a dose of 20 mg/kg body weight. When 5 ml/kg body weight of a solution containing 25 mg/ml of polyunsaturated phosphatidylcholine was administered 45 min before the induction of the mucosal damage, a dramatic reduction in the size of gastric ulcers was observed ($P < 0.001$). The average ulcer size decreased as a consequence of the treatment in the EtOH model from (untreated) 46.67 ± 1.67 mm to (when pretreated) 8.33 ± 5.43 mm and in the Ind model from (untreated) 38.33 ± 4.01 mm to (when pretreated) 5.83 ± 1.54 mm (Fig. 1).

Chronic antral ulcers were induced in starved and re-fed animals by subcutaneous administration of indomethacin (Ind) as described above. 5 ml/kg body weight of a solution containing 25 mg/ml of polyunsaturated phosphatidylcholine was administered 30 min after the Ind-injection. As a consequence a pronounced reduction in ulcer size was observed, the size of the ulcers decreased from (untreated) 5.83 ± 1.19 mm to (when treated) 0.75 ± 0.49 mm. When, instead, hydrogenated phosphatidylcholine was used, a somewhat smaller reduction was achieved, from (untreated) 9.67 ± 2.12 mm, to (when treated) 3.33 ± 1.02 mm (Fig. 2).

Some natural foods are known to contribute potentially protective compounds that might strengthen the barrier [95]. Banana, both the plaintain type and the common type, especially when unripe, belong to these foods [96, 97]. Both are rich in phospholipids with phosphatidylcholine, phosphatidylethanolamine and phosphatidylinositol as the main classes, rather similar to the composition of GI surfactants. In addition, the unripe

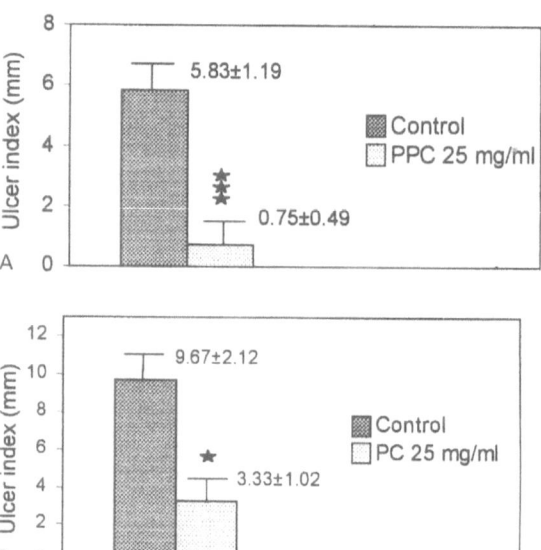

Fig. 2 Effect of a single dose of 25 mg/ml of (A) exogenous polyunsaturated phosphatidylcholine (PPC) and (B) exogenous hydrogenated phosphatidylcholine (PC) in the chronic indomethacin model. Asterisks indicate a significant difference from control value by a two-tailed unpaired t-test (* $P < 0.05$. *** $P < 0.001$)

Fig. 1 Effect of a single dose of 25 mg/ml exogenous polyunsaturated phosphatidylcholine (PPC) in acute ethanol model (A) and acute indomethacin model (B). Values are expressed in mean $\pm$ SEM for six rats. Asterisks indicate a significant difference from control value by a two-tailed unpaired t-test (*** $P < 0.001$)

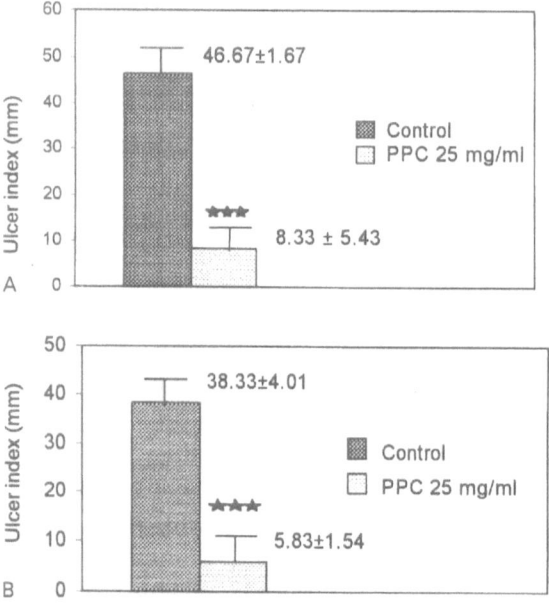

banana is also rich in pectin, a polysaccharide with a unique ability to form gels, very similar to that of guar gum and oat gum. All these fibers are candidates to be mucus substitutes. Because of its polymeric structure the natural mucus gel (secreted by the surface epithelial cells) has a gel "stickiness" and distinctive flow properties, which are unique and distinguishes mucus from all substitutes. Polysaccharidic gels such as those of pectin, guar gum and oat gum, are all more rigid in nature. Kåre Larsson and I have observed in vitro, when studying a pectin solution in water, that when the pH is reduced to the level of 1.0, that is the pH seen in the stomach, a so-called two-phase separation occurs [98], which might favor its use as a mucus substitute, especially in the stomach. One of the phases consists in a pectin gel with high pectin concentration and the other is a diluted pectin solution. It is likely that when this separation occurs at the level of the gastrointestinal mucosa, the gel phase will adhere to the mucosa layer and hereby considerably increase the protecting capacity of the mucus.

The fact that unripe banana is rich not only in membrane phospholipids and pectin, but also in cellulose, was the reason why we found it interesting to study its protecting effects in different rat models. Again, in collaboration with Kåre Larsson we induced stomach ulcers [99] with the methods described above. Banana suspension was administered 30–45 min before the induction of gastric ulcers by alcohol or indomethacin. Only administration of the banana suspension led to 50% reduction in ulcer size both in the acute alcohol and indomethacin models, but in the chronic indomethacin ulcer model a much less pronounced and only temporary protective effects were obtained. A complete protection against development of acute ulcers however, could be achieved when pectin and polyunsaturated phosphatidylcholine were used in concentrations higher than those normally occurring in the fruit.

The influence of the high content of cellulose in banana has not been studied so far; nor the effects of its natural antioxidants. It should not be disclosed that protective capacity of the banana fruit is not confined to only one or two ingredients but rather due to interaction of several active components. This should be the object of future studies.

The intestinal mucosa

The rate of turnover of intestinal mucosa is high, and most of the mucosa cells are renewed within a three-day cycle. Under normal conditions appr. 350 g of intestinal mucosa are broken down and replaced every day. In inflammatory conditions such as ulcerative colitis the amount of broken down mucosa is often doubled or more. The high activity of phospholipase A_2 induced breakdown of surfactants and membrane lipids creates a considerable need for exogenous supply of phospholipids, necessary for the regeneration process. Such a need is often not easy to meet as the Western diet is known to be deficient in polyunsaturated, and especially in omega-3, fatty acids, which is why one could anticipate positive clinical outcome from supplementing these lipids. As there is an increased turnover of mucus, supplementation of mucinous substances could be expected to further support the renewal of the mucosa and its protection layer. Together with Kåre Larsson, we, studied the effects of externally supplied phospholipids in rats with experimental ulcerative colitis [100]. An isolated colonic segment was exposed to a 4% acetic acid solution for 15 s, which resulted in a uniform mucosal inflammation and a sixfold increase in mucosal permeability. Instillation of 2.5–5 ml of either phosphatidylcholine or phosphatidylinositol in concentrations of 10 mg/ml, either immediately before or after the exposure to acetic acid, gave an almost optimal protection, leading to both reduced inflammation and reduction in the permeability defect otherwise obtained. As a matter of fact, phosphatidylcholine produced an almost complete, and the more polyunsaturated phosphatidylinositol a complete, protection. In a subsequent series of experiments [101] enteric bacterial translocation was produced by a 90% reduction of the liver size through liver resection. 1.5 ml/100 g body weight of a solution containing 10 mg/ml of either phosphatidylcholine (PC) or phosphatidylinositol (PI) was administered with a gastric tube both 12 h and 30 min before the liver reduction. PC prevented almost totally, and the more polyunsaturated (omega-3 containing) PI prevented totally, translocation of enteric bacteria to mesenteric lymph nodes, systemic circulation and portal vein. Furthermore, both PC and PI prevented the postoperative decrease in intestinal mucosal mass and enterocyte protein content, otherwise seen. In another series of experiments using the same model a nonionic surfactant, ethylhydroxyethyl cellulose (EHEC) was administered 2 h before the liver reduction [102]. Whereas bacterial overgrowth and translocation, increased bacterial adherence, diminished intestinal and mucosal mass were observed in the control animals, these changes were totally prevented after pretreatment with EHEC. In an in vitro study [102] where one ml of a 1.68% EHEC solution was added to the medium, inhibition of both bacterial growth and DNA synthesis could be observed. Simultaneously, changes in hydrobiology, decrease in bacterial surface hydrophilicity and hydrophobicity and increased surface neutrality, were observed after 1–3 h, changes which might explain a reduced bacterial ability to proliferate, to attach to and to invade enterocytes. As bacterial translocation is more likely to be depending on the characteristics of the invading bacterial species than on the immune situation of the host, this observation has the potential of being a very important step towards understanding and prevention of translocation.

Patients with inflammatory diseases like ulcerative colitis are known to have a significantly reduced mucin polymer content in the adherent colonic mucosa [103], resulting in these patients in a thinner and discontinuous colonic mucosal barrier, paralleled by an elevation in mucolytic proteinase activity. It seems logical to try a local supply of mucus-like polymers, alone or in combination with phospholipids, in order to prevent these changes. Proteoglucans like heparin are known to have anti-inflammatory effects, including inhibition of neutrophil adherence to endothelium, inhibition of neutrophil elastase (essential for penetration of the endothelium), inhibition of local formation of thrombin, and binding of chemokines; for review see ref. [104]. Intravascular supply of heparin has in recent studies, alone or in combination with sulfasalazine, led to most impressive results, with remission of the disease observed in almost all cases within 2–10 weeks [105–107]. The mechanisms for this action is not fully understood. One explanation could be that heparin inhibits the ability of human secretory type II phospholipase A2 to hydrolyze phospholipids [108], thereby supporting a maintained barrier function. Another explanation could be that heparin in itself is a part of an improved barrier function. Under all circumstances it is reasonable to assume that significant clinical effects can be obtained by local supply of bioactive, mucosa-adhesive and gelforming polymers, applied alone or in combinations.

Protection of synovia-covered surfaces

Lubrication of joints and tendons are of the greatest importance for their function. Since Meyer et al. [109] in 1939 indentified the most dominant viscous component as hyaluronic acid (HA), the function of this polymer under normal and pathological conditions has received large attention. The structure, physical chemical properties and biology of HA has recently been extensively reviewed by Laurent [110]. HA has the ability to form a three-dimensional network of entangled coils, which form a gel-like structure. It has been shown recently [111] that specific proteins, such as hyaladherins and also others, bind large numbers of single hyaluronan molecules together to form gigantic aggregates. During the last 25 yr attempts have been made to improve the lubricating ability of human and animal joints by intra-articular administration of HA, especially in cases of traumatic and degenerative arthropathies, and effects lasting for several months after a single injection have frequently been reported [112, 113]. However, the fast disappearance of intra-articularly injected HA [114–116] – only 33% remains in rabbit knee joints after 24 h – seems more or less to exclude the possibility of this agent simply replacing depolymerized endogenous HA, thereby improving synovial fluid viscoelasticity and joint lubrication [117]. Furthermore, it has become increasingly evident that the claim often made that high molecular weight HA has therapeutic advantages does not stand up to scientific scrutiny [118]. These are the reasons why other important biological functions of HA have come into focus. Among these are the ability of HA to inhibit leukocyte and mononuclear cell phagocytosis, to inhibit adherence and mitogen-induced proliferation, to inhibit degradation of phospholipids and production of prostaglandins and leucotreines, its antioxidative and its analgesic effects – for review see refs. [117, 118].

It was demonstrated by Bole [119] more than thirty years ago that lipids normally are present at articular surfaces in quantities comparable to those of the polysaccharides. Later studies showed that rinsing the articular surfaces with a lipid solvent increased joint friction by 150% while treatment with hyaluronidase seemed not to have any obvious influence – for review see ref. [120]. Hills demonstrated that a mixture of HA and liposomes offers a superior lubrication than liposomes and HA alone, suggesting a synergistic effect. It has also been shown that

appreciable amounts of phospholipids and metabolites thereof such as prostaglandins are present in arthritic joints [121]. Recent studies have shown capacity of highly purified and native HA preparations to form complexes with phospholipids, including the platelet activating factor [122]. Furthermore, interaction between these molecules alters both their configuration and rheological properties [122]. It is likely, as has been shown to be the case with mesothelium-covered membranes, that in the future also other polymers will prove to have equally good or better ability to lubricate synovia-covered surfaces.

Other applications

Matrix engineering and viscosurgery

Ever since Balazs developed and introduced the first non-inflammatory elastoviscous hyaluronan preparation in early 1970s [123] and proposed its use as a viscosurgical fluid it has been on several indication with the aim to expand tissues. Highly purified HA is used worldwide in ophthalmic surgery for appr. two decades with the aim to facilitate cataract removal and intraocular lens implantation – for review see Balazs [124]. HA does not only facilitate visualization and help manipulating the tissues, it is also claimed to build a barrier to blood, fibrin and invading cells, and help controlling postsurgical adhesions and scar formation [125]. Although considered a great progress in the past, it is obvious today that HA is far from the perfect tool as it, due to its water-soluble character, has a rather short tissue residence time. This largely limits its use in the control of postsurgical healing. Attempts to overcome this by modifying the HA molecule has so far been met with limited success. Although in the past few attempts have been made to replace HA with polymers of other origin or to combine different polymers, this is likely to be increasingly tried in the years to come.

Conclusions

Extensive research during the last two decades has convincingly demonstrated that it is possible to recondition cellular and tissue membranes in most parts of the body, thereby making them resistant to inflammation and ulceration and prevent translocation of microbes and toxins through the tissues. By doing so it has also been possible to reduce and sometimes totally eliminate unwanted sequelae of tissue injuries and inflammation, such as scar formation and adhesions. The ability to form gel and to adhere to tissues seem to be the most important characteristics of the polymers proven to be effective. Polymers with these

Progr Colloid Polym Sci (1998) 108:21–33
© Steinkopff Verlag 1998

characteristics have proven also to be useful for drug delivery to various tissues.

So far most of the interest has focused on natural polymers of proteoglucan nature. There is, however, a growing interest to also try other polymers, either natural or slightly modified. Several of the currently investigated polymers have the potential of being as effective or even more effective than hyaluronan, so far the golden standard. One should remember that nature is complex and most of the functions in the body depend on interplay between several compounds. It is likely that in the future the most effective protection will be obtained from combining substances with different biological functions. Not only additive but potentiating effects can be expected.

It should also be remembered that the body harbors appr. 1.2 kg of commensal – "protective" microbial flora. This flora has been shown to be an important part in the barrier function of the gastrointestinal tract [126–131], but are also known to be of importance for the integrity of the respiratory organs and the skin. Some microbes are known to produce surfactants, others to produce polymers. Future research will show if endogenous or supplied microbes can be used to further build barrier and membrane protection. These novel thoughts are shared with Kåre Larsson.

References

1. Marriott C, Gregory NP (1990) Mucus physiology and pathology. In Lenaerts V, Gurny R (eds) Bioadhesive Drug Delivery Systems, pp 1–23. CRC Press, Boca Raton, FL
2. Peppas NA, Buri PA (1985) J Contr Rel 2:257–275
3. Gu JM, Robinson JR, Leung SHS (1988) CRC Crit Rev Ther Drug Carrier Systems 5:21–67
4. Tsuchida E, Abe K (1982) Advances in Polymer Science 45:1–130
5. Mortazavi SA, Smart JD (1994) J Pharm Pharmacol 46:86–90
6. Allen A, Cunliffe WJ, Pearson JP, Sellers LA, Ward R (1984) Scand J Gastroenterol 19(suppl):101–113
7. Kerr LJ, Kellaway IW, Rowlands C, Parr GD (1990) The influence of poly(acrylic) acids on the rheology of glycoprotein gels. In Proc Int Symp Cont Rel Bioact Mater 17:122–123
8. Montazavi SA, Smart JD (1994) J Pharm Pharmacol 46:86–90
9. Morishita M, Morishita I, Takayama K, Machida Y, Nagai T (1993) Biol Pharm Bull 16:319–321
10. Tokomura T, Tsushima Y, Machida M, Kayano M, Nagai T (1993) Biol Pharm Bull 16:319–321
11. Balant LP, Doelker E, Buri P (1990) Eur J Drug Metab Pharmacokinet 15:143–153
12. Levin MS (1993) J Biol Chem 268:8267–8276
13. Artursson P, Lindmark T, Davis SS, Illum L (1994) Pharm Res 11:1358–1361
14. Schipper NGM, Vårum KM, Artursson P (1996) Pharmaceutical Research 13:1686–1692
15. Smith PL, Wall DA, Gochoco CH, Wilson G (1992) Adv Drug Deliv Rev 8:253–290
16. Swensson ES, Curatola WJ (1992) Adv Drug Deliv Rev 8:39–92
17. Kurihari-Bergstrom T, Good WR, Feisullin S, Signor C (1991) J Control Rel 15:271–278
18. Vermeer BJ (1991) J Contr Rel 15:261–266
19. McBurney EJ, Noel SB, Collins JH (1989) J Am Acad Dermatol 20:508–510
20. Nigalaye AG, Adusumilli P, Bolton S (1990) Drug Dev Ind Pharm 16:449–467
21. Nakatsuka S, Andrady AL (1992) J Appl Polym Sci 44:17–28
22. Hassan EE, Parish RC, Gallo JM (1992) Pharm Res 9:390–397
23. Chandy T, Sharma CP (1993) Biomaterials 14:939–944
24. Thacharodi D, Panduranga Rao K (1993) Chem Technol Biotechnol 58:177–81
25. Thacharodi D, Panduranga Rao K (1993) Int J Pharm 96:33–39
26. Thacharodi D, Panduranga Rao K (1995) Biomaterials 16:145–148
27. Ebel JP (1990) Pharm Res 7:848–851
28. Eldridge JH, Hammond CJ, Muelbroek JA, Staas JK, Gilley RM, Tice TR (1990) J Control Res 11:205–214
29. Regen SL, Czech B, Singh A (1980) J Am Chem Soc 102:6638–6640
30. Hub HH, Hupfer B, Koch H, Ringsdorf H (1980) Angew Chem Int Ed Engl 19:938–940
31. Okada J, Cohen S, Langer R (1997) Pharmaceutical Research 12:576–582
32. Lehr CM, Bouwstra JA, Schacht EH, Junginger HE (1992) Int J Pharm 78:43–48
33. Illum L, Farraj NF, Davis SS (1994) Pharm Res 11:1186–1189
34. Rentel CO, Lehr CM, Bouwstra JA, Leussen HL, Junginger HE (1993) Proc Control Release Soc 20:446–447
35. Artursson P, Lindmark T, Davis SS, Illum L (1994) Pharmaceutical Research 11:1358–1361
36. Smedsröd B, Pertoft H, Gustafson S, Laurent TC (1990) Biochem J 266:313–327
37. Bengmark S (Ed) (1989) The Peritoneum and Peritoneal Access. Wright, London
38. Miserocchi G, Agostino E (1971) J Appl Physiol 30:208–213
39. McGowen L, Pitkow HS, Davis RH (1976) Experientia 32:314–315
40. Ziegler C, Torchia M, Grahame GR, Ferguson IA (1989) Perit Dial Int 9:47–49
41. Castor CW, Naylor B (1967) Cancer 20:462–466
42. Yung S, Coles GA, Williams JD, Davies M (1994) Kidney Internat 46:527–533
43. Heldin P, Laurent TC, Heldin CH (1989) Biochem J 258:919–922
44. Honda A, Iwai T, Mori Y (1989) Res Commun Chem Pathol Pharmacol 1014:305–312
45. Ellis H (1983) Br J Surg 69:241–243
46. Weibel MA, Majno G (1973) Am J Surg 126:345–355
47. Christen D, Buchmann P (1991) Hepatogastoenterology 38:283–286
48. diZerega GS (1994) Fertil Steril To 61:219–235
49. Grahame GR, Torchia MG, Dankewich KA, Ferguson IA (1985) Perit Dial Bull 5:109–111
50. DiPaolo N, Bouncristiani U, Capotondo L, Gaggiotti E, DeMia M, Rossi P et al (1986) Nephron 44:365–370
51. Rozga J, Andersson R, Srinivas U, Ahrén B, Bengmark S (1989) Nephron 52:134–38
52. Ar'Rajab A, Ahrén B, Rozga J, Bengmark S (1991) J Surg Res 50:212–215

53. Snoj M, Ar'Rajab A, Ahrén B, Bengmark S (1992) Br J Surg 79: 427–429

54. Snoj M, Ar'Rajab A, Ahrén B, Larsson K, Bengmark S (1993) Res Exp Med 193:117–22

55. Ar'Rajab A, Snoj M, Larsson K, Bengmark S (1995) Eur J Surg 161: 341–344

56. Bengmark S (1997) Peritoneal conditioning: The role of the supply of natural membrane lipids. In diZerega GS et al (eds) Pelvic Surgery, Adhesion Formation and Prevention, pp 103–113. Springer, New York

57. Kappas AM, Barsoum GH, Ortiz JB, Keighley MRB (1992) Eur J Surg 158:33–35

58. Treutner KH, Klimaszewski M, Bertram P, Schumpelick V (1994) Adhäsionsprophylaxe mit Lipidverbindungen. Langenbecks Arch Chir (Suppl), Springer, Berlin Heidelberg New York, pp 45–48

59. Treutner KH, Bertram P, Klimaszewski M, Winkeltau G, Schumpelick V (1995) Eur Surg Res 27(Suppl 1): 89–90

60. Treutner KH, Bertram P, Schumpelick V (1997) Experimental prevention of peritoneal adhesions in general surgery. In diZeraga GS et al (eds) Pelvic Surgery, Adhesion Formation and Prevention. Springer New York, pp 71–78

61. Urman B, Gomel, Jetha N (1991) Fertil Steril 56:563–567

62. Urman B, Gomel V. Fertil Steril 56:568–570

63. Burns JW, Skinner K, Colt J, Sheidlin A, Bronson R, Yaacobi Y et al (1995) J Surg Res 59:644–652

64. Mitchell JD, Lee R, Hodakowski GT, Neya K, Harringer W, Valeri R, Vlahakes GJ (1994) J Thor Cardiovasc Surg 107:1481–1488

65. Hagberg L, Gerdin B (1992) J Hand Surg 17A:935–941

66. Elkins TE, Bury RJ, Ritter JL, Ling FW, Ahokas RA, Homsey CA, Malinak LR (1984) Fertil Steril 41: 926–928

67. Elkins TE, Ling FW, Ahokas RA, Abdella TN, Homsey CA, Malinak LR (1984) Fertil Steril 1984:929–932

68. Peck LS, Quigg JM, Fossum GT, Goldberg EP (1995) J Invest Surg 8:337–348

69. Wurster SH, Bonet V, Mayberry A, Hoddinott M, Williams T, Chaudry IH (1995) J Surg Res 59:97–02

70. Becker JM, Dayton MT, Fazio VW, Beck DE, Stryker' SJ, Wexner SD et al (1996) J Am Coll Surg 183: 297–306

71. Matsuda T, Mogh addam MJ, Miwa H, Sakurai K, Iida F (1992) ASAIO J 38:M154–M157

72. Hingham PA (1990) Composition containing derivates of chitin for preventing adhesions. European patent EP 426 368 A2

73. Kennedy R, Costain DJ, McAlister VC, Lee TDG (1996) Surgery 120:866–870

74. Tanford C (1980) The Hydrophobic Effect: Formation of Micelles and Biological Membranes, 2nd ed. Wiley, New York

75. Lindman B, Wennerström H (1980) Top Curr Chem 87:1–83

76. Thuresson K (1996) Solution properties of hydrophobically modified polymers, Thesis, Lund University

77. Falk K, Holmdahl L, Halvarsson M, Larsson K, Lindman B, Bengmark S. Polymers that reduce adhesion formation. Patent application number: Sweden 9702698-3

78. Goldberg EP (1997) Tissue-protective solutions and films for adhesion prevention. In diZerega GS et al (eds) Pelvic Surgery, Adhesion Formation and Prevention. Springer, Heidelberg, pp 79–92

79. diZerega GS (1997) Use of adhesion prevention barriers in pelvic reconstructive and gynecological surgery. In diZerega GS et al (eds) Pelvic Surgery, Adhesion Formation and Prevention. Springer, Heidelberg, pp 188–209

80. Slomiani BL, Sarosiek J, Slomiany A (1987) Dig Dis 5:125–145

81. Sarosiek J, Slomiani A, Takagi A, Slomiany BL (1984) Biochem Biophys Res Commun 118:523–531

82. Murty VLN, Sarosiek J, Slomiany A, Slomiani BL (1984) Biochem Biophys Res Commun 121:521–529

83. Slomiany A, Jozwiak Z, Takagi A, Slomiany BL (1984) Arch Biochem Biophys 229:560–567

84. Spychal RT, Marrero JM, Saverymuttu SH, YuCW, Corbishley CM, Northfield TC (1990) Gastroenterology 98:1250–1254

85. Goggin PM, Northfield TC, Spychal RT (1991) Scand J Gastroenterol 26(suppl 181):65–73

86. Orchard JL, Bickerstaff CA (1986) Abstract Gastroenterology 90:1573

87. Eliakim R, Karmeli F, Razin E, Rachmilewitz D (1979) Gastroenterology 95:1167–1172

88. Wardle TD, Hall L, Turnberg LA (1996) GUT 38:355–361

89. Raab Y, Sundberg Ch, Hällgren R, Knutsson L, Gerdin B (1995) Am J Gastroenterol 90:614–620

90. Hillingsö J, Kjeldsen J, Laursen LS et al. (1995) Clin Pharm Ther 57:335–341

91. Ferraris I, Karmeli F, Eliakim R, Klein J, Fiocchi C, Rachmilewitz D (1993) Gut 34:665–668

92. Butler BD, Lichtenberger LM, Hills BA (1983) Am J Physiol 244: G645–G651

93. Karlsson KA (1986) Chem Phys Lipids 42:153–172

94. Dunjic BS, Axelsson J, Ar'Rajab A, Larsson K, Bengmark S (1993) Scand J Gastroenterol 28:89–94

95. Tovey FI (1990) Curr Opin Gastroenterol 6:891–893

96. Best R, Lewis DA, Nasser N (1984) Br J Pharmac 82:107–116

97. Hills BA, Kirwood CA (1989) Gastroenterology 97:294–303

98. Bengmark S, Larsson K, to be published

99. Dunjic BS, Svensson I, Axelsson J, Ar'Rajab A, Larsson K, Bengmark S (1993) Scand J Gastroenterol 28: 894–898

100. Fabia R, Ar'Rajab A, Willén R, Andersson R, Ahrén B, Larsson K et al (1992) Digestion 53:35–44

101. Wang XD, Andersson R, Soltesz V, Wang WQ, Ar'Rajab A, Bengmark S (1994) Scand J Gastroenterol 29: 1117–1121

102. Wang XD, Guo W, Wang Q, Soltesz, Andersson R (1995) J Invest Surg 8:65–84

103. Rankin BJ, Srivastava D, Record CO, Pearson JP, Allen A (1995) Biochemical Society Transactions 23:104S

104. Zavgorodniy LG, Mustyats AP (1982) Klin Med (Moskva) 60:74–80

105. Zahernakova TV, Kashmenskaya NA, Maltseva IV et al (1984) Soviet Med (Moskva) 1984:110–113

106. Gaffney PR, O'Leary JJ, Doylé CT, Gaffney A, Hogan J, Smew F et al (1991) Lancet 337:238–239

107. Gaffney PR, Doyle CT, Gattney A, Hogan J, Hayes DP, Annis P (1995) Am J Gastroenterol 90:220–223

108. Dua R, Cho W (1994) Eur J Biochem 221:481–490

109. Meyer K, Smyth EM, Dawson MH (1939) J Biol Chem 12:319–327

110. Laurent TC, Fraser JRE (1992) FASEB J 6:2397–2404

111. Toole BP (1990) Curr Opin Cell Biol 2:839–844

112. Grecomoro G, Martoran U, Dimarco C (1987) Pharmatherapeutica 5:137–141

113. Dixon ASJ, Jacoby RK, Berry H, Hamilton EB (1988) Curr Med M 11: 205–213

114. Brown TJ, Laurent UBG, Fraser JRE (1991) Exp Physiol 76:125–134

115. Fraser JRE, Kimpton WG, Pierscionek BK, Cahill RNP (1993) Semin Arthritits Theum 22(Suppl 1):9–17

116. Laurent UBG, Fraser JRE, Engström-Laurent A, Reed RK, Dahl LB, Laurent TC (1992) Matrix 12:130–136
117. Ghosh P (1994) Clin Exp Rheum 12:75–82
118. Aviad AD, Houpt JB (1994) J Rheumatol 21:297–301
119. Bole GG (1962) Arthritis Rheum 5:589–601
120. Hills BA (1989) J Rheumatol 16:82–91
121. Wise CM, White RE, Agudelo CA (1987) Semin Arthritis Rheum 16:222–230
122. Ghosh P, Hutadilok N, Adam N, Lentini A (1994) Int J Biological Macromolecules 16(5):237–244

123. Balazs EA (1971) Hyaluronic Acid and Matrix Implantation. Biotrics Inc, Arlington MA
124. Balazs EA (1989) The introduction of elastoviscous hyaluronan for viscosurgery. In Rosen ES (ed) Viscoelastic Materials: Basic Science and Clinical Applications. Proc 2nd Int Symp North Eye Inst, Pergamon Press, Oxford
125. Weiss C, Band P (1995) Clin Pod Med Surg 12:497–517

126. Bengmark S, Larsson K, Molin G (1994-95) Biotechnol Therapeut 5:171–194
127. Bengmark S, Jeppsson B (1995) J Parent Enter Nutr JPEN 19:410–415
128. Bengmark S (1995) Clin Nutr 15:1–10
129. Bengmark S (1998) GUT 42:2–7
130. Bengmark S, Nutrition, to be published
131. Bengmark S, Nutrition, to be published

Progr Colloid Polym Sci (1998) 108:34–39
© Steinkopff Verlag 1998

T. Nylander
T. Arnebrant
R.E. Baier
P.-O. Glantz

Interactions between layers of salivary acidic proline rich protein 1 (PRP 1) adsorbed on mica surfaces

T. Nylander (✉)
Physical Chemistry 1
Center for Chemistry
and Chemical Engineering
Lund University
P.O. Box 124
S-22100 Lund
Sweden
E-mail: Tommy.Nylander@fkem1.lu.se

T. Arnebrant · P.-O. Glantz
Department of Prosthetic Dentistry
Centre for Oral Health Sciences
Carl Gustafs Väg 34
S-21421 Malmö
Sweden

R.E. Baier
Department of Oral Diagnostic Sciences
University at Buffalo
Buffalo, NY
USA

Abstract Interactions between layers of acidic proline rich protein 1 (PRP 1) films adsorbed on mica surfaces were investigated using the interferometric surface force technique. The influence of the type of electrolyte present (NaCl and $CaCl_2$) as well as the effect of the sequential addition of 10% human mixed saliva was explored. In the presence of NaCl very weak adsorption was observed and the force curves resembled the ones in pure NaCl. Only after long equilibration a thin layer was adsorbed and the forces became entirely repulsive. In the presence of $CaCl_2$ a substantially thicker layer was adsorbed. After short adsorption times the long-range repulsive forces were replaced by an attractive force below a surface separation of about 120 Å, which caused the surfaces to slide into contact (approximately 40 Å from mica–mica contact). As the layer built up, the attractive force was reduced and the force curve showed an inflection point instead of a jump. Finally, the force became entirely repulsive and no adhesion was observed. The introduction of mixed whole saliva drastically changed the interaction, whose range became longer than expected for electrostatic repulsive forces under the experimental conditions. This indicates that steric forces then prevailed as the force curves were almost identical to those recorded between mica surfaces after adsorption from mixed whole saliva alone.

Key words Surface force measurements – mica surfaces – adsorption – proline rich proteins – saliva

Introduction

The ability of saliva to form oral films on solids is an important factor for the maintenance of oral health and surface integrity cf. [1]. Salivary proteinaceous macromolecules, ranging from short peptides to high molecular weight mucins and differing in physicochemical properties, are essential in this respect. Fractions worth noting are the so-called acidic proline rich proteins (PRPs) (molecular weights 11–16 kD) [2–5] shown to promote the binding of certain oral microorganisms [6, 7] and the high molecular weight glycoproteins (MG1 and MG2) (>1000 kD and 200–250 kD, respectively) [8, 9] which are associated with bacterial binding as well as protection and lubrication. The lubricating effect of salivary proteins is a relevant factor in the development of salivary substitutes for patients suffering from Sjögren's syndrome and similar diseases.

Late stages of films formed on tooth surfaces, referred to as dental plaque, can give rise to diseases such as caries and peridontal disease. The buildup of dental plaque has been suggested to be a consequence of specific (short-range) interactions between microorganisms and adsorbed salivary proteins and at a further stage between microorganisms within the biofilm [6, 7, 10, 11]. Another view is that the process is less specific and controlled by the system striving to minimize its free energy through adsorption/adhesion of appropriate surface active components from the liquid phase (saliva) [12–15].

Characterization of salivary protein adsorption onto oral surfaces is a subject frequently addressed in the literature cf. [2, 4, 15–19]. Regarding the mechanism of film buildup, informative parameters would, apart from amounts and isotherms, be the interaction between surfaces (particles) covered by adsorbed films of salivary proteins and its distance dependence. We have a previous study reported a purely repulsive interaction between films adsorbed from whole mixed human saliva [20].

The so-called acidic proline rich proteins constitute a major (20–30% of the total salivary protein) fraction of salivary proteins which, as the name implies, have a high content of proline. In fact this amino acid along with glycine and glutamine make up 70–88% of the total amino acid content [3, 5]. Usually, four major acidic PRPs are discussed and these are denominated PRP 1–4. PRP 1 and PRP 2 both consist of 150 residues (molecular weight 16 300) the only difference is that the residue in position 4 is aspargine in PRP 1 and aspartate in PRP 2. PRP 3 and 4 correspond exactly to the first 106 residues (molecular weight 11 000) of PRP 1 and 2, respectively. Alternatively PRP 1 and PRP 3 may be denoted protein C and protein A. The N-terminal end, in particular the first 30 residues, contains basically all negatively charged residues including two Ca-binding phosphoserines per molecule. Except for the C-terminal carboxy group the C-terminal end contains the positively charged amino acids and most of the proline, glutamine and glycine [3]. Acidic PRPs bind calcium with a strength that indicates that they are involved in the maintenance of an appropriate level of ionic Ca^{2+} in saliva. They also inhibit the formation of hydroxyapatite and adhere strongly to this mineral [2–4]. Furthermore, they are a major part of the acquired dental pellicle [21] and have been shown to selectively bind oral microorganisms such as *Streptococcus mutans* and *Actinomyces viscosus* [6, 7]. Consequently, this group of proteins is considered to be of prime importance in the buildup process of dental plaque. They have therefore attracted substantial interest in studies of the mechanism behind this process [4, 21] and in attempts to reduce its rate [10, 22].

Bearing this in mind it should be worthwhile to measure the interaction between films adsorbed from purified PRP 1 to determine the character of this interaction and whether it is affected by addition of whole saliva. In addition, the influence of the electrolyte present (NaCl or $CaCl_2$) was explored.

Materials and methods

Materials

PRP 1, a gift from Dr. Donald Hay of the Forsyth Dental Center, Boston, was isolated from human saliva and purified according to Hay and Moreno [23]. The protein was diluted in 1 mM NaCl or $CaCl_2$ to a final concentration of 17 µg/ml (1.05×10^{-6} moles/l). The pH of these solutions was 6.8.

10 ml of unstimulated whole saliva was donated during 10–15 min periods between 8 and 9 a.m. on the day of the measurement by a subjectively healthy 40 year old male found to be free from oral diseases at clinical examination. In a range of previously performed surface chemical studies saliva from the used test subject had demonstrated interfacial properties at air–liquid and solid–liquid interfaces well within the ranges of those of a large group of healthy subjects [17, 19, 24, 25]. No food or drink was consumed during a 2 h period prior to saliva collection. The collected saliva was immediately diluted in 1 mM NaCl or $CaCl_2$ to a final concentration of 10% (v/v), centrifuged for 30 min at 4000 g and injected in the measuring chamber of the surface force apparatus. The pH values of the solutions were determined to be 6.9 in NaCl and 6.5 in $CaCl_2$.

The water used was ion exchanged, distilled, and passed through a Milli Q water purification system (Millipore Corporation), giving water with a conductivity of 0.7 µS/cm and showing no bubble persistence. All glasswares were cleaned in a mixture of concentrated sulfuric and nitric acid, 1:1 (v/v), and thoroughly rinsed with pure water.

All chemicals used were of analytical grade and the sodium chloride was further purified by roasting for 4 h at 600 °C.

Methods

The interferometric surface force apparatus (SFA), Mark IV [26], was used to study the interactions between salivary films. The technique as been described in detail elsewhere [26–28]. Before each experiment, the surface force apparatus (fitted with a small volume chamber of about 30 ml) was dismantled and all inner parts were thoroughly cleaned in distilled water, ethanol and finally sprayed with ethanol and blown dry with ultra pure nitrogen. All operations (cleaning, assembly of the instrument, mica cleaving and gluing) were performed in a laminar flow cabinet under essentially dust-free conditions.

The mica–mica contact position was first measured both in dry nitrogen and in 1 mM sodium or calcium chloride solution and at least two reproducible force curves were recorded in salt solution. The surfaces were separated by about 0.2 mm and the chamber was emptied, making sure that a drop of liquid was left between the mica surfaces. The PRP 1-solution (~25 ml) was injected into the chamber and the surfaces were separated and brought together several times to ensure that the liquid residing between the surfaces was properly mixed. The same procedure was used when the PRP solution was replaced with salt solution and in all occasions involving replacement of the solution within the measuring chamber. Experiments were carried out according to the following sequences:

Sequence no. I:
1) Adsorption of PRP 1 from 1 mM NaCl solution.
2) Addition of 1 mM $CaCl_2$ to the PRP solution after 28 h of adsorption.*
3) Replacement of the PRP solution by pure 1 mM NaCl after 33 h of adsorption.
4) Introduction of 10% whole human saliva in 1 mM NaCl after 50 h.

Sequence no. II:
1) Adsorption of PRP 1 from 1 mM $CaCl_2$ solution.
2) Replacement of PRP solution with pure 1 mM $CaCl_2$ after 28 h.
3) Introduction of 10% whole human saliva in 1 mM $CaCl_2$ after 30 h.

All measurements were performed at $22.0 \pm 0.2\,°C$.

Results and discussion

The build up of the PRP layers on the mica surfaces in the presence of 1 mM NaCl and 1 mM $CaCl_2$ is illustrated in Figs. 1 and 2, respectively. It should be noted that the deviation between similar experiments was less than 10% and each force curve was repeated at least twice at two different positions.

As illustrated in Fig. 1, the force curve recorded in 1 mM NaCl is the one expected for clean mica surfaces in a 1:1 electrolyte [29, 30]. It demonstrates a good agreement with the DLVO theory [31, 32]. The introduction of PRP solution in the measuring chamber initially (after 30 min) led to a shift of the force curve towards larger surface separations. However, on further compression the surfaces moved into mica–mica contact. After 2 h, less force was needed to bring the surfaces into contact and the magnitude of the force had decreased somewhat, suggesting a decrease in the surface charge. It cannot be ruled out that the PRP preparation might contain some residual

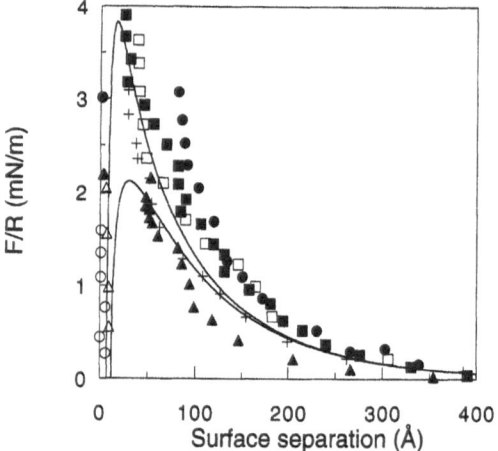

Fig. 1 The normalized force versus distance between mica surfaces in 1 mM NaCl (+) and 30 min (●), 2 h (▲), 20 h (■) after the addition of 17 μg/ml (1.05×10^{-6} M) PRP in 1 mM NaCl. The force curves recorded on decompression are also inserted and are indicated by the corresponding unfilled symbols. The solid line represent the DLVO fit using a Debye length, κ^{-1}, of 97 Å and a surface potential, Ψ, of 80 mV

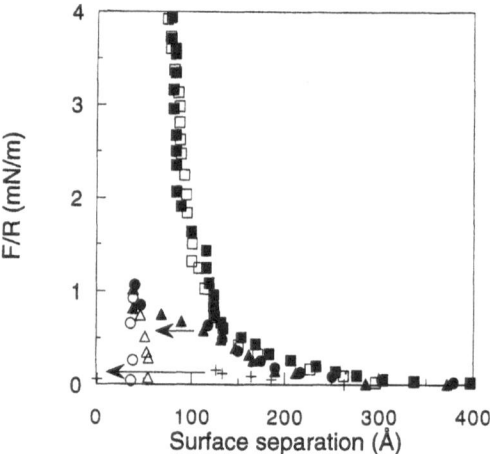

Fig. 2 The normalized force vs. distance between mica surfaces in 1 mM NaCl (+) and 30 min (●), 2 h (▲), 20 h (■) after the addition of 17 μg/ml (1.05×10^{-6} M) PRP in 1 mM NaCl. The force curves recorded on decompression are also inserted and are indicated by the corresponding unfilled symbols. The arrows indicate when the surfaces jump into contact (adhesive force)

calcium ions which can interact with the surface and decrease the net-charge. The adhesive force recorded on decomposition after 30 min and 2 h were, however, similar to the one observed in pure NaCl. This behavior indicates that the adsorbed layer was squeezed out from the confinement between the surfaces or that no

---

* In order to minimize concentration gradients of Ca^{2+} in the vicinity of the mica surface, this was carried out by emptying the chamber according to the procedure described above and adding an appropriate amount of $CaCl_2$ solution. The solution was then mixed and reinjected into the measuring chamber.

adsorption had taken place. The force curve recorded 20 h after injection of PRP, indicates some adsorption as the force now showed almost no hysteresis, i.e. the force curve recorded at decompression was similar to the one recorded at compression. The onset of the steric wall seems to be located at a surface separation of 20 Å from mica–mica contact. This layer can hardly be more than a protein monolayer thick and therefore points to a low affinity of PRP for the mica surface under the experimental conditions. An extremely flat, high density adsorbed film was observed by Christersson and Dunford for Human Parotid Saliva (HPS) on hydrophobed germanium [33] and by Baier et al. for this preparation of PRP on both hydrophilic and hydrophobic germanium [34].

The force curves recorded in 1 mM $CaCl_2$ were markedly different (Fig. 2). First, the recorded double layer force was weak and the surface carried a low surface charge, as expected and shown experimentally. Addition of PRP caused, however, a substantial increase in the repulsive force. At short equilibrium times (30 min), the repulsive force was replaced by an attractive force at about 120 Å surface separation which made the surfaces slide into contact at about 40 Å from mica–mica contact and this distance did not change upon further compression. This clearly indicates the presence of an adsorbed layer. Since the long-range repulsive forces increased in the presence of the adsorbed layer it is tempting to suggest that the adsorbed layer increased the charge of the mica surface. Upon further increase in the equilibrium time (2 h) the surfaces were no longer sliding into contact spontaneously. Instead a plateau was obvious at a surface separation of about 120 Å. Furthermore, the contact was still adhesive. After about 20 h the surface force curve was purely repulsive and without noticeable hysteresis. The onset of the steric wall had moved outwards to about 100 Å and did not move further at increased adsorption times. This steady state or equilibrium conditions are judged to represent formation of a complete adsorbed layer.

As described above, two different methods were used to form the adsorbate in the presence of calcium. In the first experiment the PRP was adsorbed from a 1 mM NaCl solution for 20 h before addition of $CaCl_2$ solution to a concentration of 1 mM, whilst in the second experiment PRP was directly adsorbed from a 1 mM $CaCl_2$ solution. As shown in Fig. 3 almost the same force curve was observed for PRP irrespective of the procedure. Further, the force curves upon approach were reminescent of the one recorded on the first approach in NaCl but different in that they were entirely repulsive and showed no hystereses (Fig. 3).

The effect of Ca^{2+} can be related to partial neutralization of the negatively charged mica surfaces and/or to Ca^{2+} binding to the protein. As discussed above, PRP 1 is known to possess two phosphoserines per molecule which

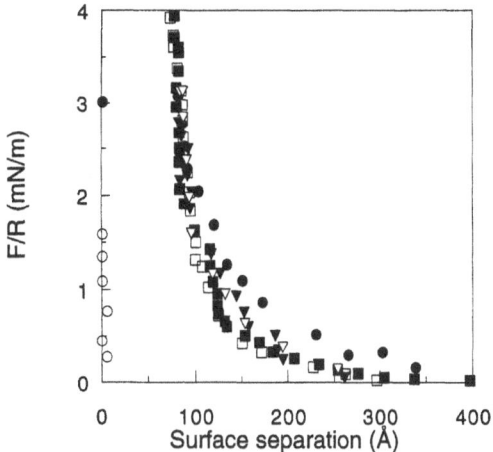

Fig. 3 The normalized force vs. distance between mica surfaces in 1 mM NaCl after 30 min (●) of adsorption from a 17 μg/ml $(1.05 \times 10^{-6}$ M) PRP solution and after 28 h under the same conditions followed by 30 min adsorption from a PRP solution of the same concentration in 1 mM $CaCl_2$ (▼). For comparison the curve after 20 h of adsorption from 17 μg/ml $(1.05 \times 10^{-6}$ M) PRP in 1 mM $CaCl_2$ (■) is inserted. The force curves recorded on decomposition are also inserted and are indicated by the corresponding unfilled symbols

are probably responsible for the reported binding capacity of two moles of Ca^{2+} per mole protein and located in the N-terminal, apatitic binding, part of the protein [3, 5]. With respect to the reported affinity of PRP to hydroxy and fluoroapatite [2–4] it should be mentioned that mica does not contain Ca but has a surface energy $(130–170$ mJ/m^2) [35] in the same order of magnitude as fluoroapatite $(95$ mJ/m^2) [36] and enamel $(77 \pm 10–87 \pm 6$ mJ/m^2) [37, 38]. Incorporation of Ca^{2+} in the adsorbed layer usually has a screening effect on the electrostatic repulsion between the surface and the protein as well as between the adsorbed molecules and thus allows the molecules to pack more densely at the interface.

In the presence of Ca^{2+}, the initial steep steric force sets in at a separation of about 120 Å after short adsorption times (30 min and 2 hours) but the layers could be compressed further to a separation of approximately 40 Å where the hard wall steric contact sets in. This indicates the presence of a biphasic structure with an inner dense phase close to the mica surface and a more open compressible fraction present closer to the aqueous phase. With time, the hard wall moved outwards, indicating an increased density in the outer fraction. Similar structural features of adsorbed layers have been observed for an amphiphilic milk protein, β-casein [39], which possess many common structural features with the PRP 1, for example the presence of phosphoserines and a highly segregated distribution of the charged amino acids. It is also interesting to note that the calcium phosphate nanoclusters of casein micelles are stabilized by interaction with the phosphorylated sequences of the calcium sensitive caseins, as β-casein [40]. Thus, it might be speculated that the PRP-fractions have a similar stabilizing function for the micelle-like structures reported to occur in saliva [41–44].

The interaction after sequential addition of 10% saliva in a 1 mM NaCl solution to a layer of PRP adsorbed from a 1 mM CaCl$_2$ solution is shown in Fig. 4A. For comparison the force curves recorded in pure PRP solution containing 1 mM NaCl or 1 mM CaCl$_2$ are also inserted. The addition of the saliva solution gave rise to a steric wall at a substantially larger surface separation as compared with the original PRP solution. This force curve in fact resembles the one recorded after adsorption from pure salivary solutions (Fig. 4A, [20]). The change in the interaction upon saliva addition might result from the formation of a mixed layer of PRP and other salivary components or even a replacement of PRP. The replacement of initially adsorbed protein components by other more surface active, and usually higher molecular weight proteins is a well-known phenomenon frequently discussed in the literature cf. [45, 46].

When comparing the long-range repulsive-interactions one has to bear in mind that they are recorded at different ionic strengths, which based on the salt concentrations for PRP in 1 mM NaCl are 0.001, 0.0025 and 0.009, respectively. Pure electrostatic repulsive forces are expected to be proportional to $e^{-\kappa D}$, when κ^{-1} is the classical Debye length. The Debye lengths decrease in the order PRP in 1 mM NaCl ($\kappa^{-1} = 97$ Å), PRP in 1 mM CaCl$_2$ ($\kappa^{-1} = 56$ Å) and 10% saliva in 1 mM NaCl (in total 9 mM 1:1 electrolyte, $\kappa^{-1} = 32$ Å) assuming that the present salt contribute as 1:1 (NaCl) or 2:1 (CaCl$_2$) electrolytes. It should be noted that this calculation is valid only if the protein concentration is low compared to that of the added electrolyte. In other cases the proteins contribute significantly to the measured electrostatic decay length as described by the theoretical model for asymmetric electrolytes derived by Ninham and Mitchell [47, 48]. In these cases the obtained decay lengths can be considerably shorter than the classical Debye length [49]. The net-charge of PRP1 is about -8 at physiological pH [23]. Thus under the given conditions, 1 μM PRP and 1 mM NaCl, the expected decay length will be 94 Å when the protein charge is taken into account. This is very close to the value for pure 1 mM NaCl indicating that the effect of the protein can be neglected in this case. For a complex mixture of proteins as saliva, a similar calculation is very difficult, but the real value of the decay length is expected to be lower than the one estimated from the concentration of simple electrolyte given above. In Fig. 4B, the force curves are plotted on a logarithmic scale and the linear part of the curves was fitted to an exponential force. The obtained slopes of the linear fits correspond to the follow-

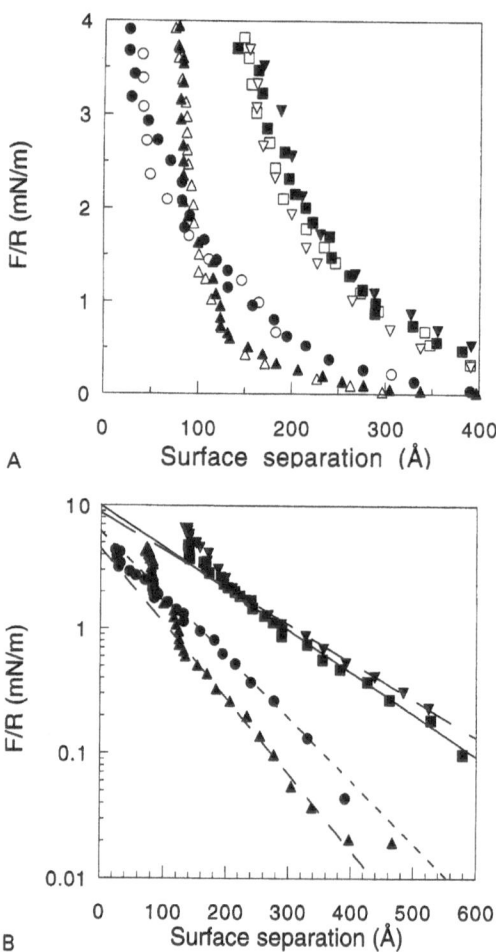

Fig. 4 The normalized force versus distance between mica surfaces 20 h after the addition of 17 μg/ml (1.05×10^{-6} M) PRP in 1 mM NaCl (●), and in 1 mM CaCl$_2$ (▲), as well as the curve recorded 20 h after the subsequent addition of 10% saliva to a PRP layer adsorbed from a 1 mM CaCl$_2$ solution (■) and in pure 10% saliva solution (▼). (A) shows the force curves in linear scale. The force curves recorded on decompression are also inserted and are indicated by the corresponding unfilled symbols. (B) The force curves recorded on compression are shown in logarithmic scale. A linear fit was performed on the long-range part of the force and the resulting lines are inserted. An electrostatic repulsive force is roughly proportional to $e^{-\kappa D}$, where κ^{-1} is the Debye screening length. The obtained slopes of the linear fits correspond to the following values of κ^{-1}: 86 Å (- - -) for PRP in NaCl, 72 Å (– – –) for PRP in CaCl$_2$, 129 Å (——) for saliva added after PRP adsorption and 143 Å (— — —) for pure saliva solution

ing values of κ^{-1}: 86 Å for PRP in NaCl, 72 Å for PRP in CaCl$_2$, 129 Å for saliva when added for PRP adsorption and 143 Å for pure saliva solution. Bearing in mind the low magnitude of the force, which gives some uncertainty in the fitting, the values obtained for the PRP films roughly corresponds to an electrostatic repulsive force. However, as the slope of the long-range force in the presence of saliva was much smaller than expected, it can be concluded that the force was not purely electrostatic. As our earlier study demonstrates that the decay length did not change significantly with increased ionic strength, it can be concluded that steric forces prevail in salivary solutions. The interaction between such films is discussed in detail elsewhere [20] but essentially resembles the one obtained for interacting layers of mucins adsorbed on mica [50, 51] indicating a strong contribution of high molecular weight glycoproteins to the interaction.

In conclusion, our results indeed indicate that in the presence of Ca^{2+} the PRP-1 layer consists of a bottom densely packed fraction attached to the surface and a less dense (compressible) outer portion. These fractions may consist of the N- and C-terminal portions of the protein, respectively, as previously suggested for adsorbed layers on apatitic surfaces. With respect to the reported microorganisms binding capacity of the C-terminal portion it is of course interesting to speculate about the accessibility of the binding sequence to species entering the adsorbed layer from solution. The orientation featuring a lower density outer layer should allow orientational freedom and be more ideal for receptor–ligand type interactions. In the presence of whole saliva, however, it is still a challenge to envisage a mechanism for interaction between receptor ligands of approaching microorganisms and binding sites of adsorbed PRP given the long-range nature of the steric repulsion.

Acknowledgements Financial support from the Swedish Council for Forestry and Agricultural Research (SJFR), the Swedish Research Council for Engineering Sciences (TFR) and the Swedish Medical Research Council (MFR) (Grant no. B 96-25X-05712-17C) is gratefully acknowledged as also the support from the U.S. National Institute for Dental Research (Grant no. DE 07760 "Conformation of Dental Salivary Molecules").

References

1. Mandel ID (1986) J Dent Res 66:623–627
2. Hay DI (1967) Archs Oral Biol 12:937–946
3. Hay DI (1983) In Lazzari EP (ed) CRC Handbook of Experimental Aspects of Oral Biochemistry, pp 319–355
4. Moreno EC, Kresak M, Hay DI (1982) J Biol Chem 257:2981–2989
5. Bennick A (1987) J Dent Res 66:457–461
6. Gibbons RJ, Hay DI (1988) Infect Immun 56:439–445
7. Gibbons RJ, Hay DI (1989) J Dent Res 68:1303–1307
8. Tabak LA (1990) Crit Reviews in Oral Biology and Medicine 1:229–234
9. Loomis RE, Prakobphol A, Levine MJ, Reddy MS, Jones PC (1987) Arch Biochem Biophys 258:452–464
10. Gibbons RJ, Hay DI (1989) In: Switalski L, Höök M, Beachey E (eds) Molecular Mechanisms of Microbial Adhesion, Springer, New York, pp 143–163
11. Gibbons RJ (1996) J Dent Res 75:866–870
12. Baier RE (1970) In Manly RS (ed) Adhesion in Biological Systems. Academic Press, New York, pp 15–48
13. Baier RE, Meyer AE (1981) In Mittal KL (ed) Surface Energetics and Biological Adhesion, Int Symp on Physicochemical Aspects of Polymer Surfaces. Plenum Press, New York
14. Baier RE, Meyer AE, Natiella JR, Natiella RR, Carter JM (1984) J Biomed Mat Res 18:337–355
15. Glanto P-O, Baier RE (1986) J Adhesion 20:227–244
16. Baier RE, Glantz P-O (1978) Acta Odontol Scand 36:289–301
17. Vassilakos N, Arnebrant T, Glantz P-O (1992) Scand J Dent Res 100:346–353
18. Vassilakos N, Rundegren J, Arnebrant T, Glantz P-O (1992) Arch Oral Biology 37:549–557
19. Vassilakos N, Glantz P-O, Arnebrant T (1993) Scand J Dent Res 101:339–343
20. Nylander T, Arnebrant T, Glantz P-O (1997) Colloids Surfaces A, Physicochemical and Engineering Aspects 129–130:339–344
21. Bennick A, Chau G, Goodlin R, Abrams S, Tustian D, Madapallimattam G (1983) Archs Oral Biol 28:19–27
22. Perinpanayagam HER, van Wuyckhuyse BC, Ji ZS, Tabak LA (1995) J Dent Res 74:345–350
23. Hay DI, Moreno EC (1989) In Tenovou JO (ed) Human Saliva: Clinical Chemistry and Microbiology. CRC Press, Boca Raton, FL, pp 131–150
24. Adamczyk E, Arnebrant T, Glantz P-O (1997) Acta Odont Scand, in press
25. Sefton J, Arnebrant T, Glantz P-O (1992) Acta Odont Scand 50:221–226
26. Parker JL, Christenson HK, Ninham BW (1989) Rev Sci Instrum 60:3135–3138
27. Israelachvili JN (1973) J Colloid Interface Sci 44:259–272
28. Israelachvili JN, Adams GE (1978) J Chem Soc Faraday Trans 74:975–1001
29. Pashley RM (1981) J Colloid Interface Sci 85:531–546
30. Shubin VE, Kékicheff P (1993) J Colloid Interface Sci 155:108–123
31. Derjaguin BV, Landau L (1941) Acta Phys Chim USSR 14:633–662
32. Verwey EJW, Overbeek JTG (1948) Theory of the Stability of Lyophobic Colloids. Elsevier, New York
33. Christersson C, Dunford R (1991) Biofouling 3:237–250
34. Baier RE, Meyer AE, Dombroski DM, Nassar U, Merrick JM, Hay DI, Olivieri MP (1994) Surfaces in Biomaterials '94 Symposium Notebook, Surfaces in Biomaterials Foundation, Minneapolis, pp 17–22
35. Christenson HK (1993) J Phys Chem 97:12034–12041
36. Aning M, Welch DO, Royce BSH (1971) Phys Lett 37A:253–254
37. Busscher HJ, Retief DH, Arends J (1987) Dent Mater 3:60–63
38. Weerkamp AH, Uyen HM, Busscher HJ (1988) J Dent Res 67:1483–1487
39. Nylander T, Wahlgren NM (1997) Langmuir 13:6219–6225
40. Holt C, Sawyer L (1988) Protein Eng 2:251–259
41. Glantz P-O, Natiella JR, Vaughan CD, Meyer AE, Baier RE (1989) Acta Odont Scand 47:17–24
42. Glantz P-O, Wirth SM, Baier RE, Wirth JE (1989) Acta Odont Scand 47:7–15
43. Glantz P-O, Friberg SE, Wirth SM, Baier RE (1989) Acta Odont Scand 47:111–116
44. Rykke M, Smistad G, Rölla G, Karlsen J (1995) Colloids Surf B: Biointerfaces 4:33–44
45. Vroman L, Adams AL (1969) Surf Sci 16:438–446
46. Horbett TA, Brash JL (1987) In Brash JL, Horbett TA (eds) Proteins at Interfaces – Physicochemical and Biochemical Studies, Vol 343. American Chemical Society, Washington DC, pp 1–37
47. Mitchell DJ, Ninham BW (1978) Chem Phys Lett 53:397–399
48. Nylander T, Kékicheff P, Ninham BW (1994) J Colloid Interface Sci 164:136–150
49. Kékicheff P, Ninham BW (1990) Europhys Lett 12:471–477
50. Malmsten M, Blomberg E, Claesson PM, Carlstedt I, Ljusegren I (1992) J Colloid Interface Sci 151:579–590
51. Perez E, Proust JE (1987) J Colloid Interface Sci 118:182–191

Progr Colloid Polym Sci (1998) 108:40–46
© Steinkopff Verlag 1998

B. Forslind
L. Norlén
J. Engblom

A structural model for the human skin barrier

B. Forslind (✉)
Associate Professor
L. Norlén
EDRG, Medical Biophysics, MBB
Karolinska Institute
S-171 77 Stockholm
Sweden
E-mail: bosse@mango.mef.ki.se

J. Engblom
Food Technology
Lund University, Lund
and Bioglan AB, Malmö
Sweden

Abstract Based on the composition of lipids extracted from the horny layer of human skin a two phase structure organization has been proposed for the skin barrier. The dominant part represents lipids in crystalline (gel) state envisioned as domains surrounded by lipids in liquid crystalline state. On theoretical grounds this *domain mosaic model* can be shown to provide an effectively water tight skin barrier, still allowing a minute leakage of water necessary to keep the keratin of the corneocytes hydrated to a level that ensures plasticity of the horny layer. Perturbing the liquid crystalline phase, e.g. by introduction of penetration enhancers, can be shown to cause an increase in transdermal transport by several orders of magnitude.

Key words Lipid structure – skin barrier model

Skin barrier function and perspiratio insensibilis

The most important task for the human skin barrier is to create a water tight enclosure of the body to prevent water loss. Water homeostasis is a strict requirement for normal physiological function of the body as an uncontrolled loss of water results in drastic increase in salt concentrations with consequent harmful effects on cells and tissues.

The tissues of the body contain 60–80% water and the gross water content is mainly regulated by the kidneys. Normal water loss *at rest* amounts to approximately 1.5–2.5 l/d of which 0.5 l are lost via respiration, 1.0–1.5 l via the urinary tract, roughly 300 ml through the *perspiratio insensibilis* (the unnoticed water loss through the skin excluding the sweat glands). The rest is excreted via feces and sweat. If we focus on the perspiratio insensibilis and relate this to the surface area we find that it represents $<7 \text{ ml/m}^2/\text{h}$, i.e. a negligible amount, indeed unnoticeable without the aid of sensitive instruments [1]. Consequently, the normal skin can be regarded as effectively water tight. However, loss of barrier function in an area of 20×20 cm, e.g. the size of two palms, will put survival to hazard. This is the main concern at burn injuries. Due to the fact that man has a comparative large surface area in relation to the volume enclosed by the integument (Fig. 1) compared to most mammals the resulting volume-to-surface ratio is unfavorable.

The horny layer – stratum corneum

The horny layer represents the barrier proper since once this part of the skin has been removed substances diffuse freely into or out of the body system. A detailed study of the morphology of this layer reveals that the corneocytes provide a mechanical scaffold for the stacked bilayers of lipids which occupy the intercorneocyte spaces (Fig. 2a). The approximate dimensions of a roughly hexagonal corneocyte represent a thickness of 0.3 μm and a diameter of

Progr Colloid Polym Sci (1998) 108:40–46
© Steinkopff Verlag 1998

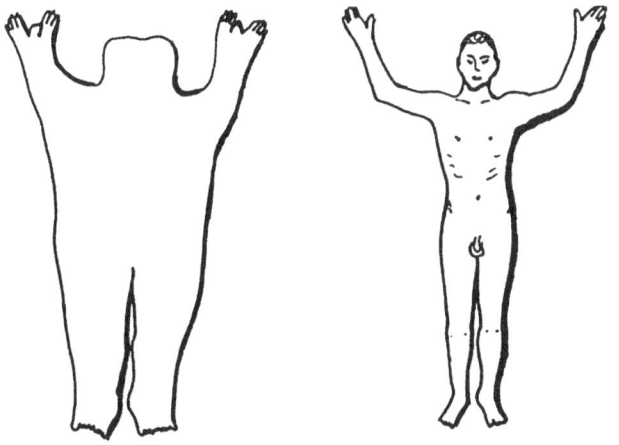

SKIN SURFACE OF MAN

SKIN SURFACE OF MICE

Fig. 1 Surface of man compared to that of mouse. About 50% of man's body surface resides over the limbs compared to approximately 10% for the mouse that can attain a practically spheroid form at rest thus optimizing surface-to-volume ratio

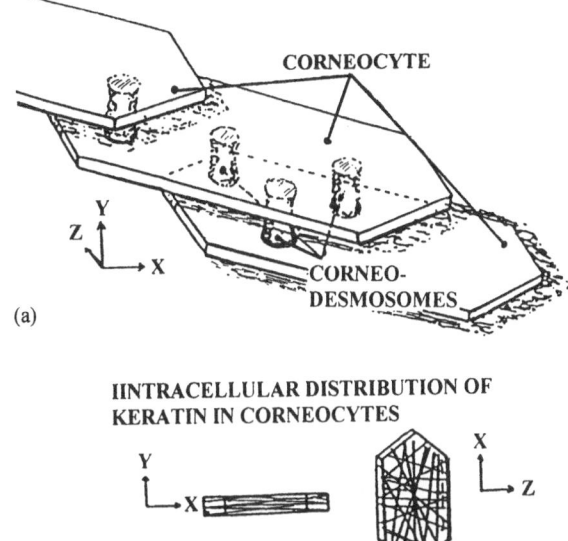

Fig. 2 (a) The organization of the corneocytes provides a scaffold for stacked bilayers of lipids in the intercellular space between corneocytes. (b) The dispersion of keratin fibrils in the plane of the skin surface results in an internal reinforcement that stabilizes the cell form in this plane

30 μm. The corneocytes maintain a mechanically stable form due to an internal reinforcement provided by keratin filaments that are oriented mainly in the plane of the very flat cell (Fig. 2b), these straight keratin filaments being anchored in the protein envelope of the corneocyte. Each corneocyte is mechanically coupled to its neighbors by special protein "rivets", desmosomes. When the horny layer swells this will take place effectively only in the vertical direction. In fact, swelling corneocytes Norlén et al. [2] have shown that the horizontal extension is about 1–3% whereas in the vertical direction the corneocytes may swell >25%. This ensures that the intercellular compartment where the lipid bilayers reside will suffer a minimum of distortion in the horizontal dimensions at swelling and also minimizes surface roughness on swelling which would impair the mechanical properties of the skin.

In the transmission electron microscope the stratum corneum is found to present a diversified image. The topmost part contains corneocytes with varying degrees of translucency that represents different degrees of mass loss. The intercorneocyte spaces contains patches of electron dense material representing the remains of the extracellular lipids. Loss of desmosome structure in the stratum corneum parallels these findings which are the result of enzymatic digestion by proteases and lipases present in the stratum corneum. The integrity of the corneocyte envelope is compromised in this process as was shown in an experimental penetration study by Boddé et al. [3] who demonstrated that the interior of the corneocytes in the topmost part of the stratum corneum contained the metal tracer used. This finding was not at hand in the lowermost layers!

In the deepest part of the stratum corneum the electron contrast is more or less homogeneous both with regard to the corneocytes as well as the intercellular lipids. As an approximation one may state that the stratum corneum thus can be divided into three layers, where the bottom third represents the intact barrier, the topmost third prepares the corneocytes for shedding and the intermediate part is a transient zone between the two extremes (cf. [4, 5]). Even if the intermediate layer thus is expected to have impaired barrier properties it still represents the two-fold lipid–water compartment and will therefore impede the penetration of hydrophilic and hydrophobic substances by partitioning effects.

Skin barrier research of the past

Contact allergy and irritative reactions in the skin as well as an interest in understanding how topically administered drugs penetrate the skin, focused the interest on the barrier properties of its outmost layer, the stratum corneum or the horny layer. Early research suggested that lipids played an important role for the barrier function, however, the first barrier model did not appear until 1975 when Michaels et al. presented what has been denominated the "brick and mortar" model [6]. This model envisioned the stratum corneum as a two-compartment structure where the lipids represented a hydrophobic pathway whereas the corneocytes with their content of hydrophilic keratin represented a hydrophilic one.

Based on this model Elias and his coworkers [7, 8] did pioneering work on skin barrier function which has had important effects on our present view on the barrier and barrier function. Eventually, it has been established that the intercorneocyte space is the true pathway for essentially all substances with the major exception for water which passes more or less freely through corneocytes [3]. However, the *brick and mortar model* is not a structural model but a conceptual one and does not take lipid structure into account as a means for explaining the properties of the human skin barrier. This lack *of a structure-function concept* may be one reason for the fact that numerous studies on penetration of substances through the skin barrier has been performed on animal models albeit the lipid composition of the horny layer from these animals is vastly different from that of the human skin barrier [9]. Thus, in recent years it has become clear that the only substitute for human skin in penetration studies is that of pig [10], an animal that in many respects is closely related to man, e.g. omnivore, similar basic immunology, naked skin with a lipid composition that closely mirrors that of human skin, etc.)

The human skin is unique in many aspects but the lack of a fur is a crucial characteristic. The isolation properties of a fur achieved by an effective trapping of thermally low conductive air gives physical and mechanical protection to the skin of mammals preventing loss of water and direct access to the skin by chemical substances. The presence or absence of a fur respectively is likely to be responsible for the conspicuous differences in lipid composition and content shown by furry mammals as compared to humans, "the naked ape".

A normal cell membrane which has to exhibit mechanically plastic properties contains lipids with an average chain length of 16 carbons or 18 carbons with a double bond. At normal physiological temperatures (approx. 20–40 °C) these carbon chains may be in a *liquid state* [11]. Double layers of such lipids in the liquid crystalline state

will allow water to pass through the membrane more or less freely [12] and a cell membrane is therefore not a barrier to water. On the other hand, bilayers in the close packed *crystalline (gel) state* will effectively bar the penetration of water molecules [13, 14]. The composition of the barrier lipids (Table 1) makes it obvious that the majority of the stratum corneum lipids will form crystal-(gel-) structures, hence provide a water tight enclosure.

There is a continuous shedding of corneocytes from the horny layer surface in response to the need of ensuring an intact barrier in spite of an always persisting environmental load on the integument. The process is supported by a flow of cells from the differentiating, metabolically active epidermis (Fig. 3), stratum basale to stratum

Fig. 3 Cross section of the cellular part of human skin, the epidermis. The size of the cross section is approximately 120 μm of which the stratum corneum takes 10 μm. However, the intact lipid barrier is found in the lower 1/3 of the stratum corneum. The cellular part of the skin is separated from the connective tissue part, the dermis, by a mechanical support, the basal lamina which allows free diffusion of nutrients and waste products from the epidermis

CROSS SECTION OF HUMAN EPIDERMIS

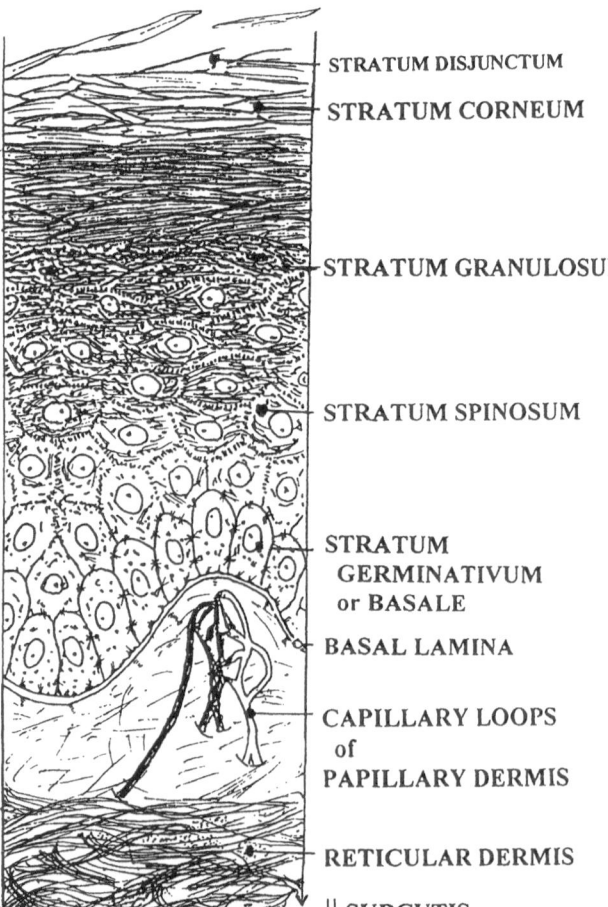

STRATUM DISJUNCTUM

STRATUM CORNEUM

STRATUM GRANULOSUM

STRATUM SPINOSUM

STRATUM GERMINATIVUM or BASALE

BASAL LAMINA

CAPILLARY LOOPS of PAPILLARY DERMIS

RETICULAR DERMIS

⇓ SUBCUTIS

Progr Colloid Polym Sci (1998) 108:40–46
© Steinkopff Verlag 1998

granulosum. The basal cells are standing on a very special collagenous sheet which is secreted by these keratinocytes. This sheet, the *basal lamina*, is anchored to the fine collagen fibrils and ground substance material of the supporting papillary dermis (Fig. 3). The basal cells divide in an orderly manner through a control of the cell division by presumably Ca^{+2}-triggered channels, so-called "gap junctions" that form interconnections between the basal cells. The progeny will subsequently be filled with fibrous keratin during the differentiation process that carry the cells towards the final destiny, the horny layer, the *stratum corneum*. In addition the epidermal keratinocytes synthesize glycosylated ceramides [15] which together with long chain fatty acids and cholesterol esters are stored in membrane enclosed bodies in the form of bilamellar sheets (Fig. 4). Eventually, at the border between the topmost cell layer of the so called viable epidermis, the *stratum granulosum*, and the lowermost stratum corneum cells, these lamellar bodies fuse with the cell membrane and their content is extruded into the extracellular space. Apparently, immediately after this process the cell membrane is exchanged for a protein envelope, the nucleus and the cytoplasmic nucleic acids are disintegrated and the resulting corneocyte contains essentially only the fibrous keratin. In fact, this process is so fast that it has not yet been recorded by the transmission electron microscope.

As stated before, the corneocytes are bound to their neighbors by strong "rivets", corneodesmosomes. Through these means the corneocytes of the horny layer form a scaffold into the intercellular spaces of which lipid bilayers are inserted as a continuous lipid phase. Between the hydrophilic head groups of apposed lipid bilayers there is a thin water sheath in direct contact with these head groups (Fig. 5). Thus the intercellular space of the stratum corneum contains both a hydrophilic and a hydrophobic (lipid) pathway [16].

The domain mosaic model

Our work has been based on the knowledge of the unique composition of the barrier lipids of the naked ape as compared to that of furry animals or genetically "manipulated" naked animal species [17–19]. Chromatographic analyses of the lipids extracted from human and mammal skins have revealed that the human skin is unique with regard to its high content of long chain ceramides [15, 20, 21]. In addition to the ceramides (mostly saturated) free fatty acids and cholesterol constitute the bulk of the barrier lipids.

The dominance of the long chain ceramides and fatty acids suggest that the bulk of barrier lipids reside in crystalline (gel) bilayer domains which are separated by an

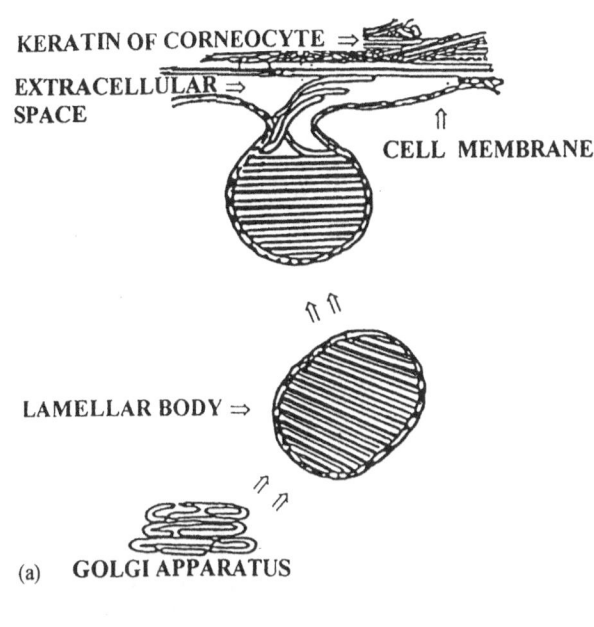

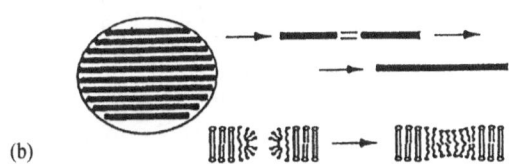

Fig. 4 (a) The lamellar bodies are produced by the Golgi apparatus of the cell machinery. Glycosylated ceramides dominate the content of these bodies and are organized in bilayers enclosed by a bilayer membrane that in the final stage will fuse with the stratum granulosum cell membrane to extrude the bilayer sheaths into the intercellular compartment. (b) It is plausible that these lipid units subsequently are joined by lipids in the liquid crystalline state (cf. [32]) (Fig. 4b taken from Ref. [32])

interdomain phase in the liquid crystalline state. Bearing in mind that transmission electron microscopy demonstrates that the intercorneocyte bilayers are stacked in multiple layers [4, 5] such mosaics of domains will constitute an essentially water tight structure. Any water molecule escaping from the body will have to suffer a meandering way out of the system. A random diffusion path in the interbilayer water sheath will occupy a comparatively long time compared to the vertical passage through a liquid crystalline area to reach the water sheath separating the passed bilayer from the next one. In fact Engström has estimated that if just a few percent of the interdomain areas are transformed, e.g. by a penetration enhancer to form a vertical channel, the over-all effect on penetration may increase one or more orders of magnitude [16].

With the present state of art the evidence for the domain mosaic model for the human skin barrier remains

Fig. 5 Bilayer organization of the stratum corneum lipids. In the corneocyte a protein envelope has replaced the lipid cell membrane of the cell of the viable epidermis and some lipids are covalently bound to this envelope (not represented here) Between the hydrophilic head groups of the stacked bilayers resides a water tablet within which diffusion transport may take place parallel to the skin surface

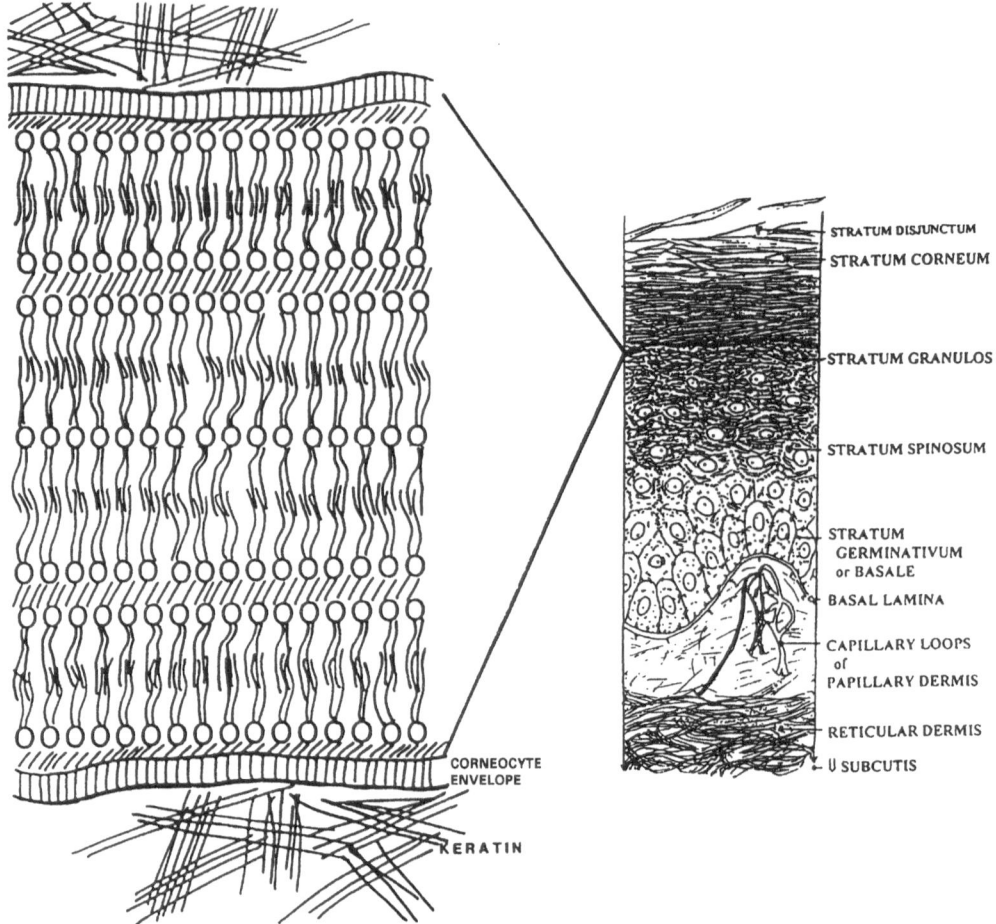

circumstantial. However, there are several reports giving evidence for the presence of a high degree of ordering of the humans skin barrier lipids, i.e. by NMR [22, 23], by differential scanning calorimetry [24–26] by infrared spectroscopy [27, 28], and by X-ray diffraction [29–31]. The X-ray diffraction studies give no direct support to the presence of a liquid crystalline phase or a swelling of the lipid phase on excessive hydration of stratum corneum. It should be born in mind that this crystallographic method actually is insensitive to minor perturbations in a mixed system, e.g. inclusions of low crystallographic order such as a liquid crystalline phase mixed with highly ordered crystalline (gel) domains in the present case. This means that if the liquid crystalline phase occupies less than 20% of the lipid part only a very high degree of ordering would appear in the diffractograms. Engblom has addressed this problem in his thesis [32] and finds it unlikely that changes in the interdomain phase will appear in the X-ray diffractograms. Correspondingly, it is likely that the water adsorbed in the lipid compartment is confined to the interdomain phase, thus not detected by X-ray diffraction although present. The bulk of water absorbed in the stratum corneum will be found in the corneocytes [2]. In

this context it is appropriate to underline that the morphological information provided by transmission electron microscopy on stained specimens must be regarded with caution since the preparation involves dehydration using organic solvents and substitution of the water phase with plastic. The structural organisation of the barrier lipids must therefore be compromised to a significant degree.

The morphology of the stratum corneum is also a factor hindering a perceptible swelling in the lipid compartment. The corneocytes are, as stated above, mutually coupled to all neighbors by means of the rigid corneodesmosomes and this arrangement will not allow an increased size of the intercorneocyte compartment. Rather the corneocyte swelling up to >25% in the vertical direction is expected to compress the lipid bilayer stacks. An increased pressure in the intercorneocyte compartment may force at least some of the lipids in the liquid crystalline state into the gel state [33], thus making transport through the lipid barrier less likely than normal.

In this context it is of interest to consider the recent results of a quantitative HPLC analysis of lipids from the deepest part of the stratum corneum, stratum compactum. This study shows that the lipid yield after stripping of the

Table 1 Lipid composition (wt%) of human and porcine stratum corneum

	Swartzendruber et al. (1988)	Wertz et al. (1987)	Norlén et al. (1997)*	Gray et al. (1982)	Wertz et al. (1992)
SC origin	Human	Human	Human	Porcine	Porcine
Ceramides	41.1	41	47	52.4	39.7
Free fatty acids	9.1	9	11	22.0	11.2
– unsaturated		0.7			
– saturated		8.3	11		
Cholesterol	?	27	24	20.5	28.4
Cholest esters	10.0	10	18	1.4	12.8
Cholest sulfate	1.9	1.9			2.0
Triglycerides	0	—		1.9	1.3

*Data to be published.

skin in vivo (Table 1) results in a lipid composition pattern [34] corresponding to that found in epidermal cysts [35], i.e. with a very low content of contaminating triglycerides (cf. Table 1). Special notice should be given the fact that Norlén's data [34] show a very high content of saturated long chain ($\geq$ C20) free fatty acids (11 wt%).

The role of the cholesterol is a somewhat enigmatic one and a problem that has not been outspokenly addressed in the past. In ordinary cell membranes the unsaturated phospholipids are compacted by introduction of cholesterol. However, in lipid films consisting of saturated lipids, cholesterol may have fluidizing properties (Friberg, personal communication, 1997). In more generalized terms cholesterol fluidizes lipid gel phases and stabilizes lamellar liquid crystals. In view of the fact that our recent data confirm that the free fatty acid part of human skin lipid extracts contain long chain saturated species (>C20) one role of cholesterol may therefore be to ensure a certain degree of fluidity in the interdomain areas. Futhermore, Wennerström (Personal communication) holds that cholesterol may act as a lineactant between crystalline/gel and liquid crystalline phases (cf surfactant).

Conclusions

We have proposed a two phase structure organization for the skin barrier, the *domain mosaic model*, based on the composition of lipids extracted from the horny layer of human skin. The model which assumes that all barrier lipids are organized in bilayer configuration represents the dominant part of the lipids in the crystalline (gel) state envisioned as domains. Such crystalline domains are surrounded by lipids in liquid crystalline state. On theoretical grounds our model can be shown to provide a water tight skin barrier, still allowing a minute leakage of water necessary to keep the keratin of the corneocytes hydrated to a level that ensures plasticity of the horny layer. A lipid construct of this kind will allow penetration only in the liquid crystalline phase, and perturbing this phase, e.g. by introduction of a penetration enhancer, may result in an increase in transdermal transport by several orders of magnitude. The model also satisfies mechanical/physical requirements on a tight barrier allowing mechanically satisfactory responses from the lipid structures under tensional stress.

Acknowledgments Professor Kåre Larsson's generous support, advice, and kind interest during the development of the domain mosaic model cannot be underestimated. We have also enjoyed constructive discussions with Professor Sven Engström which have resulted in crucial developments of the model. Professor Stig Friberg has generously shared his vast knowledge in the field with us.

For generous financial support we are indebted to the Swedish Work Environmental Foundation (#94-0414, #95-0289), the Swedish Council for Work life research (#96-0486, #96-0110) (BF), the Edvard Welander foundation and the Karolinska Institute funds (BF, LN).

References

1. Nilsson GE: Linköping University Medical Dissertations No 48, Linköping (1977)
2. Norlén L, Emilson A, Forslind B (1997) Arch Exp Derm 00:000-000
3. Boddé H, van den Brink I, Koerten HK, de Haan FHN (1990) J Controlled Released 15:227-236
4. Fartasch M, Bassuskas ID, Diepgen TL (1993) Br J Dermatol 128:1-9
5. Menon G, Ghadially R (1997) Microsc Res Technol 37:180-192
6. Michaels, AS, Chandrasekaran SK, Shaw JE (1975) AIChE J 21:985-996
7. Elias PM (1983) J Invest Dermatol 80:44S-49S
8. Williams ML, Elias PM (1987) CRC Crit Rev Therap Drug Carrier Systems 3:95-122
9. Hotchkiss S (1994) New Scientist 1910:24-27
10. Swartzendruber DC, Wertz PW, Kitko DJ, Madison KC, Downing DT (1989) J Invest Derm 92:251-257

46

B. Forslind et al.
A structural model for the human skin barrier

11. Singer SJ, Nicholson GL (1972) Science 175:720–731
12. Alberts B, Bray D, Lewis J, Raff M, Roberto K, Watson JD (1989) The Molecular Biology of the Cell, 2nd ed, Ch 6. Garland Publications, New York
13. Iraelachvili JN, Marcelja S, Horn RG (1980) Quart Rev Biophys 13:121–200
14. Larsson K (1994) Lipids – Molecular Organization, Physical Function and Technical Application, Vol 5. Oily Press Lipid Library, Oily Press, Dundee, UK
15. Gray GM, White RJ (1978) J Invest Dermatol 70:336–341
16. Engström S, Engblom J, Forslind B (1995) In: Brain KR, James VJ, Walters KA (eds) Proc Prediction of Percutaneous Penetration, Vol 4b. STS Publishing, Cardiff C59, 1995
17. Forslind B (1994) Acta Dermato Venereol 74:1–6
18. Forslind B (1995) Thrombosis Res 80:1–22
19. Forslind B, Engström S, Engblom J, Norlén L (1997) J Derm Sci 14:115–125
20. Yardley HJ, Summerly R (1981) Pharm Ther 13:357–383
21. Gray GM, Yardley HJ (1975) J Lipid Res 16:434–440
22. Thewalt J, Kitson N, Araujo C, MacKay A, Bloom M (1992) Biochem Biophys Res Communications 188:1247–1252
23. Fenske DB, Thewald JL, Bloom M, Kitson N (1994) Biophys J 67:1562–1573
24. Guy CL, Guy RH, Golden GM, Mak VHW, Francoeur ML (1994) J Invest Derm 103:233–239
25. Cornwell PA, Barry BW, Bouwstra JA, Gooris GS (1996) J Pharm 127:9–26
26. Ongpipanattanakul B, Francoeur ML, Potts RO (1994) Biochem Biophys Acta 1190:115–122
27. Mantsch HH, McElhaney RN (1991) Chem Phys Lipids 57:213–226
28. Moon DJ, Rerek ME, Mendelsohn R (1997) Biochem Biophys Res Commun 231:797–801
29. Bowstra JA, de Vries MA, Gooris GS, Bras W, Brusse J, Ponec M (1991) J Controlled Release 15:209–220
30. Bowstra JA, Gooris GS, Salmons de Vries MA, van der Spek JA, Bras W (1992) Int J Pharm 84:205–216.
31. Bouwstra JA, Gooris GS, Bras W, Downing DT (1995) J Lipid Res 36:685–695
32. Engblom J (1996) Thesis, Lund University, 1996
33. Cheng A, Hummel B, Mencke A, Caffrey M (1994) Science 67:293–303
34. Norlén L, Forslind B (1997) Manuscript to be submitted
35. Wertz PW, Schwartzendruber DC, Madison KC, Downing DT(1987) J Invest Dermatol 89:419–4125

Progr Colloid Polym Sci (1998) 108:47–57
© Steinkopff Verlag 1998

P. Skagerlind
B. Folmer
B.K. Jha
M. Svensson
K. Holmberg

Lipase–surfactant interactions

P. Skagerlind[1] · B. Folmer · B.K. Jha
M. Svensson · K. Holmberg (✉)
Institute for Surface Chemistry
P.O. Box 5607
SE 114 86 Stockholm
Sweden

[1]*Present address*
Kemira Kemi AB
P.O. Box 902
SE 251 09 Helsingborg
Sweden

Abstract Interactions between different amphiphiles and *Rhizomucor miehei* lipase were investigated by a variety of techniques. Complex formation in aqueous bulk solution was studied using surface tension measurements. Interactions at the oil–water and the solid–water interfaces were investigated by measuring mobility of emulsion droplets and by ellipsometry, respectively. The results from the different methods were coherent and indicated that cationic surfactants form complex with the lipase over a broad pH range, also below the isoelectric point of the lipase. No such interactions were found for neither anionic or nonionic surfactants. It is postulated that the interaction between cationic surfactant and lipase is due to a combination of electrostatic attraction and hydrophobic interaction and that no such combined interaction occurs with anionic surfactants. The interaction between cationic surfactant and lipase leads to a reduction of reaction rate in lipase-catalyzed hydrolysis of a palm oil. It is also shown that in the same model reaction a normal straight chain alcohol ethoxylate is a substrate for the lipase. An appreciable amount of fatty acid ester of the surfactant is formed as biproduct of the reaction. Branched-tail alcohol ethoxylates are not substrates and appear not to be competitive inhibitors for the enzyme. Likewise, the double-tailed ester surfactant sodium bis(2-ethylhexyl) sulfosuccinate (AOT) seems not to interact with the enzyme active site. Thus, anionic and nonionic surfactants with bulky hydrophobic tails are the preferred surfactants for microemulsion-based reactions with *Rhizomucor miehei* lipase as catalyst.

Key words Lipase – *Rhizomucor miehei* – enzyme – surfactant – cationic surfactant – interaction – complex – surface tension – ellipsometry – microemulsion

Introduction

Lipases are surface active proteins known to exert their action at the interface between a hydrophobic and a hydrophilic region. In biological systems lipases operate in membranes which represent an interfacial milieu with the oil–water boundary being stabilized by a pallisade layer of polar lipids, mainly phospholipids. From a surface chemistry point of view such lipid mono- or bilayers are a mixture of zwitterionic, cationic and anionic surfactants, usually with a slight net positive charge [1].

During the last 15 yr there has been considerable research interest in the use of lipases in microemulsions, mainly of the water-in-oil (W/O) type containing only a few percent water. In the literature such systems are sometimes referred to as reversed micellar systems. The interest stems partly from the fact that such systems are biomimetic, i.e. they resemble, from a physico-chemical point of view, the natural environment of lipases [2, 3]. Biological membranes need not be composed of flat bilayers of lipid molecules. Nonbilayer lipid structures seem to be essential for many processes occurring in the living cell, such as fusion and compartmentalization of membranes [4]. So called "lipid particles" can be seen as reversed micelles sandwiched between monolayers of polar lipids. It is also known that many enzymes induce formation of such bilayer structures upon incorporation into both model and biological membranes. Hence, studies of enzymes in W/O microemulsions are of relevance to biology in a wider sense than biocatalysis.

Another driving force for research on enzyme catalyzed reactions in water-poor media is the preparative interest in the use of lipases as catalysts for stereo- and regioselective esterifications and transesterifications [5–9]. Applications described in the literature include lipid transformations, such as preparation of synthetic cocoa butter [10], surfactant synthesis [11], enantioselective ester synthesis [12, 13] and formation of macrocyclic lactones of interest as perfume ingredients [14].

In the W/O microemulsions the lipase is believed to be present in the water droplets but extend out into the continuous oil domain [15, 16]. Thus, the enzyme competes with the surfactant for a place at the oil–water interface. In principle, the surfactant may either repel the enzyme from the interface as an effect of its greater surface activity or it may promote access of the enzyme to the interface by electrostatic attraction. In the latter case the presence of enzyme in the interfacial region should be very dependent on the sign of charge of the surfactant and on the net charge of the lipase, which, in turn, is governed by the pH in the aqueous domains.

Against this background it is obvious that studies of surfactant–lipase interaction are motivated in order to understand the behavior of lipases in microemulsions. Knowledge about the interaction between lipase and surfactant is also vital for formulation of detergents and other cleaning products which contain the two ingredients. This communication summarizes recent work from our laboratory on lipase–surfactant interaction and the paper contains new findings as well as previously published results.

Materials and methods

Chemicals

Purified lipase from *Rhizomucor miehei* and the structurally similar *Humicola lanuginosa* were kindly provided by Novo Nordisk, Denmark. The *Rhizomucor miehei* lipase is a single chain protein consisting of 269 amino acids with a total molecular weight of 29 472 [17] and an isoelectric point of 3.5 [18]. Further details about this enzyme can be found in an earlier publication [18].

An amphoteric surfactant, sodium *N*-(2-hydroxy-dodecyl)sarcosinate, as well as three nonionic surfactants, octa(ethylene glycol)mono(2-butyloctyl) ether (branched $C_{12}E_8$), tetra(ethylene glycol)monododecyl ether (linear $C_{12}E_4$) and hexa(ethylene glycol) mono-2-[1-(2-ethyl-butyl)-3-(2-ethylhexyl)]glyceryl ether (branched $C_{14}E_6$) were kindly provided in a purified form by Akzo Nobel Surface Chemistry, Sweden. Two homolog pure alcohol ethoxylates, penta(ethylene glycol)monododecyl ether (pure linear $C_{12}E_5$) and hepta(ethylene glycol) mono-dodecyl ether (pure linear $C_{12}E_7$) were bought from Nikko Chemicals, Japan. The anionic surfactant sodium bis(2-ethylhexyl)sulfosuccinate (AOT), the cationic surfactant didodecyldimethylammonium bromide (DDDMAB) and the nonionic surfactant β-1-octylglucoside were purchased from Sigma, Germany.

Glyceryl tri[1-^{14}C]oleate was from Amersham and palm oil was from Aarhus Oliefabrik, Denmark. All other reagents were reagent grade and used as received.

Reactions

Microemulsions were prepared by mixing isooctane, palm oil, phosphate buffer pH 7.5 and surfactant as has been described before [19]. Reactions in microemulsions were performed and the yields were determined as given in the same reference.

The radiochromatographic procedure to quantify the yield of reaction between alcohol ethoxylate and fatty acid has been described in ref. [20].

Emulsions

Emulsions were prepared with an Ultra Turrax T25 for two minutes at 24 000 rpm. Surfactant-stabilized emulsions consisted of aqueous buffer, isooctane and surfactant in a weight ratio of 95.00:4.95:0.05. Enzyme-stabilized emulsions contained *R. miehei* lipase instead of surfactant and in the same amount.

Mobility measurements were performed with freshly prepared emulsions after dilution 100 times with buffer. Lipase or surfactant was then added stepwise and after each addition the electrophoretic mobility of the droplets was determined. The instrument used was a Malvern Zeta Zizer II from Malvern Instruments, UK. The emulsions, after each addition of surfactant or lipase, were stirred for

30 s before the measurement. The temperature used was 22 °C.

NMR measurements

Self-diffusion measurements were obtained using the ^{1}H-NMR Fourier transform pulsed gradient spin echo technique with a standard Jeol FX-100 NMR spectrometer operating at a frequency of 99.6 MHz and at a temperature of 37 ± 0.5 °C [21]. The measurements were performed by varying the duration of the gradient probe (1 G/cm) at a constant measuring time of 140 ms.

The structure of the fatty acid ester of the alcohol ethoxylate $C_{12}E_5$ was determined by ^{1}H-NMR as described in an earlier communication [20].

Surface tension measurements

Surface tension was measured with a KSV Sigma 70 instrument using a du Noüy ring. The Zuidema-Waters correction method for the ring was used. Surfactant addition was done with a Methrom dosimat titration unit. Measurements were performed in glass beakers cleaned with bichromate sulfuric acid. The surface tension measurements of surfactant only were carried out starting with a 1 mM phosphate buffer, to which the surfactant in 1 mM phosphate buffer was added. Measurements in the presence of enzyme were made with a solution of 0.05 g/dm^3 lipase in 1 mM phosphate buffer to which a concentrated solution of surfactant in buffer containing the same concentration of enzyme was added. Measurements were performed in basic, neutral and acidic environment.

Ellipsometry

Ellipsometry was used to measure adsorbed amount and thickness of the adsorbed film on hydrophobized silica plates. The method is based on the measurement of changes of ellipticity of polarized light upon reflection at an interface. The instrument used was a Rudolph thin-film ellipsometer, type 436, controlled by a computer. A xenon lamp, filtered to 4015 Å, was used as light source.

The hydrophobized silica plates were made by treatment of thoroughly cleaned plates with dichlorodimethylsilane by a procedure described before [18]. A cuvette was carefully cleaned and filled with buffer. Before the plates were immersed in the cuvette, they were rinsed in ethanol and water and blown dry with nitrogen gas. A zero line was recorded from the buffer solution before the surfactant was added. Surfactant was then added into the cuvette

solution until a concentration of 1.5 times the cmc had been reached. The adsorption of surfactant on the hydrophobized silica plate was recorded. After 60 min, lipase solution was added, giving a concentration of 0.05 g/dm^3 in the cuvette. The adsorption of enzyme on top of the adsorbed surfactant layer was monitored. Calculation of adsorbed amount was done with de Feijters' formula, using a refractive index increment of 0.15 cm^3/g which is a common value for surfactants [22]. The refractive index increment for lipase is 0.19 cm^3/g [23].

Results

Hydrolysis of palm oil in microemulsion

Lipase catalyzed hydrolysis of a triglyceride, palm oil, was performed in L2 (water-in-oil) microemulsions based on aqueous buffer, isooctane and surfactant. The *Rhizomucor miehei* lipase used is 1,3-specific, i.e. it catalyzes hydrolysis at positions 1 and 3 of the triacylglycerol and leaves the 2-position intact. Thus, two moles of fatty acids liberated from one mole of triglyceride is the theoretical yield.

Three double-tailed surfactants, one anionic, one cationic and one nonionic were used, as well as one standard straight chain nonionic surfactant. The structures are given in Fig. 1, compounds I, II, IV and V. All four microemulsions had the same composition in terms of water : oil : surfactant weight ratio. As shown in Table 1, self-diffusion NMR indicated that all four compositions had small values of D_w/D_w^o, i.e. the ratio of the measured self-diffusion of water in the microemulsion to that of neat water [19, 24]. This is indicative of a water-in-oil microstructure. Furthermore, the self-diffusion measurements indicated that the structure, at least for the two alcohol ethoxylates, did not change much during the course of the reaction. Thus, the microemulsion structure seems not to be a governing factor for the reaction rate.

The reaction profiles are given in Fig. 2. It is evident from the figure that whereas the microemulsions based on AOT and branched $C_{12}E_8$ gave quantitative yield of fatty acid in the 24 h reaction time used, those based on DDDMAB and linear $C_{12}E_4$ did not.

Esterification of alcohol ethoxylate $C_{12}E_5$ and fatty acid

Triglyceride hydrolysis in a microemulsion based on a straight chain alcohol ethoxylate ($C_{12}E_4$) gave poor yield of free fatty acid (Fig. 2). In order to demonstrate whether or not the fatty acids formed were consumed by the surfactant under formation of fatty acid esters of the alcohol ethoxylate, reactions were performed with glyceryl

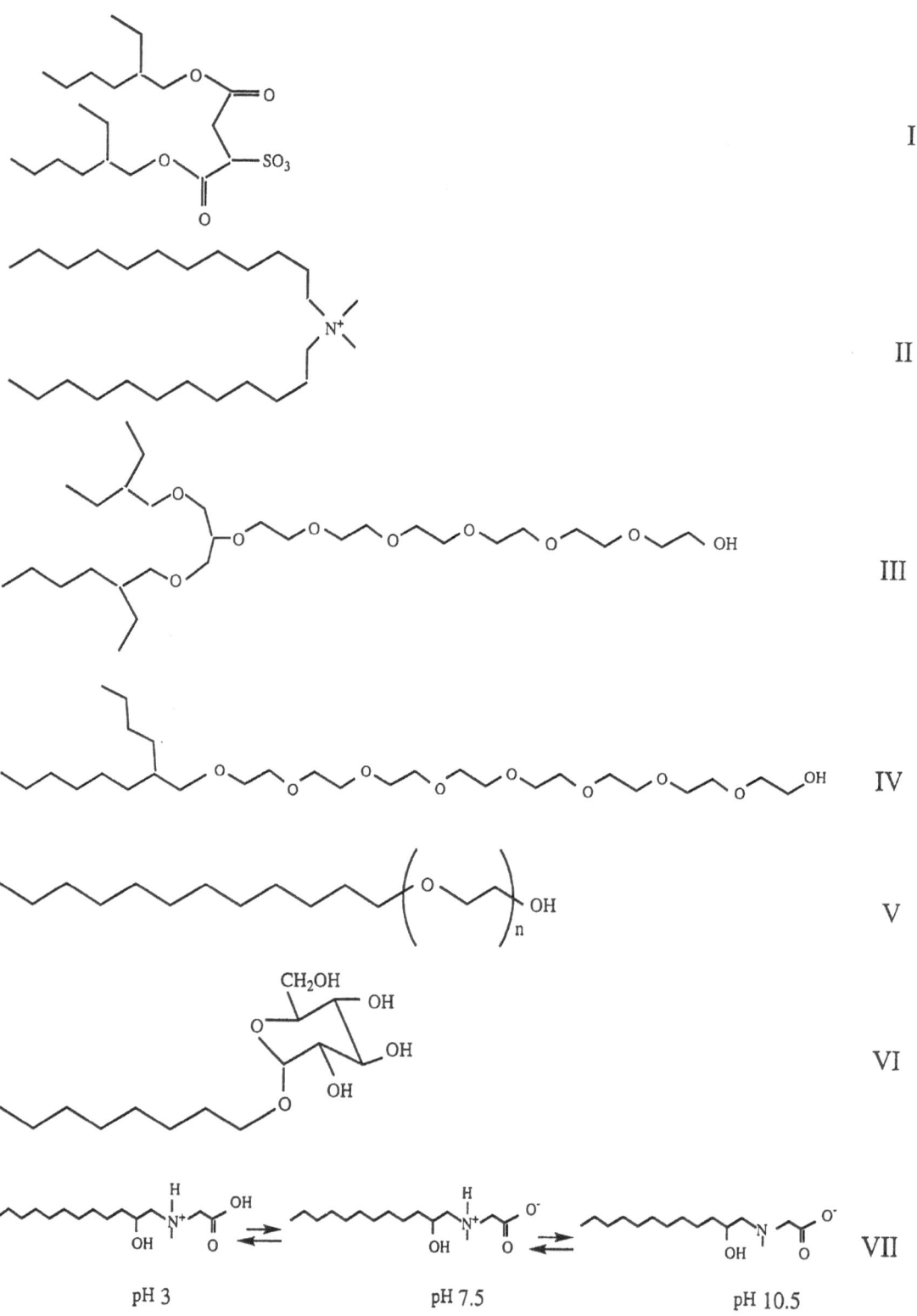

Fig. 1 Structures of surfactants used. I: sodium bis(2-ethylhexyl)sulfosuccinate (AOT), II: didodecyldimethylammonium bromide (DDDMAB), III: hexa(ethylene glycol)mono-2-[1-(2-ethylbutyl)-3-(2-ethylhexyl)]glyceryl ether (branched $C_{14}E_6$), IV: octa(ethylene glycol)mono(2-butyloctyl)ether (branched $C_{12}E_8$), V: dodecanol reacted with n moles of ethylene oxide (linear $C_{12}E_n$), VI: β-1-octylglucoside, and VII: sodium N-(2-hydroxydodecyl)sarcosinate

Table 1 Ratio of self-diffusion coefficient of water measured in the reaction systems, D_w, to that of neat water, D_w^o, at 37 °C at different reaction times. The value of D_w^o used was 3.3×10^{-9} m²/s [19]

Microemulsion surfactant	Reaction time (h)			
	0	1/2	4	24
AOT	0.03			
Branched $C_{12}E_8$	0.09	0.06	0.03	0.03
DDDMAB	0.12			
Linear $C_{12}E_4$	0.03	0.03	0.06	0.06

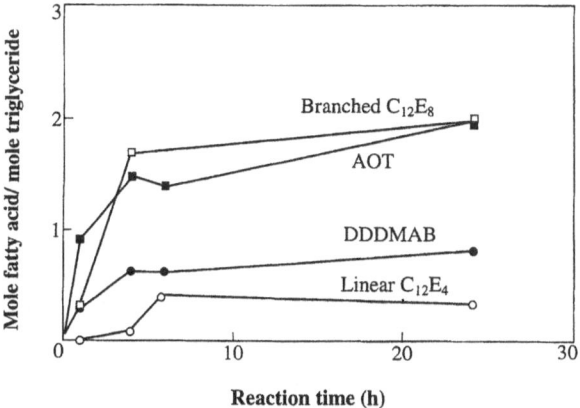

Fig. 2 Yield of palm oil hydrolysis as a function of time in microemulsions based on isooctane, palm oil, distilled water and four different surfactant in the weight proportion 50:5:8:37 [19]

Table 2 Molar distribution of ¹⁴C-labeled oleic acid at the start of the reaction and after 24 h hydrolysis in $C_{12}E_5$-based microemulsion using *Rhizomucor miehei* as catalyst [20]

Compound	Reaction time (h)	
	0	24
Tri-and diglyceride	3	0.17
Monoglyceride	0	0.72
Free fatty acid	0	1.3
Fatty acid ester of $C_{12}E_5$	0	0.81

trioleate radiolabeled in the fatty acid part [20]. In the reaction a homolog pure alcohol ethoxylate (pure linear $C_{12}E_5$) was used as microemulsion surfactant. Table 2 shows the distribution of the radiolabeled acid after completed reaction. It is evident that a substantial fraction of the fatty acid formed during the lipase catalyzed hydrolysis undergoes lipase catalyzed esterification with the surfactant alcohol.

Surface tension measurements

Surface tension plots for different surfactants, i.e. determination of surface tension as a function of surfactant concentration, was performed with and without lipase present as a way to detect surfactant-lipase complex formation in the bulk water phase. When the measurements were performed in the presence of lipase an enzyme concentration of 0.05 g/dm³ was always used. This concentration had been found to be well above the value of lipase concentration that gave a constant surface tension in surfactant-free solutions [18].

The structure of the amphoteric surfactant used, sodium *N*-(2-hydroxydodecyl)sarcosinate, is shown in Fig. 1, compound VII. The surface tension reduction obtained with the surfactant was measured at three different pH-values: 10.5, 7.5 and 3.0; at which the surfactant carried a negative, net zero and positive charge, respectively. The surface tension vs. logarithmic surfactant concentration graphs were compared with those obtained with surfactant in the presence of lipase at the same pH. At high pH the amphoteric surfactant behaves as an anionic surfactant. With increasing concentration the surface tension curve without enzyme approaches the curve with enzyme, as is shown in Fig. 3, top. At neutral pH, when the alkylsarcosinate behaves as a nonionic surfactant, the surface tension plots with and without enzyme coincide at higher concentration (Fig. 3, middle). At low pH the amphoteric surfactant is positively charged. As shown in Fig. 3, bottom, the surface tension in the presence of enzyme remains at an almost constant value until a relatively high surfactant concentration is reached before it decreases steeply. As can be seen from the latter figure, the surface tension plots with and without surfactant cross each other. This behaviour is indicative of aggregate formation between the enzyme and the positively charged surfactant at surfactant concentrations below cmc. Proper packing of surfactant at the air–water interface will not occur until the enzyme is "saturated" with surfactant molecules.

To acertain that formation of the enzyme-surfactant complex at low pH, when the surfactant is positively charged, was not due merely to charge variations of the enzyme, surface tension plots for the permanently cationic surfactant DDDMAB were also recorded in basic, neutral and acidic solutions (Fig. 4). As expected, the surface tension vs. log concentration plots of surfactant only did not show any significant pH dependence. In the curves for the systems containing both enzyme and cationic surfactant, on the other hand, there is a trend towards higher surfactant concentration before the onset of surface tension reduction with increasing pH. This can be seen from the location of the arrows which indicate the surfactant

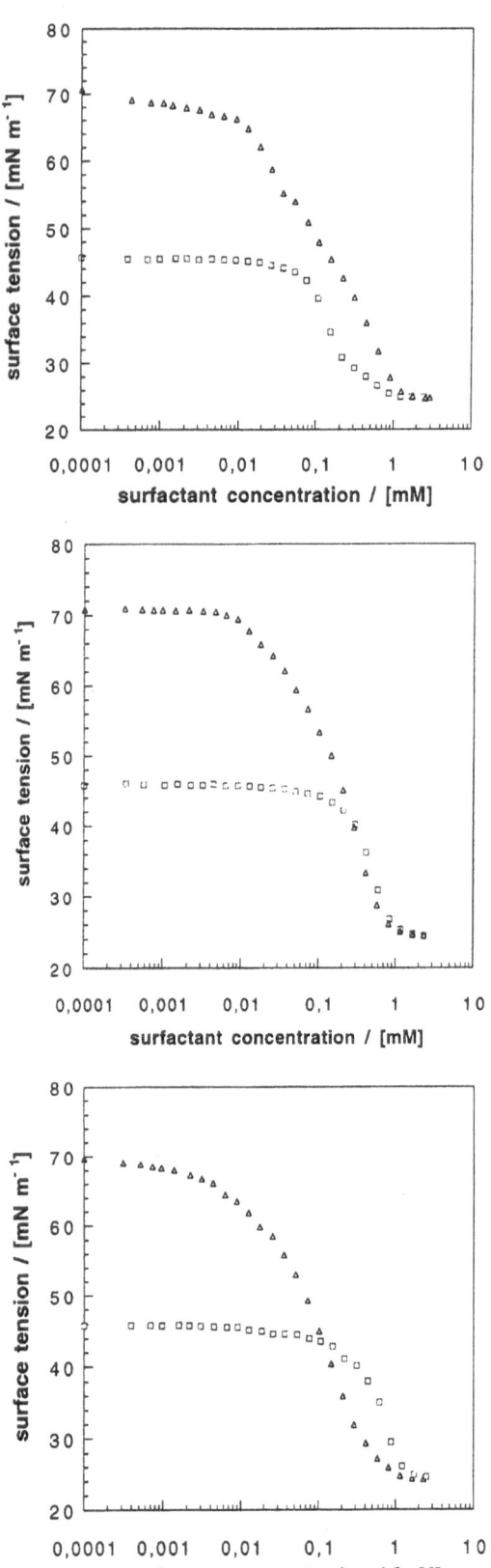

concentration at which the enzyme is saturated with surfactant at that specific pH.

The surface tension plot for the anionic surfactant AOT (Fig. 1, compound I) is shown in Fig. 5, top. Plots for two nonionic surfactants, one ethoxylated alcohol, pure linear $C_{12}E_7$, and one alkylglucoside, β-1-octylglucoside, are given in Fig. 5, middle and bottom, respectively. None of the three diagrams show any sign of surfactant–polymer interaction.

Measurements of electrophoretic mobility of emulsion droplets

Emulsions were made with the double-tailed surfactants I, II and III of Fig. 1. *Rhizomucor miehei* lipase was added in portions to the ready-made emulsion and the electrophoretic mobility of the droplets was determined. It is evident from Fig. 6 that whereas the emulsions stabilized with anionic or nonionic surfactant were unaffected by the lipase, the charge of the emulsion droplets stabilized by cationic surfactant decreased rapidly on enzyme addition. The reduction in surface charge from highly positive to slightly negative for the droplets stabilized by cationic surfactant most probably is caused by the enzyme adsorbing on top of the surfactant layer. The net charge of the lipase at the pH of 7.5 used is slightly negative (the lipase has a net zero charge at pH 3.5) which fits well with the plateau value of zeta potential obtained. These experiments indicate that no such interaction takes place with droplets stabilized by the anionic surfactant but they do not give information about whether or not the enzyme adsorbs on the droplets stabilized with nonionic surfactant.

Figure 7 shows results from the reverse experiment. The three surfactants were added in portions to an emulsion stabilized with only lipase. Also this experiment is strongly indicative of lipase adsorbing to droplets stabilized with cationic surfactant and seems to rule out interaction between the enzyme and the negatively charged droplets.

Ellipsometry

Adsorption of surfactant on hydrophobized silica was measured by ellipsometry. After stabilization, subsequent addition of *Rhizomucor miehei* lipase was recorded. As in

Fig. 3 Surface tension vs. log concentration for the amphoteric surfactant sodium *N*-(2-hydroxydodecyl) sarcosinate at pH 10.5 (top), pH 7.5 (middle), and pH 3.0 (bottom). Triangles indicate surfactant only and squares indicate surfactant + 0.05 g/dm³ lipase

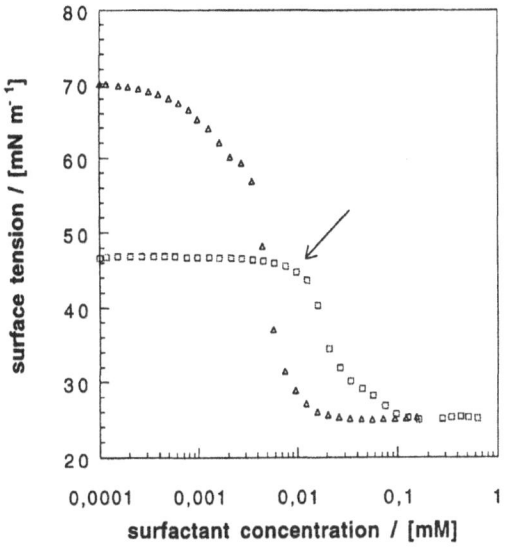

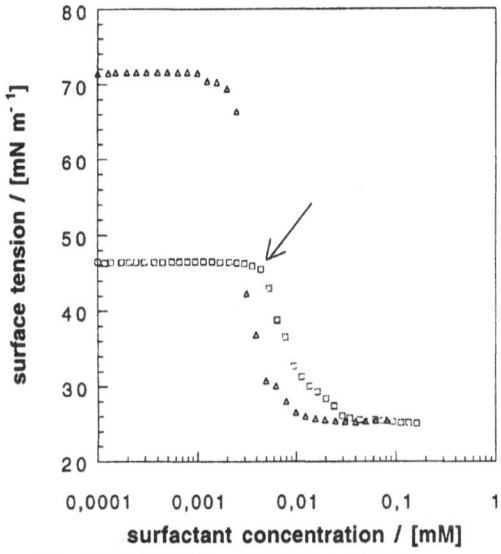

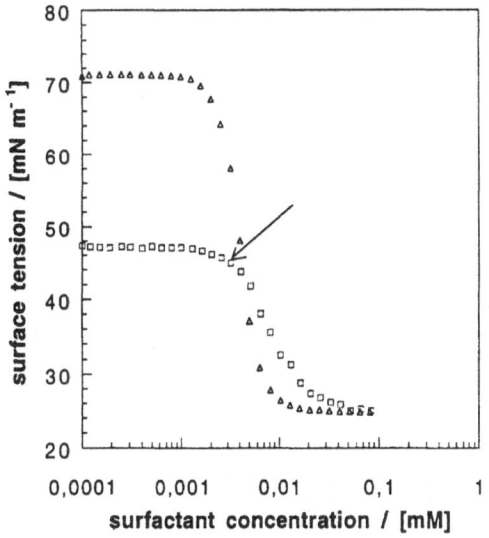

Fig. 4 Surface tension vs. log concentration for the cationic surfactant DDDMAB at pH 10.5 (top), pH 7.5 (middle), and pH 3.0 (bottom). Triangles indicate surfactant only and squares indicate surfactant + 0.05 g/dm³ lipase

the surface tension studies, described above, the amphoteric surfactant was used at three different pH values, viz., 10.5, 7.5 and 3.0. In basic and neutral environment the surfactant adsorbs in an amount of 0.75 and 0.65 mg/m², respectively (Fig. 8, top and middle). Addition of lipase after 3600 s, gives no change in adsorbed amount. Under acidic conditions the situation is strikingly different. The adsorbed amount of surfactant was 0.5 mg/m². Addition of lipase after 3600 s increased the adsorbed amount to 2.55 mg/m², as can be seen from Fig. 8, bottom. A control experiment with the permanently cationic surfactant DDDMAB was performed at pH 7.5 (Fig. 9), again to ascertain that complex formation of the amphoteric surfactant under acidic but not under neutral or alkaline conditions was not merely due to changes in the net charge of the enzyme. A plateau value of adsorption of 1.15 mg/m² was measured with only surfactant. After addition of enzyme the adsorbed amount increased to 6.3 mg/m².

Discussion

The results from triglyceride hydrolysis in microemulsions and from studies of lipase–surfactant interactions in bulk water, at the oil–water interface and at the solid–water interface are coherent and indicate that the cationic surfactants form complex with the *Rhizomucor miehei* lipase but that structurally similar anionic or nonionic surfactants do not. It is also clearly demonstrated that the enzyme–surfactant complex forms in solution at a pH both above and below the isoelectric point of the lipase. Thus, positively charged surfactants associate also with lipase carrying a net positive charge. Negatively charged surfactants, on the other hand, do not form complex with positively charged lipase.

Most likely, the complex formation between the lipase and positively charged surfactants is due to a combination of electrostatic and hydrophobic interactions. If the enzyme contains hydrophobic domains adjacent to negatively charged groups but not in the vicinity of positively charged groups, cationic but not anionic surfactants can use both types of attractive interactions for the binding to the lipase. Complex formation between cationic surfactant and lipase, also below the isoelectric point of the enzyme may be due to the presence of carboxylic groups with very low pK_a values in the enzyme situated adjacent to hydrophobic amino acid residues. Alternatively, such complex

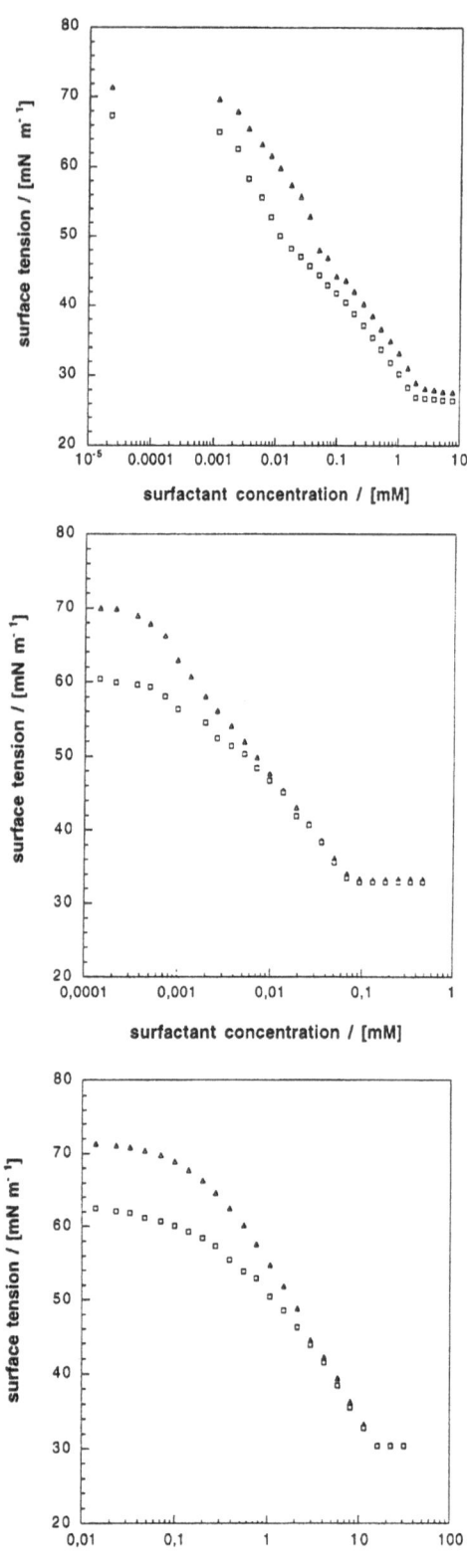

Fig. 5 Surface tension vs. log concentration for the anionic surfactant AOT (top) and the nonionic surfactants pure linear $C_{12}E_7$ (middle) and β-1-octylglucoside (bottom) at pH 7.5. Triangles indicate surfactant only and squares indicate surfactant + 0.05 g/dm^3 lipase

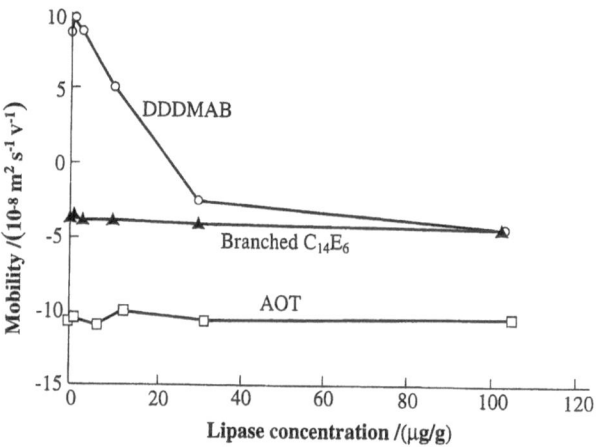

Fig. 6 Electrophoretic mobility of emulsion droplets as a function of lipase concentration. Emulsions were made with three different surfactants and lipase was post-added to the emulsion

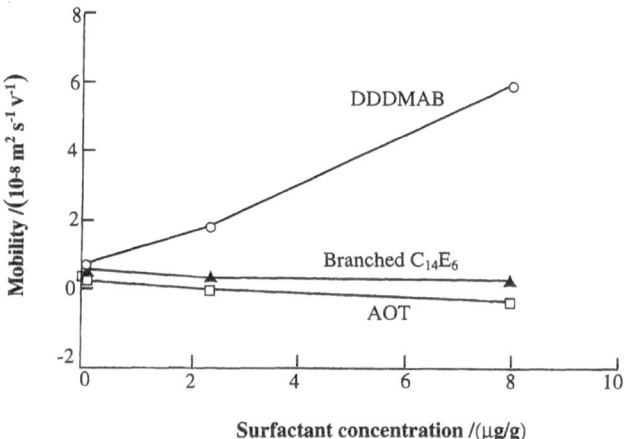

Fig. 7 Electrophoretic mobility of emulsion droplets as a function of surfactant concentration. Emulsions were made with lipase as emulsifier and three different surfactants were post-added to the emulsion

formation may be the result of protein carboxylic groups titrating during the association with oppositely charged surfactants. Complex formation between cationic surfactant and oppositely charged hydrophobically modified polyelectrolyte at pH values below the pK_a of the acid groups have been reported in the literature [25, 26].

The structural explanation to the lipase-surfactant interaction, i.e. the hypothesis that there are negatively charged groups adjacent to hydrophobic domains in the lipase, is

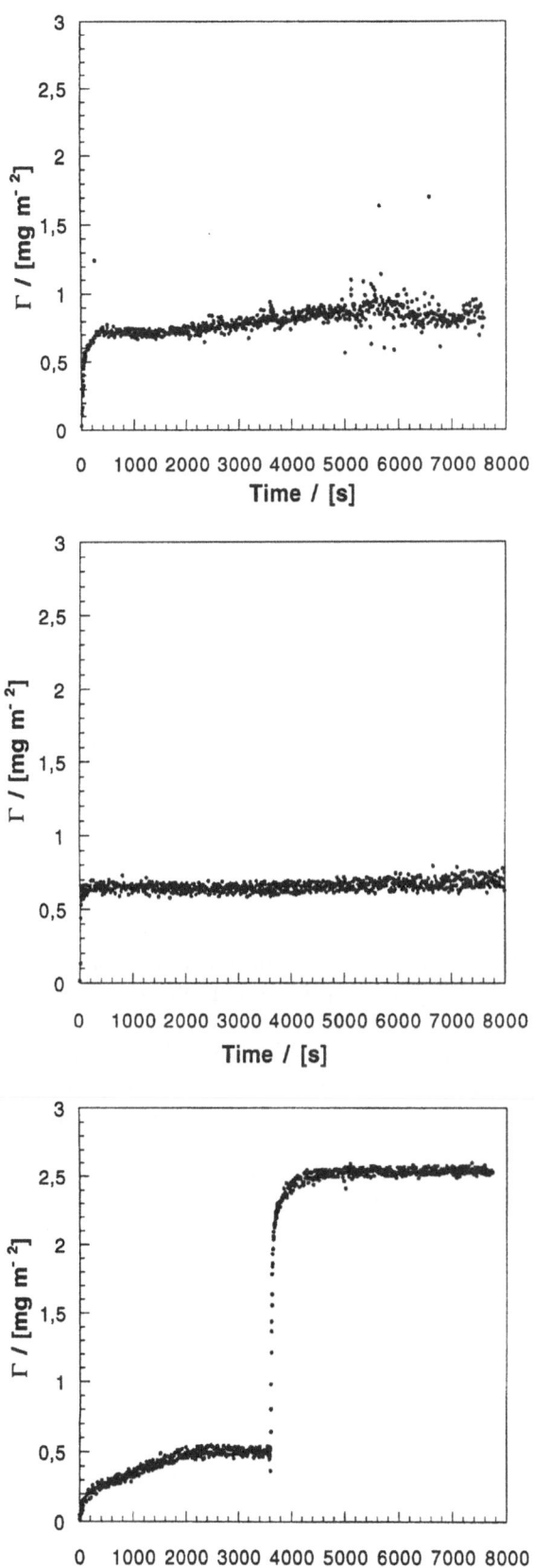

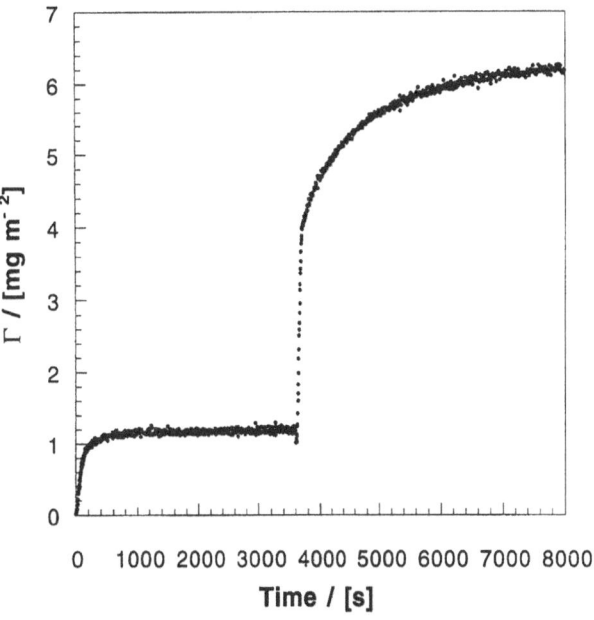

Fig. 9 Adsorption of the cationic surfactant DDDMAB on a hydrophobic surface and the effect of addition of lipase at 3600 s as monitored by ellipsometry at pH 7.5

tentative and verification by analysis of the lipase tertiary structure is in progress.

The observation that *Rhizomucor miehei* lipase interacts strongly with cationic but not with anionic and nonionic surfactants is of relevance to the much discussed issue of the position of lipase in water-in-oil microemulsions. It is of particular interest that the anionic surfactant used in this work, AOT, is the surfactant most frequently employed in the studies of microemulsion-based enzymatic reactions. It is generally believed that in water-in-oil type microemulsions the enzyme molecules are located (entrapped) in the water droplet region but the exact position may vary depending on the hydrophilic-lipophilic balance of the enzyme and to some extent on the nature of the solvent. A truly hydrophilic enzyme is likely to be located in the water core of the droplet, surrounded by a water layer. A surface active enzyme, such as lipase, has a strong driving force for the oil–water interface where it competes with the surfactant [15]. The fact that lipases need a hydrophobic surface in order to open the lid, covering the active site, could also be taken as support for a location at the interface; however the concept of "interfacial activation" has lately been questioned [27].

Fig. 8 Adsorption of the amphoteric surfactant sodium *N*-(2-hydroxydodecyl)sarcosinate on a hydrophobic surface and the effect of addition of lipase at 3600 s as monitored by ellipsometry at pH 10.5 (top), 7.5 (middle), and 3.0 (bottom)

Determination of the location of an enzyme in a micro-heterogeneous system is not straightforward, however. Pileni et al. [28] have studied the effect of chemical modifications of enzymes on the microstructure of AOT-based water-in-oil microemulsions containing an enzyme in the aqueous core. The hydrophilic enzymes ribonuclease and α-chymotrypsin were transformed into surface active proteins by derivatization with long-chain acid chlorides or chloroformates. Using small-angle X-ray scattering and conductivity measurements, no major change in droplet size and no perturbation of the droplet surface could be detected when the native enzymes were replaced by the modified ones [28–30]. Since also the present study does not indicate any attractive interaction between lipase and anionic surfactant, neither in bulk water, nor at the oil–water or the solid–water interface, it seems doubtful that lipases are allowed a place at the oil–water interface of AOT-based microemulsions.

Access to the interface has been claimed to be necessary in order for lipases to catalyze reactions with hydrophobic substrates present in the continuous hydrocarbon domain [9]. This would certainly be the case in more static systems, e.g. emulsions, but need not be true in the highly dynamic microemulsions. The rapid formation and disintegration of the microstructure of such systems may provide sufficient contact between enzyme and hydrophobic substrate even if the enzyme competes poorly with the surfactant for a place at the interface.

The results from the lipase catalyzed hydrolysis reactions show considerable differences in reaction rate depending on the choice of microemulsion surfactant. The self-diffusion measurements indicate that the structure of the microemulsion is not decisive. The reduced reaction rate obtained with the cationic surfactant is most likely due to formation of the lipase–surfactant complex discussed above. It is obvious that such interaction can be affect the catalytic activity if a surfactant binds at or in the vicinity of the enzyme active site.

The low yield of fatty acid in the hydrolysis reactions with the microemulsion based on straight-chain alcohol ethoxylate has a different explanation. As was demonstrated in the experiments with radiolabeled triglyceride, the fatty acid formed undergoes lipase catalyzed esterification with the alcohol ethoxylate. This side reaction occurs to an appreciable degree; in fact, during the conditions used more than 25% of the fatty acid in the starting triglyceride ended up as surfactant ester (Table 2). In a technical formulation a surfactant with a broad homolog distribution will be used, not a homolog pure alcohol ethoxylate as the $C_{12}E_5$ sample used in the experiments accounted for in Table 2. A $C_{12}E_5$ surfactant of technical quality can be expected to give fatty acid ester to an even higher degree since it typically contains around 20% unreacted dodecanol, which is more nucleophilic than the ethoxylated species and, thus, more reactive in esterification reactions [24].

In separate experiments, using o-nitrophenylpalmitate instead of a triglyceride as substrate, it was shown that the straight-chain alcohol ethoxylate (pure linear $C_{12}E_5$) also inhibited ester hydrolysis by competing with the ester substrate for the active site of the lipase [20]. For instance, it was found that the rate of o-nitrophenylpalmitate hydrolysis in an AOT-based micoemulsion decreased considerably upon addition of a small amount of $C_{12}E_5$ but that the reaction rate was unaffected by addition of the same amount of AOT to a $C_{12}E_5$-based microemulsion. Evidently, AOT, in spite of being an ester surfactant, is not a competitive inhibitor (and is not a substrate) for the lipase, whereas linear $C_{12}E_5$ is. As can be seen from Fig. 2, a microemulsion based on branched $C_{12}E_8$ was as good a reaction medium as the AOT-based system for palm oil hydrolysis. Branched $C_{12}E_5$ seems neither to be a competitive inhibitor, nor a substrate for the Rhizomucor miehei lipase used. The above results strongly suggest that in order to prevent lipase inhibition and/or unwanted reactions involving the amphiphile, ester and alcohol surfactants used in the microemulsion formulation should have a bulky hydrophobic tail.

Conclusions

1. The microbial lipase from Rhizomucor miehei forms complex with cationic surfactants both in aqueous bulk solution and at the oil–water and solid–water interfaces. The complexes form over a wide pH range, also below the isoelectric point of the enzyme. No such complex formation could be seen with neither anionic nor nonionic surfactants.

2. Straight chain alcohol ethoxylates are competitive inhibitors for the Rhizomucor miehei lipase. In lipase-catalyzed hydrolysis of triglycerides fatty acid esters of the surfactant are formed in appreciable amounts. Branched-tail alcohol ethoxylates, as well as the double-tailed ester surfactant AOT, are not competitive inhibitors for the enzyme and do not undergo lipase-catalyzed reactions under the conditions used.

Acknowledgement KH is grateful to Prof Kåre Larsson for valuable discussions on lipase action. Dr Karin Bergström at Akzo Nobel Surface Chemistry AB and Drs Per Falholt and Kim Borch at Novo Nordisk A/S are thanked for the gifts of surfactants and lipases, respectively.

Progr Colloid Polym Sci (1998) 108:47–57
© Steinkopff Verlag 1998

References

1. Larsson K (1994) Lipids – Molecular Organization, Physical Functions and Technical Applications. The Oily Press, Dundee, Scotland, Chap 1
2. Martinek K, Levashov AV, Khmelnitski YuL, Klyachko NL, Berezin IV (1982) Science 218:889
3. Burdette RA, Quinn DM (1986) J Biol Chem 261:12016
4. Larsson K (1984) Lipids – Molecular Organization, Physical Functions and Technical Applications. The Oily Press, Dundee, Scotland, Chap 9
5. Bello M, Thomas D, Legoy MD (1987) Biochem Biophys Res Comm 146:361
6. Hayes DG, Gulari E (1992) Biotechnol Bioeng 40:110
7. Stamatis H, Xenakis A, Provelegiou M, Kolisis FN (1993) Biotechnol Bioeng 42:103
8. Holmberg K (1997) Microemulsions in Biotechnology. In Solans C, Kunieda H (eds) Industrial Applications of Microemulsions. Marcel Dekker, New York
9. Holmberg K, Enzymatic Reactions in Microemulsions. In Kumar P, Mittal KL (eds) Microemulsions – Fundamental and Applied Aspects. Marcel Dekker, New York, in press
10. Holmberg K, Österberg E (1987) Prog Colloid Polym Sci 74:98
11. Skagerlind P, Larsson K, Barfoed M, Hult K (1997) J Am Oil Chem Soc 74:39
12. Rees GD, Robinson B (1995) Biotechnol Bioeng 45:344
13. Hedström G, Backlund S, Slotte JP (1993) Biotechnol Bioeng 42:618
14. Rees GD, Robinson BH, Stephenson GR (1995) Biochim Biophys Acta 1257:239
15. Stamatis H, Xenakis A, Kolisis FN, Malliaris A (1994) Prog Colloid Polym Sci 97:253
16. Papadimitriou V, Petit C, Cassin G, Xenakis A, Pileni MP (1995) Adv Colloid Interface Sci 54:1
17. Brady L, Brzozowski AM, Derewenda ZS, Dodson E, Tolley S, Turkenburg JP, Christiansen L, Huge-Jensen B, Norskov L, Thim L, Menge U (1990) Nature 343:767
18. Folmer B, Holmberg K, Svensson M (1997) Langmuir 13:5864
19. Skagerlind P, Holmberg K (1994) J Disp Sci Technol 15:317
20. Skagerlind P, Jansson M, Hult K (1992) J Chem Tech Biotechnol 54:277
21. Stilbs P (1987) Prog Nucl Magn Reson Spectrosc 19:1
22. van Os NM, Haak JR, Rupert LAM (1993) Physical Chemical Properties of Selected Anionic, Cationic and Nonionic Surfactants. Elsevier, Amsterdam
23. Malmsten M (1995) J Colloid Interface Sci 172:106
24. Stark M-B, Skagerlind P, Holmberg K, Carlfors J (1990) Colloid Polymer Sci 268:385
25. Shimizu T, Seki M, Kwak JCT (1986) Colloids Surfaces 20:289
26. Goddard ED, Ananthapadmanabhan KP (1993) Interactions of Surfactants with Polymers and Proteins. CRC, Boka Raton, USA, pp 203–276
27. Verger R (1997) Tibtech January 15:32
28. Pileni MP, Cassin G, Michel F, Pitré F (1995) Colloids Surfaces B 3:321
29. Pitré F, Regnaut C, Pileni MP (1993) Langmuir 9:2855
30. Michel F, Pileni MP (1994) Langmuir 10:390

Progr Colloid Polym Sci (1998) 108:58–66
© Steinkopff Verlag 1998

K. Sato

Newest understandings of molecular structures and interactions of unsaturated fats and fatty acids

K. Sato (✉)
Faculty of Applied Biological Science
Hiroshima University
Higashi-Hiroshima, 739 Japan
Fax: + 81-824-22-7062
E-mail: kyosato@ipc.hiroshima-u.ac.jp

Abstract Recent studies on molecular structures and interactions of unsaturated fatty acids, diacylglycerols and triacylglycerols have been reviewed. An emphasis was paid to elucidate thermodynamic and kinetic behavior of polymorphic transformation, molecular-level structural analysis of crystalline phases, phase behavior and structural properties of binary mixture states of the unsaturated fatty acids and triacylglycerols. It may be concluded that the most peculiar property revealed in the unsaturated fats and fatty acids is diversity in molecular structures and interactions, which are mainly caused by the diversity in the olefinic conformations and packings. In particular, the polymorphic behavior of oleic acid itself and triacylglycerols containing oleic acid moieties has shown multiple conversions in lateral molecular packing, chain length structure and olefinic conformation. It was inferred that these properties may be applied to those of polyunsaturated fats and lipids.

Key words Unsaturated fatty acid – triacylglycerols – polymorphism – binary mixtures

Introduction

Unsaturated fatty acids are important lipid molecules which are present in biomembranes and natural fats and oils [1]. In the biomembranes, they occupy about one-half of the whole acyl chains in membrane phospholipids which promote structural fluidity and permeability through conformational flexibility of the unsaturated acyl chains. They are precursors in membrane bio-syntheses. In edible fats and oils, many of vegetable fats and oils, such as palm oil, olive oil, etc., contain high amounts of the unsaturated fatty acids, and, nutritionally important poly-unsaturates are present in fish oils. In addition, the unsaturated fatty acids are utilized in surfactants and emulsifiers and in oleochemical industry as well.

There are a plenty of scientific and technological studies on the unsaturated fats and lipids. This paper mainly describes structural aspects revealed in crystalline phases. The information obtained from the crystal structures may have two meanings: (a) to relate the crystal structures with physical properties of the unsaturated fat crystals and (b) to highlight the structural nature of individual molecules through comparable studies of various specimens.

In recent years, the newest understandings of the unsaturated lipids have been brought up in such topics as diversified polymorphism in cis-monounsaturated fatty acids [2–20], di- and triacylglycerols [21–28], and gel–liquid crystal transformations in phospholipids containing saturated–unsaturated acyl moieties [29, 30]. The main causes which have brought up these new findings may be very pure samples of purity of 99%, the use of high-tech facility of micro-probe and time-scan FT-IR spectroscopy, Synchrotron radiation X-ray diffraction with precise time resolution and so on.

Structural features of unsaturated lipids

To begin with, it is necessary to take a brief look at three main features which characterize the crystal structures of the unsaturated fats and lipids.

Figure 1a shows subcell packings of the aliphatic chains. It must be stressed that greater contributions on the criteria establishment of the subcell packing and its application to polymorphic definition have been done by Swedish school [31–35].

Figure 1b shows chain length structures of double, triple and interdigitated modifications. Figure 1c shows three typical olefinic conformations. Many researchers have always thought that the twisted *skew–cis–skew'* conformation is most typical of the cis-unsaturation. However, through the recent studies, it has been obvious that other conformations such as *skew–cis–skew*, *trans–cis–trans* or even *trans–skew–cis–skew–skew–trans* can occur as the most stable conformation as a result of external and internal influences on the structural formation, as displayed in the following sections.

Polymorphism of monounsaturated fatty acids

Single phases

Table 1 summarizes the molecular properties and thermal data of phase transitions of eight monounsaturated fatty

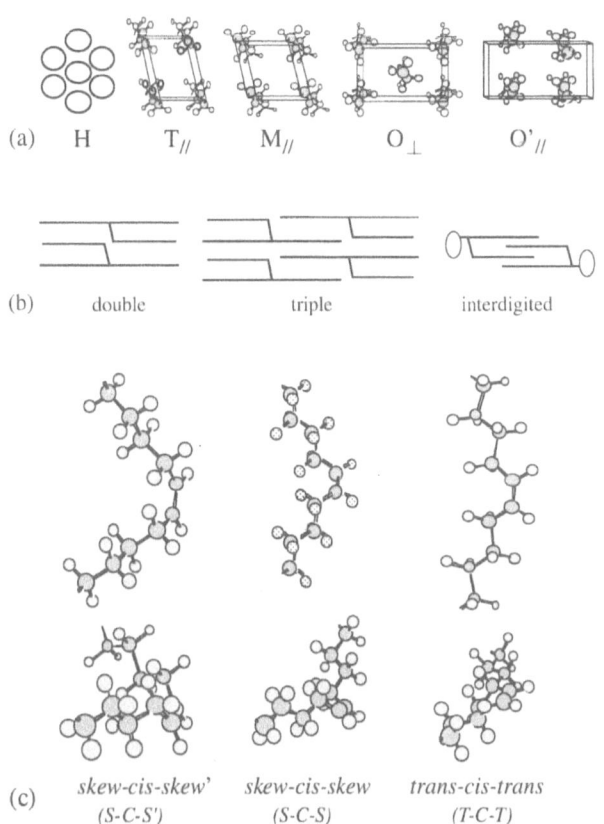

Fig. 1 (a) Subcell structures, (b) chain length structures and (c) olefinic conformations of aliphatic chains

Table 1 Polymorphism of monounsaturated fatty acids

Fatty acid	Form	Olefinic group[a] conformation	Subcell	Phase transition		
				Form	T_{tr} (°C)	ΔH_{tr} (kJ/mol)
Myristoleic (C14, ω5)		unclear	∥-type	melting	−3.8	28.9
Palmitoleic (C16, ω7)	γ	S–C–S′	O′∥	→ α	−18.4	7.5
	α	unclear	∥-type	melting	2.0	32.1
Asclepic (C18, ω7)	γ	S–C–S′	O′∥	→ α	−15.4	7.8
	α	unclear	∥-type	melting	13.8	39.8
Oleic (C18, ω9)	γ	S–C–S′	O′∥	→ α	−2.2	8.8
	α	unclear	∥-type	melting	13.3	39.6
	β₂	unclear	∥-type	melting	16.0	48.9
	β₁	T–C–T	T∥	melting	16.3	57.9
Petroselinic (C18, ω12)	LM	157°-cis-−160°	O⊥	melting[b]	–	–
	HM	S–C–S	O⊥ + M∥	melting	30.5	47.5
Gondoic (C20, ω9)	γ	S–C–S′	O′∥	→ α	−3.2	9.0
	α	unclear	∥-type	melting	23.3	49.7
Erucic (C22, ω9)	γ	S–C–S′	O′∥	→ α	−1.0	8.8
	α	unclear	∥-type	melting	31.2	5.4
	γ₁	S–C–S′	T∥	→ α₁	9.0	8.9
	α₁	unclear	∥-type	melting	34.0	54.0
Elaidic (C18, ω9)	I	*trans*	O⊥	melting	44.5	53.6

[a] S–C–S′, *skew–cis–skew'*; S–C–S, *skew–cis–skew*; T–C–T, *trans–cis–trans*.
[b] Melt-mediated transformation to HM form.

(a)

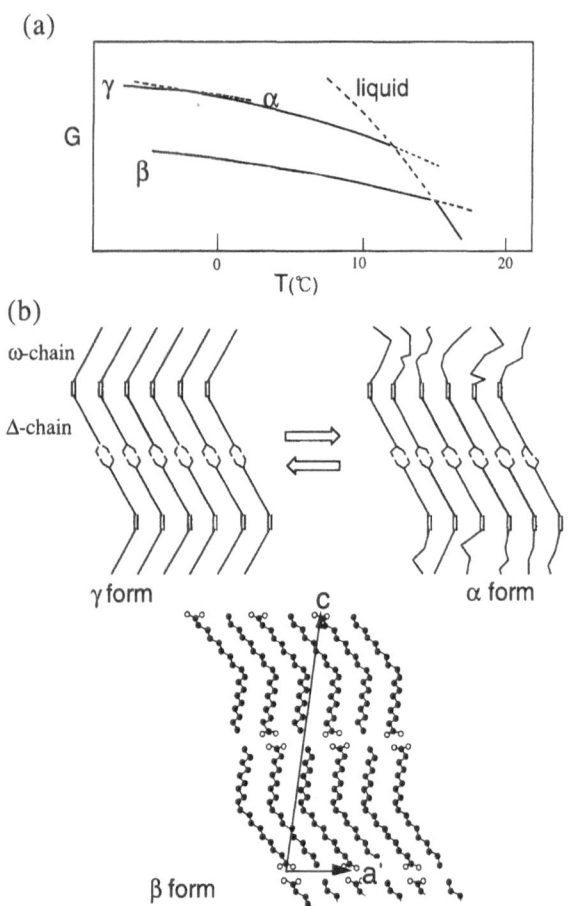

(b)

Fig. 2 (a) Gibbs energy (G) vs. temperature (T) relationship and (b) molecular structures of three forms of oleic acid

acids involving *trans*-type elaidic acid, which have been all unveiled in the past 12 years [36]. Several new properties have been found in terms of the number of individual polymorphic forms, olefinic conformation, subcell packing which is split in two even in the same molecule (petroselinic acid), and the nature of phase transformation.

Because of the limit of space, the most important findings will be highlighted on oleic acid polymorphism in this article. It partly shares common properties with other fatty acids, and partly reveals a very unique property which never occurs in other substances.

Figure 2a shows the thermodynamic stability relation of three forms of oleic acid [2], although two β forms were isolated very recently [20]. The forms of α and β have two different melting points, showing that β is most stable. In the metastable area, α transforms reversibly to γ, which were found to correspond to high-melting and low-melting forms discovered in 1962, respectively [37]. As shown in Fig. 2b, γ is the form which reveals S–C–S'

olefinic conformation keeping the all *trans* aliphatic conformation both in the ω and $\varDelta$ chain segments divided by the *cis*-double bond. However, this form transforms to α form in which the ω segment closer to the methyl end group is conformationally disordered, but the $\varDelta$ segment retains the *trans*-conformation [4]. The transformation of this type was widely observed in other cases, meaning that it is the most typical property of the *cis*-monounsaturated fatty acids.

A surprising structure was discovered in oleic acid β form as shown in Fig. 2b [20]. Two remarks are notable; one is the alternative arrangement of the methyl end and carboxyl groups in the same lamella plane, making an interdigitated chain length structure. The other is the T–C–T olefinic conformation. These properties have never been observed in the other unsaturated fatty acids. Therefore, it seems that specific cohesive interactions may be operating, such that this form becomes most structurally stable in oleic acid.

Coming back to Table 1, one may discuss the polymorphic properties revealed in eight principal monounsaturated fatty acids. The γ–α transformations were observed in palmitoleic, asclepic, oleic, gondoic and erucic acids, but not in petroselinic and elaidic acids. The latter two acids were quite similar to saturated fatty acids in terms of subcell packing and crystal morphology [39]. As for the diversity in the olefinic conformation, it is inferred that this may cause the diversity of lateral packing in acylglycerols or phospholipids, when the unsaturated fatty acid chains are forced to interact with the saturated fatty acid chains in the same lamella leaflet.

Binary mixture phases

The next interesting properties are shown in binary mixing behavior of the monounsaturated fatty acids, in which molecular interactions are exhibited remarkably.

The eutectic mixture has most commonly been observed in the all mixtures among the fatty acids displayed in Table 1. However, two exceptional mixing systems were observed in the following (Fig. 3):

(a) A miscible mixture between gondoic and asclepic acids (Fig. 3a) [39]. The two component molecules shared the two properties of the γ–α transformation and the same length of the $\varDelta$ segment.

(b) A compound formation between oleic and palmitoleic acids [12]. The two component molecules also share the γ–α transformation and the same $\varDelta$-chain length, similar to the combination of gondoic acid and asclepic acid. However, it seems that two types of the molecular compounds are formed in γ and α phases. Molecular-level

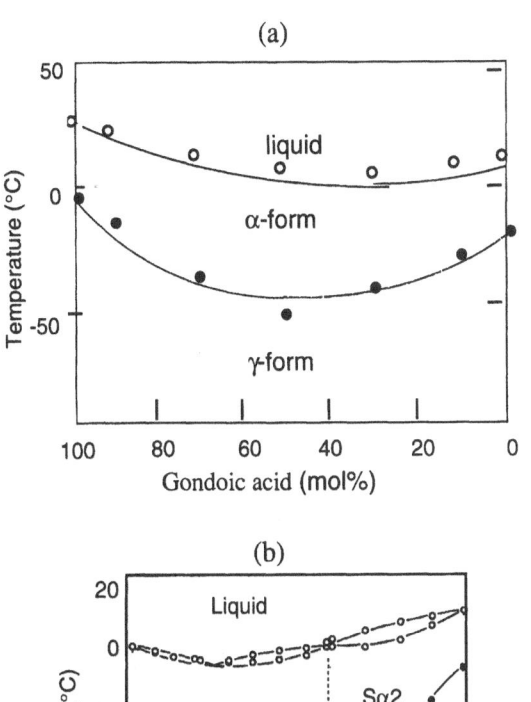

(a)

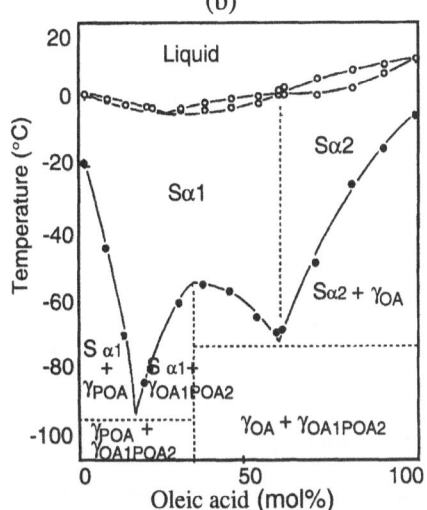

(b)

Fig. 3 Binary mixture phase diagrams of (a) asclepic acid–gondoic acid and (b) palmitoleic acid and oleic acid

understandings of the two compound structures are still open to future work.

___

Stearoyl-oleoyl-diacylglycerols (SODG)

Two types of diacylglycerols containing stearic and oleic acids (SODG) placed at the stereospecific-numbered (*sn*)-1,2 [21] and *sn*-1,3 acyl positions [22] have shown quite interesting structural behavior. Figure 4 shows the single-crystal XRD structure analysis of *sn*-1,3-SODG and a structure model of *sn*-1,2-SODG speculated by powder XRD data.

In the *sn*-1,3-SODG crystals, the stearoyl and oleoyl chains are stacked in a different leaflet in the double chain

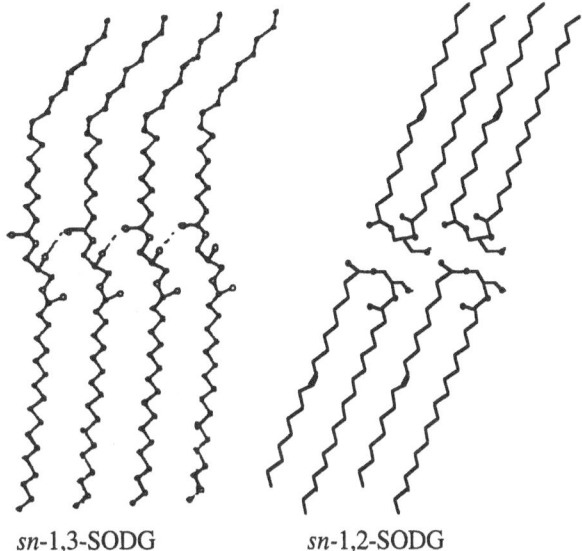

Fig. 4 Structural models of *sn*-1,3-stearoyl-oleoyl-glycerol (SODG) and *sn*-1,2-stearoyl-oleoyl-glycerol (SODG)

length structure, and subcell packing was $T_\parallel$ both for the stearoyl and oleoyl leaflets. The molecule forms an extended V-shaped conformation with the oleoyl and stearoyl chains coming off the two ends of the glycerol group. This structure is essentially equivalent to those of the di-saturated acid DG [40, 41]. The olefinic conformation was *trans–skew–cis–skew–skew–trans* (T–S–C–S–S–T), which is quite different from those revealed in the unsaturated fatty acids (Table 1).

As to *sn*-1,2-SODG in the dry state, eight forms were isolated: α, four β, β', γ_1 and γ_2 [21]. The two γ forms are metastable, occurring at low temperatures. By contrast, the six forms showed unique melting points and subcell structures: H for α, $T_\parallel$ for β and $O_\perp$ for β'. It is notable to see that the β' form having melting point of 25.7 °C was more stable than any of the β forms (20.7–23.1 °C). This makes a marked difference from the polymorphic stability in triacylglycerols, which usually show that the β forms are more stable than β' forms when the two forms occur simultaneously. Some peculiar molecular interactions would be operating in the *sn*-1,2-SODG, where the oleoyl and stearoyl chains are lying in the same leaflet. Hydration appears to partly stabilize the molecular interactions, since three forms of α_w, β_w and γ_w were obtained.

The comparison of the above two mixed acid triacylglycerols shows that, when saturated and unsaturated chains must pack side by side, marked polymorphism showing complex chain conformation, disorder and instability was induced to occur [21]. In this respect, it follows that the glycerol conformation appears to be driven by the hydrogen bond formation, which in turn determines whether the chains are interacting in the same leaflet or are segregated in separate leaflets. For example, in *sn*-1,3-SODG, the two chains point in different directions are segregated, and the acyl chains lie side by side and must interact in *sn*-1,2-SODG. This feature may be related to the lateral packing of the biomembrane lipids, which often contain the saturated and unsaturated fatty acid chains in the same lamellar leaflets [42].

Polymorphism of saturated–unsaturated mixed acid triacylglycerols

Single phases

Now let us move to triacylglycerols containing saturated and oleic acid moieties, which are major fat species in cocoa butter, palm oil and other vegetable fats.

In Fig. 5 displayed are two groups of positional isomers of saturated–oleoyl mixed acid triacylglycerols of SOS and POP, and racemic SSO and PPO. In the former group, the polymorphic transformation from the least stable α to most stable β forms undergoes through the metastable forms of γ and β'. During this transformation, the chain length structure and subcell packing convert from less stable to more stable states, although somehow different between SOS and POP. By contrast, SSO and PPO exhibit simpler transformation from α to β', maintaining the triple chain length structure. It is thought that these differences are ascribed to inter- and intra-molecular interactions through saturated acid and oleic acid moieties, glycerol groups and methyl end packings, as elaborated on SOS in the following.

Figure 6 and Table 2 summarize Gibbs free energy relationships and molecular structures of five forms of SOS examined with DSC, X-ray diffraction, FT-IR and solid-state NMR measurements [23, 27, 28]. The most important properties are irreversible conversions in the chain length structure, subcell packing and olefinic conformation, three of which may interact altogether upon the structural stabilization. For example, the least form of α is loosely packed in hexagonal subcell packing in the double chain length, in which the stearoyl and oleoyl moieties are packed in the same leaflets. This least stable form occurs by quenching neat liquid to far below the melting point of α (23.5 °C). After the formation of α, structural stabilization induces the separation of the stearoyl and oleoyl chains in different leaflets through the chain sorting to form the triple chain length structure. In more detail, the stearoyl chains convert from H in α to $T_{\parallel}$ in β_1, through parallel type and $O_{\perp}$ subcells in γ and β' forms. This conversion is

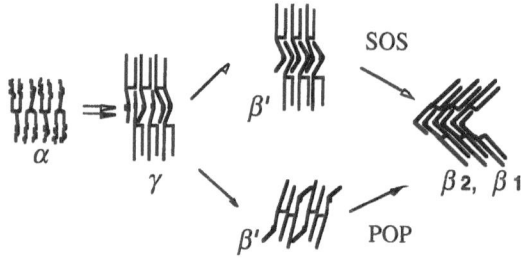

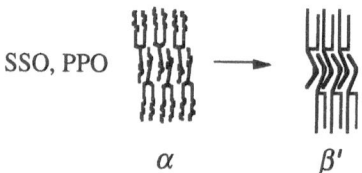

Fig. 5 Polymorphic transformations of SOS, POP, SSO and PPO

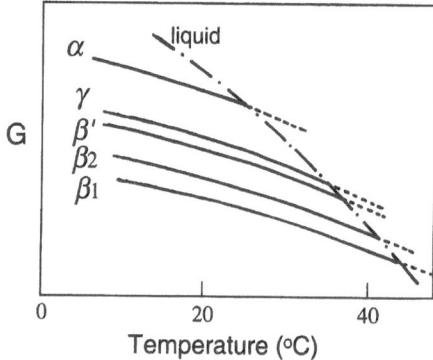

Fig. 6 Gibbs energy (G) vs. temperature (T) relationship of liquid and five forms of SOS

Table 2 Polymorphic structures of SOS

Form	Chain length	Subcell		Olefinic conformation
		Stearoyl	Oleoyl	
α	Double	H	H	Non-specific
γ	Triple	$\parallel$-type	H	Non-specific
β'	Triple	$O_{\perp}$	n.s.	Non-specific
β_2	Triple	$T_{\parallel}$	$T_{\parallel}$ or $O'_{\parallel}$	*scs'*
β_1	Triple	$T_{\parallel}$	$T_{\parallel}$	*scs'*

quite similar to those observed in saturated acids triacylglycerols [43]. By contrast, the conversion in the oleic acid chains seems to be delayed compared to the stearic acid chains, in a form that loosely packed hexagonal arrangement is maintained in γ and β' forms, showing non-specific olefinic conformation. In the stable β forms, the olefinic

Progr Colloid Polym Sci (1998) 108:58–66
© Steinkopff Verlag 1998

conformation of S–C–S′ was revealed, and the subcell packing of the oleic chains was specified. These features give rise to enrichment of the molecular structure model of SOS previously given by Larsson [44].

In addition to the polymorphic crystals displayed in Fig. 6, the most recent study using a Synchrotron radiation X-ray diffraction (SR-XRD) technique has discovered thermotropic liquid crystalline phases of SOS [45]. This discovery was done by a time-resolved analysis of transformation of SOS at 10s intervals. Furthermore, the SR-XRD study enriched the polymorphic behavior of SOS in three respects: (a) the presence of two types of thermotropic liquid crystal phases in SOS, (b) the sequential ordering process in which the lamellae formation occurred in prior to the subcell packing formation, and (c) competitive events of melt-mediated and solid-state transformations which are sensibly dependent on thermal treatments. The former two properties have not been unveiled in the previous SR-XRD studies [46–49], since their time resolution scales were too long to detect the above two events. The third is a direct proof of previously observed phenomena using DSC, conventional XRD and optical techniques [24, 50–52].

Figure 7 shows the α-mediated crystallization, in which the neat liquid was chilled at 10 °C to crystallize in α, and subjected to rapid heating to 30 °C to crystallize the β' form at the expense (melting) of the α form, as displayed in an inserted temperature profile. The rapid melting of the α form at 30 °C is shown by the disappearance of the corresponding long and short spacing XRD spectra. Soon after the melting of α, a long spacing peak of 5.1 nm appeared without the corresponding short spacing spectra. This feature is typical of smectic type liquid crystal [53], hence it was called LC1 phase. The β' form started to occur at 30 °C at the expense of LC1 phase, revealing the long spacing spectra of 7.1 nm in the first. The short spacing spectra appeared a few minutes later than the occurrence of the long spacing patterns. The presence of the liquid crystals in SOS was confirmed by direct chilling of neat liquid, and thermodynamic stability was assessed by observing the melting behavior of the LC1 phase. The thermal treatment around 33 °C revealed the second liquid crystal phase (LC2) [45]. It may be referred here that the observation of the liquid crystals may be an answer to a long-standing dispute about the presence of ordering structures in liquidus triacylglycerols [54–57].

Binary mixture phases

Peculiar molecular interactions through the oleic acid moiety of the saturated–oleoyl mixed acid triacylglycerols

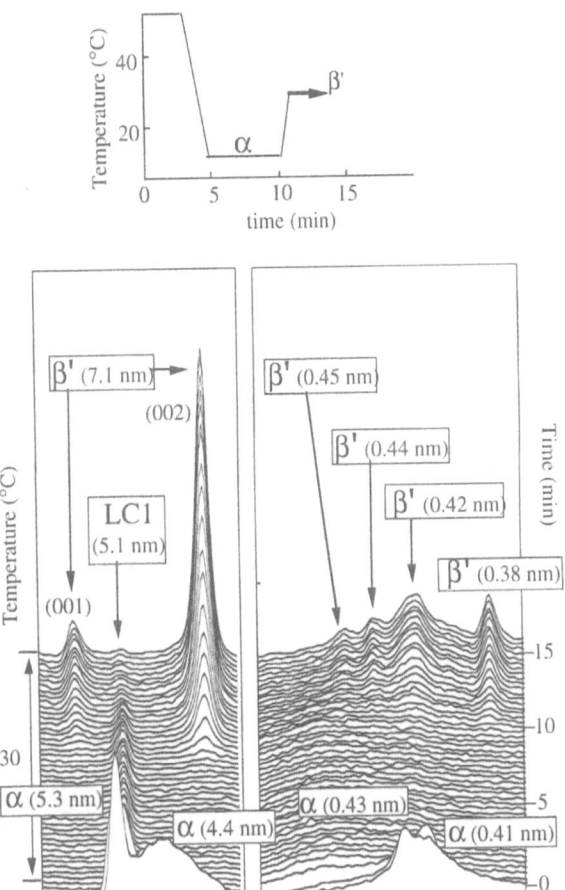

Fig. 7 Synchrotron radiation X-ray diffraction spectra of quenching and rapid heating of SOS

have been analyzed by SR-XRD and DSC on binary mixture phases of POP–PPO [58] and POP–OPO [59]. A peculiarity exhibited in these mixtures was that, as shown in Fig. 8, molecular compounds of the double chain length structure were formed at the 1:1 concentration ratios, although the component molecules are all of triple chain length.

Figure 9 shows a phase diagram in the most stable states (Fig. 9a) and a kinetic phase diagram in metastable states (Fig. 9b). It is notable that the molecular compound reveals three polymorphic transformations from α, β' and β, each of which makes monotectic phases with respect to each component material. This kinetic behavior was verified by a time-resolved SR-XRD and DSC, as shown in Fig. 10 taken for the mixture of POP:PPO = 10:90. The DSC heating record shows the melting peaks of two α forms of compound (α_c) and PPO (α_{PPO}), which were soon followed by the crystallization and transformation of

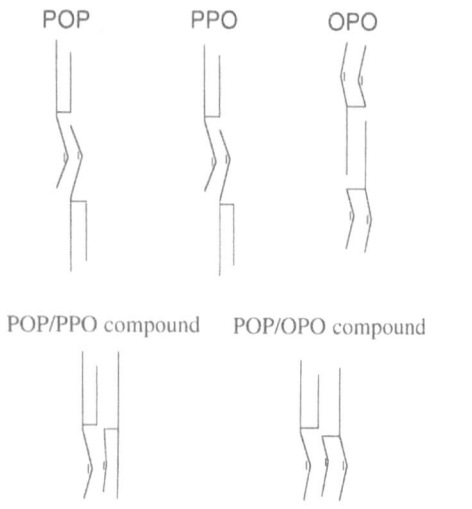

Fig. 8 Molecular models of POP, PPO, OPO and compounds of POP–PPO and POP–OPO

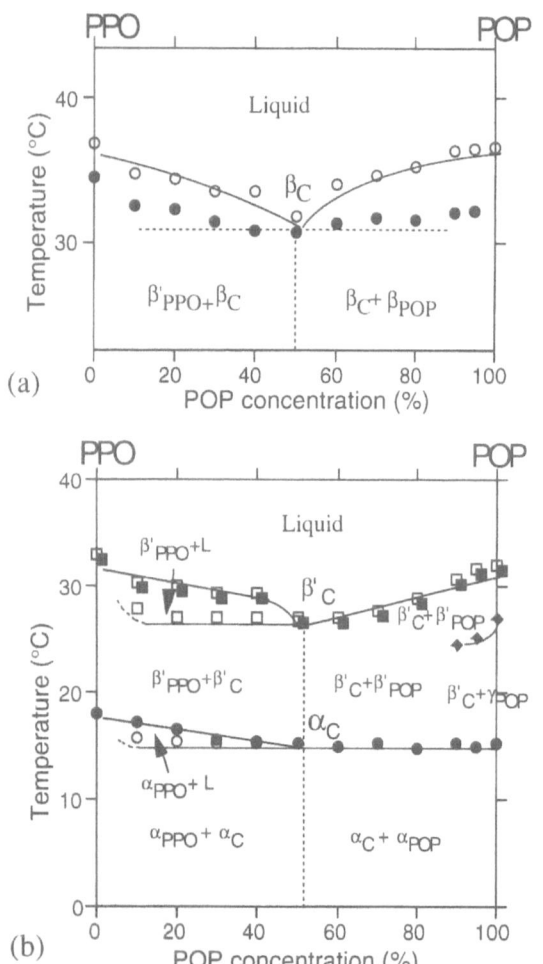

Fig. 9 Phase diagrams of (a) the most stable forms and (b) the metasable forms of binary mixtures of POP–PPO

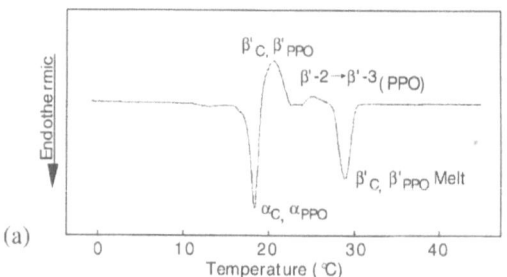

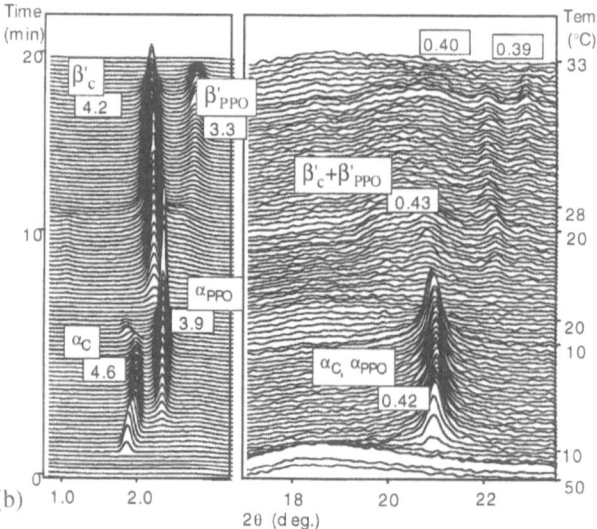

Fig. 10 Synchrotron radiation X-ray diffraction spectra and DSC recording of the binary mixture of POP–PPO = 10–90

β' of the compound (β'_c) and PPO (β'_{PPO}). The clarification of multiple endothermic and exothermic DSC peaks was perfectly done by the time-resolved SR-XRD spectra. In the same manner shown in Figs. 7 and 8, although using the conventional XRD techniques, the molecular compound formation of SOS–OSO [60] SOS–SSO [61] at the 1:1 ratio was confirmed.

It is quite interesting to determine the molecular structures of the compounds and to evaluate the interaction energies as well. FT-IR studies of the stable β_c forms of the POP–OPO and POP–PPO compounds have given the following preliminary results [62]: (a) for β_c (POP–OPO), the subcell structures of the oleic and palmitic leaflets are both $T_{\parallel}$ and the olefinic conformation is S–C–S', (b) for β_c (POP–PPO), the subcell structures are not uniquely determined, and the olefinic conformation is neither of S–C–S' type nor of S–C–S type. This means that the olefinic structures are deformed in β_c (POP–PPO), where the saturated and unsaturated chains must pack side by side in the same leaflet, similar to sn–1,2–SODG as discussed in the previous section (see Fig. 8).

Progr Colloid Polym Sci (1998) 108:58–66
© Steinkopff Verlag 1998

Conclusion

The structural properties of unsaturated fats and lipids are characterized by their diversity in polymorphism, in terms of *cis*-conformation, molecular packing, chain length structure and transformation behavior. This diversity is caused by the number and conformation of the double bonds present in a molecule as internal influence, and also by temperature, mixing and glycerol structures as external influences. Although not elaborated here, these properties have direct relations to the application; for example, (a) cocoa butter polymorphism can be basically understood in a framework of the polymorphism of SOS [36],

(b) structural control of polymorphic crystallization of cocoa butter using seed crystals could be done based on the polymorphic analysis of SOS [63], (c) granular crystals growing in palm oil blending fats were determined as POP β_1 involved in palm oil [64], and so on.

As a future work, it is interesting to point out the opportunity and challenge to polyunsaturated fatty acids and their glycerol molecules, whose structural behavior has been poorly understood. Molecular-level understanding of the physical properties of the polyunsaturated fats and lipids will be elucidated, hopefully based on the understandings on the mono-unsaturated fats and lipids reviewed above.

References

1. Hui YH (ed) (1996) Bailey's Industrial Oil and Fat Products, 5th ed. Wiley-Interscience, New York
2. Suzuki M, Ogaki T, Sato K (1985) J Am Oil Chem Soc 62:1600–1604
3. Sato K, Suzuki M (1986) J Am Oil Chem Soc 63:1356–1359
4. Kobayashi M, Kaneko F, Sato K, Suzuki M (1986) J Phys Chem 90:6371–6378
5. Suzuki M, Sato K, Yoshimoto N, Tanaka S, Kobayashi M (1988) J Am Oil Chem Soc 65:1942–1947
6. Hiramatsu N, Inoue T, Suzuki M, Sato K (1989) Chem Phys Lipids 51:47–53
7. Sato K, Yoshimoto N, Suzuki M, Kobayashi M, Kaneko F (1990) J Phys Chem 94:3180–3185
8. Hiramatsu N, Sato T, Inoue T, Suzuki M, Sato K (1990) Chem Phys Lipids 56:59–63
9. Yoshimoto N, Suzuki M, Sato K (1991) Chem Phys Lipids 57:67–73
10. Yoshimoto N, Nakamura T, Suzuki M, Sato K (1991) J Phys Chem 95:3384–3390
11. Hiramatsu N, Inoue T, Sato T, Suzuki M, Sato K (1992) Chem Phys Lipids 61:283–291
12. Inoue T, Motoda I, Hiramatsu N, Suzuki M, Sato K (1992) Chem Phys Lipids 63:243–250
13. Kaneko F, Kitagawa Y, Matsuura Y, Sato K, Suzuki M, Kobayashi M (1992) Acta Cryst C48:1054–1057
14. Kaneko F, Kobayashi M, Kitagawa Y, Matsuura Y, Sato K, Suzuki M (1992) Acta Cryst C48:1057–1060
15. Kaneko F, Kobayashi M, Kitagawa Y, Matsuura Y, Sato K, Suzuki M (1992) Acta Cryst C48:1060–1063
16. Inoue T, Motoda I, Hiramatsu N, Suzuki M, Sato K (1993) Chem Phys Lipids 66:209–214
17. Kaneko F, Yamazaki K, Kobayashi M, Sato K, Suzuki M (1994) Spectrochim Acta 50A:1589–1603
18. Ueno S, Suetake T, Yano J, Suzuki M, Sato K (1994) Chem Phys Lipids 72:27–34
19. Kaneko F, Yamazaki K, Kobayashi M, Kitagawa Y, Matsuura Y, Sato K, Suzuki M (1996) J Phys Chem 100:9138–9148
20. Kaneko F, Yamazaki K, Kitagawa K, Kikyo K, Kobayashi M, Sato K, Suzuki M (1997) J Phys Chem B 101:1803–1809
21. Li D, Small DM (1993) J Lipid Res 34:1611–1624
22. Goto M, Honda K, Li D, Small DM (1995) J Lipid Res 36:2185–2190
23. Sato K, Arishima T, Wang ZH, Ojima K, Sagi N, Mori H (1989) J Am Oil Chem Soc 66:664–674
24. Koyano T, Hachiya I, Arishima T, Sato K, Sagi, N (1989) J Am Oil Chem Soc 66:675–679
25. Arishima T, Sagi N, Mori H, Sato K (1991) J Am Oil Chem Soc 68:710–715
26. Koyano T, Hachiya I, Arishima T, Sagi N, Sato K (1991) J Am Oil Chem Soc 68:716–720
27. Yano J, Ueno S, Sato K, Arishima T, Sagi N, Kaneko F, Kobayashi M (1993) J Phys Chem 97:12967–12973
28. Arishima T, Sugimoto K, Kiwata R, Mori H, Sato K (1996) J Am Oil Chem Soc 73:1231–1236.
29. Wang Z, Lin H, Li S, Huang C (1995) J Biol Chem 270:2014–2023
30. Wang G, Lin W, Li S, Huang C (1995) J Biol Chem 270:22738–22746
31. Larsson K (1963) Acta Cryst 16:741–748
32. Larsson K (1964) Ark Kemi 23:35–56
33. Larsson K (1966) Acta Chem Scand 20:2255–2260
34. Abrahamsson S, Dahlen B, Lofgren H, Pascher I (1978) Prog Chem Fats Other Lipids 16:125–143
35. Larsson K (1994) In Lipids – Molecular Organization, Physical Functions and Technical Applications. Oily Press, Dundee, pp 75–80
36. Sato, K (1996) In: Padley F (ed), Advances in Applied Lipid Research, Vol 2. JAI Press Inc, London, pp 213–268
37. Abrahamsson S, Ryderstadt-Nahringbauer I (1962) Acta Cryst 15:1261–1264
38. Sato K, Kobayashi M (1991) In: Karl N (ed) Crystals, Vol 13, Organic Crystals I Characterization. Springer, Heidelberg, pp 65–108
39. Sato K, Yano J, Kawada I, Kawano M, Suzuki M (1997) J Am Oil Chem Soc 74:1153–1159
40. Larsson K (1963) Acta Cryst 16:741–748
41. Hybl A, Dorset D (1971) Acta Cryst B27:977–980
42. Small DM (1984) J Lipid Res 25:1490–1500
43. Hernqvist L (1988) In: Garti N, Sato K (Eds) Crystallization and Polymorphism of Fats and Fatty Acids. Marcel Dekker, New York, pp 97–137
44. Larsson K (1972) Fette Seifen Anstrichm 74:136–142
45. Ueno S, Minato A, Seto H, Amemiya Y, Sato K (1997) J Phys Chem B 101:6847–6854
46. Kellens M, Meeussen W, Riekel C, Reynaers H (1990) Chem Phys Lipids 52:79–98
47. Kellens M, Meeussen W, Gehrke R, Reynaers H (1991) Chem Phys Lipids 58:131–144
48. Kellens M, Meeussen W, Hammersley A, Reynaers H (1991) Chem Phys Lipids 58:145–158

49. Gelder RNMR, Hodgson N, Roberts KJ, Rossi A (1996) In: Myerson AS, Green DA, Meenan P (eds) Crystal Growth of Organic Materials, Am Chem Soc Washington DC, pp 209–215

50. Sato K (1993) J Phys D: Appl Phys 26:B77–B84

51. Rousset P, Rappaz M (1996) J Am Oil Chem Soc 73:1051–1057

52. Rousset P, Rappaz M (1997) J Am Oil Chem Soc 74:693–698

53. de Gennes PG (1974) In: The Physics of Liquid Crystals. Clarendon Press, Oxford, pp 1–22

54. Hernqvist L, Larsson K (1982) Fette Seifen Anstrichm 84:349–354

55. Cebula DJ, McClements DJ, Povey MJW (1990) J Am Oil Chem Soc 67:76–78

56. Cebula DJ, McClements DJ, Povey MJW, Smith PR (1992) J Am Oil Chem Soc 69:130–136

57. Larsson K (1992) J Am Oil Chem Soc 69:835–836

58. Minato A, Ueno S, Smith K, Amemiya Y, Sato K (1997) J Phys Chem B 101:3498–3505

59. Minato A, Ueno S, Yano J, Smith K, Seto H, Amemiya Y, Sato K (1997) J Am Oil Chem Soc 74:1213–1220

60. Koyano T, Hachiya I, Sato K (1992) J Phys Chem 96:10514–10520

61. Engstrom L (1992) J Fat Sci Technol 94:173–181

62. Minato A, Yano J, Ueno S, Smith K, Sato K (1997) Chem Phys Lipids 88:63–71

63. Hachiya I, Koyano T, Sato K (1989) J Am Oil Chem Soc 66:1757–1762

64. Watanabe A, Tashima I, Matsuzaki N, Kurashige J, Sato K (1992) J Am Oil Chem Soc 69:1077–1080

Progr Colloid Polym Sci (1998) 108:67–75
© Steinkopff Verlag 1998

I. Pascher
M. Lundmark
S. Sundell
H. Eibl

Conformation and packing of membrane lipids: Crystal structures of lysophosphatidylcholines

I. Pascher (✉) · M. Lundmark · S. Sundell
Department of Medical Biochemistry
University of Göteborg
Medicinaregatan 9
S-413 90 Göteborg
Sweden
E-mail: irmin.pascher@medkem.gu.se

H. Eibl
Max-Planck-Institut für
Biophysikalische Chemie
Postfach 2841
D-37018 Göttingen
Germany

Abstract The molecular conformation and packing of three lysophosphatidylcholines: 3-palmitoyl-D-glycero-1-phosphocholine (PPC), 3-hexadecyl-D-glycero-1-phosphocholine (HPC) and 3-palmitoyl-DL-glycero-1-phospho-N,N-dimethylethanolamine (PPEM$_2$) have been determined by X-ray single crystal analyses. PPC crystallizes as the monohydrate and HPC as chloroform solvate, both with triclinic unit cells (space group P1) containing two independent molecules in almost identical packing arrangements. The two molecules of PPC/HPC are mirror image conformers with respect to their head groups and pack separately in either half of a bilayer arrangement with interdigitating hydrocarbon chains and interdigitating head groups. The racemic PPEM$_2$ also crystallizes with a very similar interdigitating packing arrangement. The unit cell, however, is monoclinic (space group P2$_1$/a) and comprises four molecules, arranged as pairs of centrosymmetric D/L conformers at either side of the bilayer. In PPC and HPC the C$_{16}$-hydrocarbon chains interdigitate with an overlap of 13 carbon atoms only, leaving a cavity to accommodate the solvate molecules. The chain matrices have identical tilt (45°), but different chain packing modes (O'⊥ and O∥). In PPEM$_2$ the chain ends penetrate with 18-atoms interdigitation to the glycerol region of oppositely oriented molecules and pack in a hybrid matrix with 37° tilt. Despite the differences in hydrocarbon chain attachment (ester/ether) and degree of N-methylation the structures show great similarities, in particular with respect to the head group conformation, which apparently is favored by intrinsic energetics.

Key words Phospholipids – lysophosphatidylcholines – crystal structure – conformation – polar interactions

Introduction

Lysophosphatidylcholines constitute only a minor fraction of the lipid types occurring in biomembranes [1], primarily as intermediates in the turnover of phosphatidyl-cholines. However, different lysophosphatidylcholine species have been shown to be involved in a variety of biologically important activities [2, 3]. In particular the ether analogues, alkyl-2-acetyl-glycero-phosphocholines and alkyl-2-methyl-glycero-phosphocholines or related alkyl-phosphocholines were found to function as platelet activating factor (PAF) [4, 5] and anti-cancer drugs [3, 6–8], respectively. In aqueous dispersions lysophosphatidylcholines exhibit a complex structural behavior forming lamellar, micellar, cubic and hexagonal (H$_I$) phases [8, 9]. Below the chain melting temperature a lamellar L$_{bi}$ phase with very thin layer thickness is observed, indicating that a bilayer structure with inter-digitating hydrocarbon chains is formed [10, 11].

We have earlier reported on the crystal structure of different glycerophosphocholine compounds such as D-glycero-1-phosphocholine[1] [12], 3-lauroyl-2-deoxy-glycero-1-phosphocholine [13], 3-octadecyl-2-methyl-D-glycero-1-phosphocholine [14] and 2,3-dimyristoyl-D-glycero-3-phosphocholine [15]. In all these compounds the phosphocholine head groups show very similar conformational features. The packing pattern of the phosphocholine dipoles, however, display great variations.

In the present paper we report on the crystal structure of 3-palmitoyl-D-glycero-1-phosphocholine (PPC), 3-hexadecyl-D-glycero-1-phosphocholine (HPC) and 3-palmitoyl-DL-glycero-1-phospho-N,N-dimethylethanolamine (PPEM$_2$). These structure determinations were made to provide further information with atomic resolution on the structure of phosphatidylcholines, in particular on packing features of ester and ether derivatives and on effects on head group conformation and interactions arising from changes in the degree of N-methylation at the ethanolamine nitrogen. With respect to head group interactions

lyso-compounds are advantageous as their head group arrangement is usually more relaxed and less affected by chain packing requirements than in corresponding double-chain lipids.

Experimental

3-Palmitoyl-D-glycero-1-phosphocholine (= 1-palmitoyl-sn-glycero-3-phosphocholine) (PPC), 3-Hexadecyl-D-glycero-1-phosphocholine (= 1-hexadecyl-sn-glycero-3-phosphocholine) (HPC) and 3-palmitoyl-DL-glycero-1-phospho-N,N-dimethylethanolamine (= 1-palmitoyl-rac-glycero-3-phospho-N,N-dimethylethanolamine) (PPEM$_2$) were sythesized according to Eibl and Woolly [16]. In all compounds homologfree fatty acid or fatty alcohol constituents and reagents of highest purity were used.

Before crystallization the lipids were dried thoroughly in vacuum over P_2O_5. Crystals suitable for X-ray single crystal analysis were grown from 1–2% lipid solutions in anhydrous solvents at 18 °C. For solvent and crystal data see Table 1. The obtained crystal phases of PPC and HPC,

Table 1

Crystal data	PPC	HPC	PPEM$_2$
Molecular formula	$C_{24}H_{50}NO_7P \cdot H_2O$	$C_{24}H_{52}NO_6P \cdot CHCl_3$	$C_{23}H_{48}NO_7P$
Molecular weight	513.65	601.03	481.61
Crystallization solvent	Ether:ethanol 5:2*	Ether:chloroform 5:3*	Ether:ethanol 5:2.5*
Crystal dimensions (mm)	$0.02 \times 0.36 \times 0.48$	$0.05 \times 0.10 \times 0.29$	$0.02 \times 0.24 \times 0.43$
recorded θ-range (°)	$1 < \theta < 65$	$1 < \theta < 50$	$1 < \theta < 45$
Unique/observed reflexions ($I > 3\sigma[I]$)	2921/1004	3351/1159	2238/1243
Absorption coefficient $\mu_{CuK\alpha}$ (cm^{-1})	10.87	32.70	11.90
R-values R/R_w	0.106/0.149	0.127/0.164	0.106/0.146
Crystal class	Triclinic	Triclinic	Monoclinic
Space group	P1	P1	P2$_1$/a
Molecules/unit cell Z (independent)	2(2)	2(2)	4(1)
Unit cell parameters (with std's):			
a, b, c (Å)	5.824(3), 9.857(4), 28.877(15)	5.914(4), 9.864(3), 30.813(7)	11.238(3), 8.586(1), 28.967(6)
α, β, γ (°)	91.89(4), 90.34(4), 105.67(4)	98.84(2), 102.02(4), 106.95(4)	90.00, 96.27(2), 90.00
Unit cell volume (Å^3)	1595.07	1636.94	2778.03
Density$_{calc}$ (g cm^{-3})	1.0694	1.2186	1.1513
Bilayer thickness, l (Å)	28.86	29.33	28.80
Molecular area, S (Å^2)	55.3	55.7	48.2
Hydrocarbon matrix:			
chain packing mode	O'⊥	O‖	HS (monoclinic)
subcell: a_s, b_s, c_s (Å)	7.67(6), 5.12(3), 2.52(4)	8.86(6), 8.98(6), 2.54(4)	9.05(9), 8.58(1), 2.54(1)
$\alpha_s, \beta_s, \gamma_s$ (°)		86(1), 92(1), 90(1)	89.2(6), 101.6(1.4), 89.7(1)
chain cross-section Σ (Å^2)	19.5	19.4	19.2
chain tilt angle ϕ (°)	45	45	37
matrix thickness (Å)	10.3	9.3	16.5

* Anhydrous conditions.

---

[1] In structural-conformational studies the sn-nomenclature is unrational and preferably replaced by the more general Fischer convention, which allows identical, comparable atom numbering in chiral antipodes, see Ref. [21].

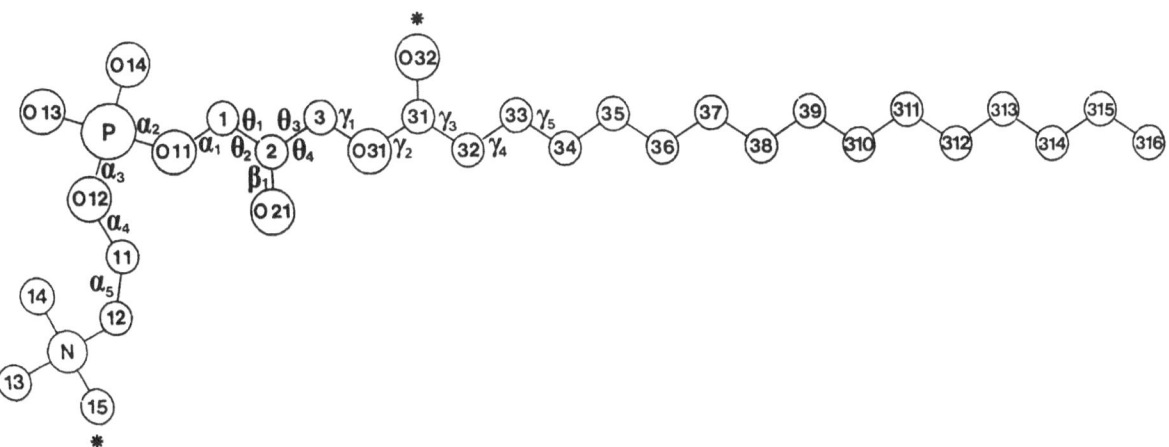

Fig. 1 Atom numbering and notations of torsion angles for palmitoyl-glycero-phosphocholine (PPC). In hexadecyl-glycero-phosphocholine (HPC) and palmitoyl-glycero-phospho-N,N-dimethylethanolamine (PPEM$_2$) the marked atoms O32* and C15*, respectively, are replaced by hydrogens. For numbering conventions of glycerolipids in structural-conformational studies see Ref. [21]

but not of PPEM$_2$, were found to accommodate one solvent molecule (water and chloroform, respectively) per lipid molecule. The crystals of PPC and HPC were unstable in moist air. They were handled in a dry atmosphere and mounted in sealed lithium glass tubes.

X-ray reflections were recorded on an Enraf-Nonius CAD4F-11 diffractometer. The angular settings of 25 reflections were measured to calculate the lattice parameters. Intensity data for one hemisphere of reflection were collected by the $\omega/2\theta$ scan method using monochromatized CuKα-radiation. Three intensity control reflections were measured every 2 h to control crystal decay. The intensities were scaled to account for this decay (10–30%). The numbers of independent reflections and observed reflections ($I > 3\sigma[I]$) for each compound are given in Table 1. All intensities were corrected for Lorentz and polarization effects but not for absorption or extinction.

The structures were solved by a combination of Patterson heavy atom method and direct methods using the program DIRDIF [17] which provided the non-hydrogen atom positions. All hydrogen atom positions except those of the water molecule in PPC were obtained from Fourier difference synthesis maps. Refinement was carried out by the full-matrix least-squares method using anisotropic temperature factors for the non-hydrogen atoms. The hydrogen atoms were assigned a common temperature factor ($B = 5 \text{Å}^2$) but their atomic parameters were not refined. The structures were refined to R- and R_w-values of about 0.10 and 0.15, respectively (see Table 1). All calculations have been performed using mainly the XTAL2.2 program system [18].

Description of the structure

Atomic coordinates and equivalent isotropic temperature factors (U_{eq}) of the non-hydrogen atoms have been deposited at the Cambridge Structure Database. Observed and calculated structure factors and anisotropic temperature factors can be obtained from this Department. The atom numbering and torsion angle notation, according to the convention of Sundaralingam [19], are shown in Fig. 1.

Molecular packing

The packing arrangements of the three lyso-PC compounds and the extension of the unit cells are shown in Fig. 2 in projections along the two short unit cell axes. In all three structures the molecules pack in a double-layer structure with interdigitating hydrocarbon chains and interdigitating head groups.

The unit cells of the enantiomeric PPC and HPC contain two different conformers, A and B, which are orientated with their heads in opposite directions and pack separately in either half of the double-layer. As will be discussed below the phosphocholine groups of molecules A and B are related by a non-crystallographic centro-symmetry and are practically mirror images with respect to conformation and interaction pattern.

The unit cell of the racemic PPEM$_2$ contains four molecules, which, however, are identical mirror image conformers related by crystallographic centro-symmetry. D- and L-enantiomers pack alternatingly within each

Fig. 2 Molecular packing of lysophosphatidylcholines in views along the unit cell *a*-axis (left) and *b*-axis (right). The unit cells of the enantiomeric PPC and HPC lipids contain two conformers, A and B, which are arranged in separate layers on either side of the double-layer structure. PPC accommodates a water molecule and HPC a chloroform molecule of solvation per lipid molecule. The unit cell of the racemic PPEM$_2$ lipid contains four molecules which are related by centro-symmetry and arranged in pairs of D/L mirror images on each side of the bilayer

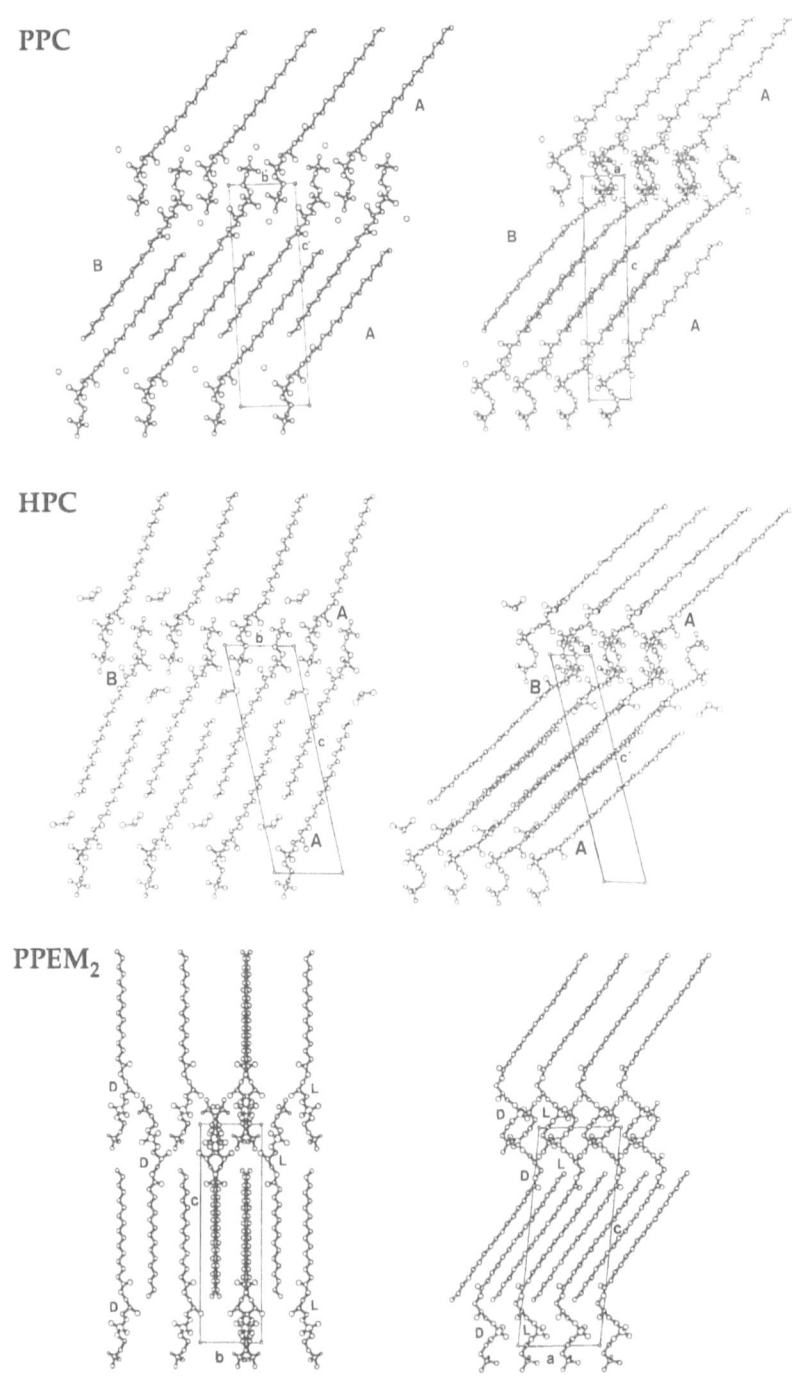

PPC

HPC

PPEM$_2$

molecular layer and the unit cell contains two pairs of D/L-conformers, one pair at each side of the bilayer.

As shown in Fig. 2 the phosphoammonio head group dipoles extend perpendiculary to the layer plane and inter-digitate with the head group dipoles of the adjacent lipid double-layer. This results in a single, about 7.5 Å thick matrix of head group dipoles at the interface between two lipid double-layers. In these packing arrangements the molecular area in the layer plane (*ab*-plane) is determined by the packing requirements of two antiparallel head group dipoles. For PPC and HPC the molecular packing cross-section (*S*) is practically identical (55.3 and 55.7 Å^2), while for PPEM$_2$ it is distinctly smaller (48.2 Å^2). To this rather large molecular packing area the hydrocarbon

Progr Colloid Polym Sci (1998) 108:67–75
© Steinkopff Verlag 1998

chains accommodate by interdigitation with the chains of oppositely oriented molecules and, additionally, by a tilt (ϕ) with respect to the layer normal of 45° in PPC/HPC and 37° in PPEM$_2$. In PPC/HPC the C$_{16}$-hydrocarbon chains thereby interdigitate by 13 carbon atoms only, leaving in extension of the chain ends a large cavity in which solvent molecules (H$_2$O and CHCl$_3$) are accommodated. In PPEM$_2$, on the other hand, the methyl chain ends interdigitate to the level of the glycerol carbon C3 of the opposite molecules, corresponding to a chain overlap of 18 atoms, not forming such an internal cavity. This partial or full interdigitation gives rise to a chain matrix layer which in PPC/HPC is only 10.3/9.3 Å thick, while in PPEM$_2$ it is 16.5 Å thick.

In the three compounds the hydrocarbon chains are arranged according to different chain packing modes [20]: orthorhombic perpendicular (O'⊥) in PPC, orthorhombic parallel (O∥) in HPC and a hybrid packing mode (HS) in PPEM$_2$ (for subcell parameters see Table 1). The packing cross-sections of the chains (Σ), perpendicular to their long axis are 19.5/19.4 Å^2 in PPC/HPC and 19.2 Å^2 in PPEM$_2$. With reference to the rather thin chain matrix in PPC/HPC their hydrocarbon chains must be considered as well packed.

As already mentioned, the incomplete chain interdigitation in PPC/HPC creates a rather large cavity in exension of each hydrocarbon chain, which is 8.9/9.9 Å in length and about 180 Å^3 in volume. In PPC this cavity is only partially filled up by a water molecule of solvation. The distance between the methyl chain end and the water molecule (5.7 Å) is considerably larger than a van der Waals contact. In an ordered chain matrix, however, the chains can slide past each other by integral zigzag units only. An interdigitation of the chains by another ethylene unit ($c_s = 2.52$ Å) would bring the methyl chain end too close to the water molecule. This empty space in extension of each chain end causes a rather low specific density ($D_c = 1.069$ g cm^{-3}) of the PPC crystal.

In HPC the cavity is, compared to PPC, one Å longer but more effectively filled by the much larger chloroform molecule. Judging from the high temperature factors of the chloroform atoms the fit of the solvate molecule is still not optimal. The heavy chlorine atoms, however, substantially increase the crystal density ($D_c = 1.219$ g cm^{-3}) of HPC.

Polar interactions

The packing patterns and interactions of the phospho-ammonio groups of the three lyso-lipids are shown in Fig. 3. As it becomes obvious from the view parallel to the double-layer interface (Figs. 2A and 3A) the P–N dipoles

extend almost perpendicular to the layer plane and interdigitate with those of the next double layer. Thereby the (+) ammonio groups of one layer protrude to the level of the (−) phosphate group of the adjacent head group layer. In this layer of interdigitating head groups two layers of (+) and (−) charges are formed, 4.2 Å apart.

In PPC/HPC the pseudo-centrosymmetry of the head groups entails that the packing and interaction pattern of one layer of phosphate (A) and choline (B) groups is a mirror image of the corresponding phosphate (B) and choline (A) pattern of the other half of the interdigitating headgroup layer. Note, however, that this mirror symmetry is not perfect and minor differences in the length of intermolecular contacts exists for molecules A and B.

In PPEM$_2$ the head groups display a rather similar interdigitating packing arrangement, however, the D- and L-enantiomers are true configurational/conformational mirror images and have identical centro-symmetry related packing contacts and interaction patterns.

Interestingly, in all three lysolipids short electrostatic contacts are established only between distinct pairs of the (+) ammonio and (−) phospho groups, rather than towards several surrounding neighbor molecules. The ammonio nitrogen in PPC/HPC thus makes two short lateral contacts (<4.0 Å) towards O12 and O13 of a neighboring phosphate group and a third short contact (3.7 Å) to the glycerol O21 of another neighbor molecule.

What is most significant for all phospholipids, however, is that the unesterified phosphate oxygens O13 and O14 strongly attract hydrogen donor groups. In all three lyso-compounds one hydrogen donor bond is provided by the fri glycerol hydroxyl group O21 → O14 ($\approx$ 2.7 Å), not intramolecularly, but intermolecularly from an a-translated neighbor molecule. The other hydrogen bond towards O13, however, comes in the three lipids from different donors. In PPC the water molecule of hydration (W1) directs one hydrogen bond (2.9 Å) to O13 (but does not engage its second hydrogen). Note, that this water molecule was retained or attracted, although PPC was dried over P$_2$O$_5$ and crystallized under anhydrous conditions.

Surprisingly, in HPC the polarized C–H bond of chloroform acts as a hydrogen donor ($\approx$ 3.1 Å) towards O13. Finally in PPEM$_2$ the not methyl-substituted ammonio hydrogen interacts with O13 of a neighboring interdigitating phosphate group. This N–H → O13 hydrogen bond (2.7 Å) together with an N(+) → O11 ionic bond (3.4 Å) directed from the ammonio group towards one neighboring phosphate group results for PPEM$_2$ in a closer contact between head group pairs and a substantial reduction (13%) of the molecular packing cross-section compared to PPC/HPC (see Table 1).

Fig. 3 Packing and interactions of the zwitter-ionic head group dipoles of PPC, HPC and PPEM$_2$ in views parallel (left) and perpendicular (right) to the bilayer interface. Hydrogen bonds are indicated by dotted lines, electrostatic and polar contacts by broken lines. Contact distances are given in Å

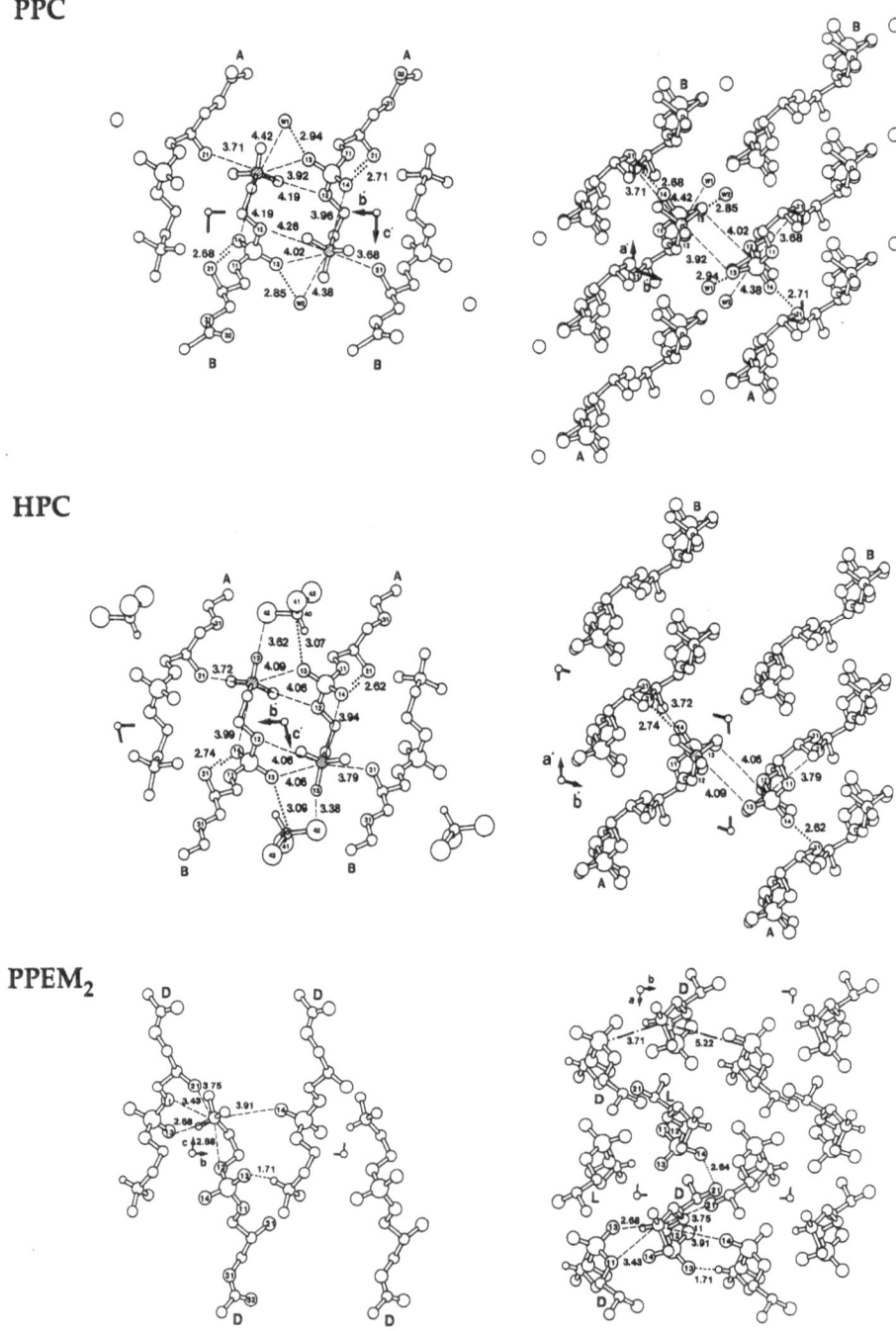

PPC

HPC

PPEM$_2$

Molecular conformation

The conformations of molecules A and B of PPC/HPC and of the D- and L-enantiomer of PPEM$_2$ are shown in Fig. 4. Torsion angles of the molecules in comparison to corresponding diacyl derivatives are given in Table 2. For notations of the torsion angles see Fig. 1 and Refs. [19, 21]. The terminology of Klyne and Prelog [22] is used to describe conformational ranges.

PPC/HPC: Although the PPC/HPC-molecules have the natural *D*-configuration the polar part of molecules A and B adopts a quasi-symmetric mirror image conformation up to glycerol carbon C2. The two glycerol oxygens O11 and O21 of molecules A and B still retain mirror image geometry ($\theta_2 = +$ sc (A) and $-$ sc (B)), since O21 is involved in the pseudo-centrosymmetric interaction pattern of the head group dipoles. Due to chirality the centro-symmetry is broken at glycerol carbon C2 and the glycerol carbons C3 in molecules A and B are oriented differently ($\theta_1 = -$ sc and ap) with respect to the polar group. On the other hand, the glycerol-attached hydrocarbon chains have from C3 onwards a rather identical antiplanar conformation but different orientations of their chain planes with respect to their head groups. In PPC A the orientation of the chain plane is additionally effected by torsion $\gamma_1 = $ ac, to allow an accommodation of the chains in the O'$\perp$ chain matrix.

PPEM$_2$: As evident from Fig. 4 and torsion angles in Table 2 the D- and L-enantiomers of PPEM$_2$ are true mirror images and resemble with respect to their polar group conformation the molecules B and A of PPC/HPC. The conformational similarity of the PPEM$_2$ L-enantiomer with the D-enantiomeric PPC/HPC molecule A comprises the α-chain only, while for the PPEM$_2$ D-enantiomer and the D-enantiomeric PPC/HPC molecule B it also includes the glycerol torsions θ_1/θ_2. In PPEM$_2$, a significant difference exists in the torsions $\theta_3/\theta_4 = -$ sc/ap (D) and sc/ap (L), which turn the glycerol oxygen O3 and the rest of the γ-chain perpendicularly to the direction of the glycerol chain. Compared to PPC/HPC this bend at the C2–C3 glycerol bond reverses the direction of the chain tilt in PPEM$_2$.

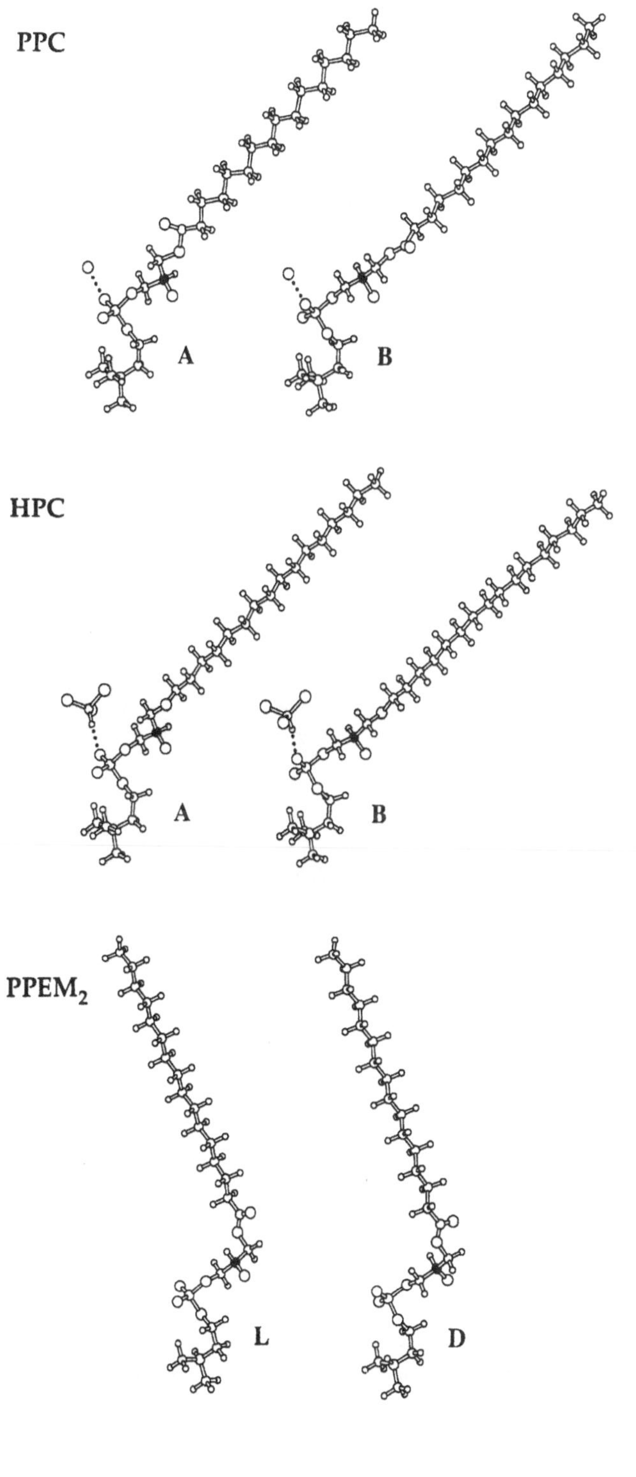

Discussion

As mentioned earlier, in the crystal stuctures of phosphatidylethanolamines and phosphatidylcholines solved so far the head groups exhibit similar conformations but very different packing patterns [21]. Among the PC-compounds only 3-lauroyl-2-deoxy-glycero-1-phosphocholine [13] shows the "expected" layer-parallel orientation of the P–N dipoles, while in 2,3-dimyristoyl-*DL*-glycero-1-phosphocholine [15] the head group layer is folded into a space-saving saw-tooth arrangement with a packing area of 38 Å^2 per molecule. In 3-octadecyl-2-methyl-*DL*-glycero-1-phosphocholine [14], on the other hand, the P–N dipoles are aligned layer-parallelly, but interdigitate with those of the adjacent bilayer. This results in a very large packing cross-section per molecule (74 Å^2) corresponding to a packing area of almost two phosphocholine groups.

The presented structures of the three lyso-PCs show still another head group arrangement with extended,

Fig. 4 Molecular conformation of conformer A (left) and conformer B (right) of PPC and HPC together with their hydrogen bonded molecules of solvation (water and chloroform, respectively), and of the two mirror image enantiomers D and L of PPEM$_2$. The molecules are shown with their polar heads in an identical projection, perpendicular to their (pseudo)-mirror plane, in order to demonstrate the mirror image conformation of the headgroups including the glycerol atoms C1–C2–O21. For PPC and HPC the pseudo-centro-symmetry is broken at the chiral glycerol carbon C2 (in black). The hydrophobic part of the molecules beyond C2 adopts an anti-planar conformation, however, the orientation of the chain planes with respect to the polar part differs in the different conformers. The mirror image conformers D and L of PPEM$_2$ have head group conformations identical with molecules B and A of PPC/HPC, but different torsions (θ_3/θ_4) about the glycerol C2–C3 bond, producing a reverse tilt of the hydrocarbon chains

Table 2 Torsion angles of PPC, HPC and PPEM$_2$. For comparison the torsion angles of the corresponding double-chain lipids DMPC[a] [15] and DLPEM$_2$[b] [23] are given (in italics). The letters A or B after the abbreviated compound name refer to the two conformationally unique molecules in the unit cells of PPC and HPC and the letters D and L to the configurational/conformational mirror image molecules in the racemic PPEM$_2$ structure. For notation of torsion angles see Fig. 1

	α_1	α_2	α_3	α_4	α_5	θ_1	θ_2	θ_3	θ_4	β_1	β_2	β_3	β_4	γ_1	γ_2	γ_3	γ_4
PPC A	−156	53	70	144	−77	−58	62	−167	73					166	169	170	−179
HPC A	−161	61	64	138	−69	−68	58	−170	61					83	174	178	179
DMPC A	*163*	*62*	*68*	*143*	*−64*	*58*	*177*	*−178*	*63*	*82*	*172*	*−81*	*45*	*−177*	*168*	*−173*	*178*
PPC B	168	−76	−59	−149	72	−179	−63	−166	77					−169	−170	175	−174
HPC B	167	−69	−56	−141	69	169	−73	−170	70					176	−179	−179	−179
DMPC B	*177*	*−74*	*−47*	*−150*	*54*	*168*	*−80*	*166*	*51*	*120*	*179*	*−134*	*67*	*102*	*176*	*180*	*180*
PPEM$_2$ D	174	−61	−64	−163	55	166	−62	−68	162					−113	179	−153	−178
PPEM$_2$ L	−174	61	64	163	−55	−166	62	68	−162					113	−179	153	178
DLPEM$_2$	*179*	*65*	*54*	*144*	*−96*	*176*	*−66*	*56*	*−60*	*148*	*173*	*−57*	*176*	*129*	*−167*	*166*	*175*

[a] DMPC = 2,3-dimyristoyl-D-glycero-1-phosphocholine dihydrate.
[b] DLPEM$_2$ = 2,3-dilauroyl-DL-glycero-1-phospho-N,N-dimethyletanolamine.

Fig. 5 Comparison of the molecular packing of PPEM$_2$ and the corresponding double-chain lipid dilauroyl-DL-glycero-phospho-N,N-dimethylethanolamine [23]. The monoacyl- and diacyl-lipid show practically identical packing and interaction pattern, but the hydrocarbon chains of the double-chain lipid do not interdigitate

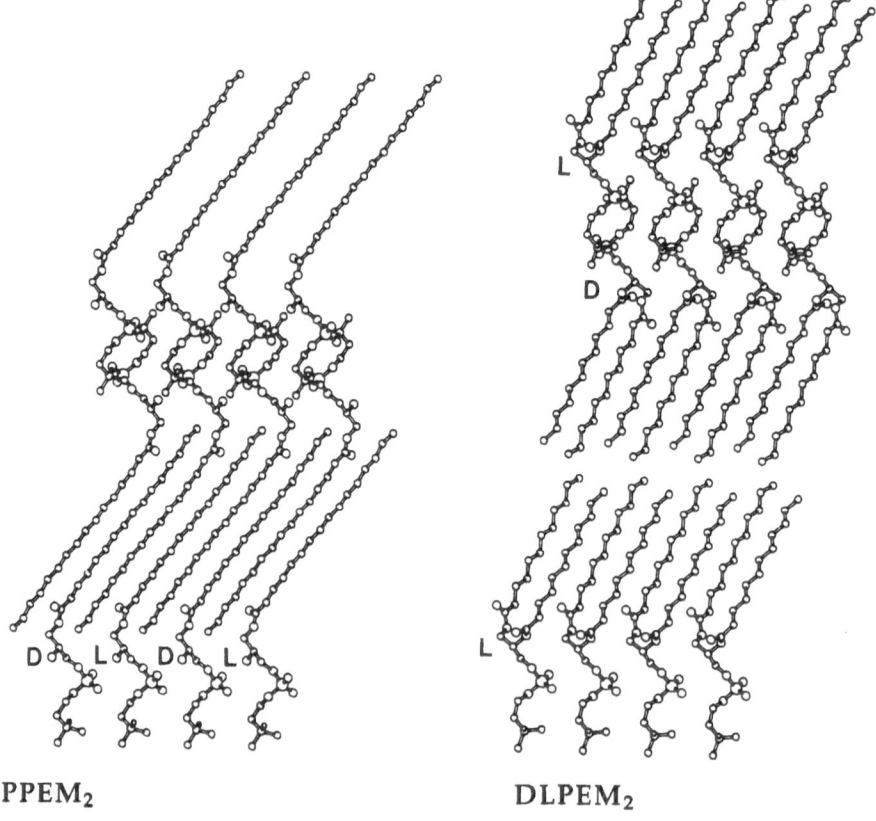

PPEM$_2$ DLPEM$_2$

layer-perpendicular P–N dipoles which interdigitate with those of the adjacent bilayer. This type of interdigitating head group dipoles, which can only arise in stacked multi-lamellar structures, is not specific for the reported lyso-lipids. The same packing arrangement with interdigitating head group dipoles has also been found in the crystal structures of 2,3-dilauroyl-DL-glycero-1-phospho-N,N-di-methyl-ethanolamine (DLPEM$_2$) [23] and 2,3-dilauroyl-DL-glycero-1-phospho-N-monomethylethanolamine (DLPEM$_1$) (Pascher and Sundell, unpublished). Interest-

Progr Colloid Polym Sci (1998) 108:67–75
© Steinkopff Verlag 1998

ingly, the double-chain lipid DLPEM$_2$ exhibits the same packing arrangement as PPEM$_2$, with the exception that the chain matrix of DLPEM$_2$ does not interdigitate (Fig 5).

It appears that in multilamellar structures the interdigitation of P–N dipoles is energetically favorable, as it gives rise to more effective lateral head group interaction at the bilayer interface. In structures with layer-parallel head groups the corresponding interactions become distributed to two separate head group layers. The two types of head group arrangements, however, do not significantly effect the molecular packing cross-section. In PPC/HPC molecular area (55 Å^2) is rather similar to that of 3-lauroyl-2-deoxy-glycero-1-phosphocholine (52 Å^2) with layer-parallel dipoles, as in the interdigitating phosphate/ choline pattern the choline groups of one lipid layer are replaced by those protruding from the adjacent lipid layer.

An interesting feature that emerges from these crystal structure analyses is the fact that, despite the great variety of observed arrangements of head group dipoles, the zwitterionic phosphoethanolamine/choline head groups always adopt a typical preferred conformation, that usually occurs in coexisting, energetically equivalent mirror image conformers. This characteristic conformation, which according to *ab initio* calculations [24, 25] is favored by intrinsic energetics is not effected by hydration, and, as confirmed by this study, not by the degree of N-methylation or the type and attachment of the hydrocarbon chains.

Acknowledgments This work is supported by the Swedish Medical Research Council (grant 0006), the Alice and Knut Wallenberg Foundation and the Skandinaviska Enskilda Banken Foundation.

References

1. White DA (1973) In: Ansell GB, Hawthorn JN, Dawson RMC (eds) Form and Function of Phospholipids. Elsevier, Amsterdam
2. Weltzien HU, Munder PG (1983) In: Mangold HK, Paltauf F (eds) Etherlipids. Academic Press, New York
3. Eibl H (1984) Angew Chem Int Ed 23:257–271
4. Pinckard RN, McManus LM, Hanahan DJ (1982) In: Wiessman G (ed) Advances in Inflammation Research, Vol 4. Raven Press, New York, pp 147–180
5. Shukla SD, Hanahan DJ (1984) Arch Biochem Biophys 232:458–466
6. Berger MR, Muschiol C, Schmähl D, Unger C, Eibl H (1985) In: Unger C, Eibl H, Nagel G (eds) Die Zellmembran als Angriffspunkt in der Tumortherapie, Aktuelle Onkologie 34. Zuckschwert Verlag, München-Bern-Wien, pp 27–36
7. Eibl H, Kaufmann-Kolle P (1995) J Liposome Res 5:131–148
8. Arvidson C, Brentel I, Kahn A, Lindblom G, Fontell K (1985) Eur J Biochem 152:753–759
9. Larsson K (1994) In: Larsson K (ed) Lipids – Molecular Organization, Physical Functions and Technical Applications, Chs 2 & 3. The Oily Press, Dundee
10. Mattai J, Shipley GG (1986) Biochim Biophys Acta 859:257–265
11. Jain MK, Crecely RW, Hille JDR, de Haas GH, Gruner SM (1985) Biochim Biophys Acta 813:68–76
12. Abrahamsson S, Pascher I (1966) Acta Crystallogr 21:79–87
13. Hauser H, Pascher I, Sundell S (1980) J Mol Biol 137:249–264
14. I. Pascher I, Sundell S, Eibl H, Harlos K (1986) Chem Phys Lipids 39:53–64
15. Pearson RH, Pascher I (1979) Nature (London) 281:499–501
16. Eibl H, Woolley P (1988) Chem Phys Lipids 47:63–68
17. Beurskens PT, Bosman WPJH, Gould RO, Van den Hark TEM, Prick PAJ (1978) Technical Report 1978/1, Crystallography Laboratory; Toernvooiveld, 6525 ED Nijmegen, The Netherlands
18. Hall SR, Stewart JM (eds) (1988) XTAL2.2 User's Manual, Universities of Western Australia and Maryland
19. Sundaralingam M (1972) Ann NY Acad Sci USA 195:324–355
20. Abrahamsson S, Dahlén B, Löfgren H, Pascher I (1978) Prog Chem Fats other Lipids 16:125–143
21. Pascher I, Lundmark M, Nyholm P-G, Sundell S (1992) Biochim Biophys Acta 1113:339–373
22. Klyne W, Prelog V (1960) Experientia 16:512–523
23. Pascher I, Sundell S (1986) Biochim Biophys Acta 855:68–78
24. Landin J, Pascher I, Cremer D (1995) J Phys Chem 99:4471–4485
25. Landin J, Pascher I, Cremer D (1997) J Phys Chem 101:2996–3004

Progr Colloid Polym Sci (1998) 108:76–82
© Steinkopff Verlag 1998

V. Razumas
Z. Talaikytė
J. Barauskas
T. Nylander
Y. Meizis

Structural characteristics and redox activity of the cubic monoolein/ubiquinone-10/water phase

In honour of Professor K. Larsson on the occasion of his 60th birthday

V. Razumas (✉) · Z. Talaikytė · J. Barauskas
Department of Bioelectrochemistry
Institute of Biochemistry
Mokslininku 12
LT-2600 Vilnius
Lithuania
E-mail: vrazumas@ktl.mii.lt

T. Nylander
Department of Physical Chemistry 1
Center for Chemistry
and Chemical Engineering
Lund University
Lund
Sweden

Y. Miezis
Department of Food Technology
Center for Chemistry
and Chemical Engineering
Lund University
Lund
Sweden

Abstract Ubiquinone-10 (UQ_{10}), a mitochondrial carrier of electrons and protons, has been entrapped in a bicontinuous Pn3m cubic phase of monoolein (MO) and water (W) at a concentration of 0.34 mol% with respect to MO. X-ray diffraction, FT-IR and cyclic voltammetry techniques have been used to study the crystallographic structure, molecular features and redox activity of the cubic $MO/UQ_{10}/W$ phase of 60.8:0.5:38.7 wt% composition compared to the cubic MO/W phase of 61:39 wt% composition. The X-ray data indicate that this low amount of UQ_{10} leaves the crystallographic structure of the Pn3m cubic phase unaltered (the lattice parameters were 96.8 ± 2.2 and 98.1 ± 0.5 Å for the binary and ternary systems, respectively). According to the FT-IR spectroscopy measurements, the acyl chains of MO in the both cubic phases contain similar types and concentrations of nonplanar (*gauche*) conformers. The headgroup-sensitive features of the FT-IR spectra also indicate that the MO bilayer arrangement is similar in the binary and ternary cubic phases. The cyclic voltammograms of the gold electrode coated by the cubic $MO/UQ_{10}/W$ phase show that the entrapped UQ_{10} undergoes electrochemically irreversible $UQ_{10}/UQ_{10}H^-$ transition over the pH range from 6 to 8. The diffusion coefficient of the coenzyme was determined to 1.9×10^{-8} cm²/s by the same technique. The location and redox function of UQ_{10} in the cubic phase are discussed on the basis of the obtained electrochemical results.

Key words Cubic phase – monoolein – ubiquinone-10 – X-ray diffraction – FT-IR – electrochemistry

Introduction

There is no doubt that the cubic liquid-crystalline phases represent the most remarkable and complicated example of the phase polymorphism of lipids. On the basis of investigations performed to date (see e.g. refs. [1–9]), two distinct types of cubic phases have been discovered: (i) three-dimensional structures based on micelles or rod-like structures and (ii) bicontinuous cubic phases where the curved lipid layers form infinite periodic minimal surfaces (IPMS).

Once the bicontinuous cubic phases of lipids are considered, it should be noted that our present knowledge of their structure, biological relevance and practical applications is largely due to the investigations conducted by Professor K. Larsson and his colleagues (for a review of their work, see refs. [4, 5, 10]). For instance, when it was realized that the bicontinuous cubic phases of the "water-in-oil" type consist of open water channel system separated

Progr Colloid Polym Sci (1998) 108:76–82
© Steinkopff Verlag 1998

by the infinite lipid bilayer, Larsson and co-workers started pioneering studies of the cubic lipid/protein/water phases [11–13]. At present there is convincing evidence that proteins and enzymes with molecular weights of up to 590 kDa can be entrapped and stabilized in the lipid-based cubic phases [14–17]. An important point is that these studies have culminated in the generation of the electrochemical biosensors [14, 17] and protein-containing bioelectrodes [15]. In our opinion, these recent developments open up two new fields of research: (i) application of the cubic phases in electroanalytical chemistry and (ii) electrochemical studies of the cubic phases which contain membranous redox active biocompounds. In this report we shall concentrate on the latter area.

A recent analysis of published electron micrographs of cell membranes from a large number of ultrastructural studies has given conclusive evidence of the occurrence of bicontinuous cubic phases in many cell types and their organelles [18]. As pointed out in [18], among the most frequent sites of identification of cubic membranes in the cell organelles are mitochondria. Although the mitochondrial respiratory chain consists of several components which catalyse the reduction of molecular oxygen in vivo, ubiquinone-10 (UQ_{10}, Fig. 1) plays a central role in this multistep process. UQ_{10}, which is a constituent of the mitochondrial inner membrane, functions as a mobile carrier of electrons and protons. Therefore, there has been intense interest in investigating the redox properties of UQ_{10}. The electrochemical behavior of the solubilized UQ_{10} has been studied in aprotic solvents [19], whereas the redox characteristics of the adsorbed coenzyme have been obtained in aqueous solutions [20–22]. More recent investigations have demonstrated redox activity of UQ_{10} within different lipidic matrices [23–25]. Although the latter systems mimic the natural environment of the co-

enzyme to some extent, the structures of the used UQ_{10}-containing matrices have not been characterized in detail.

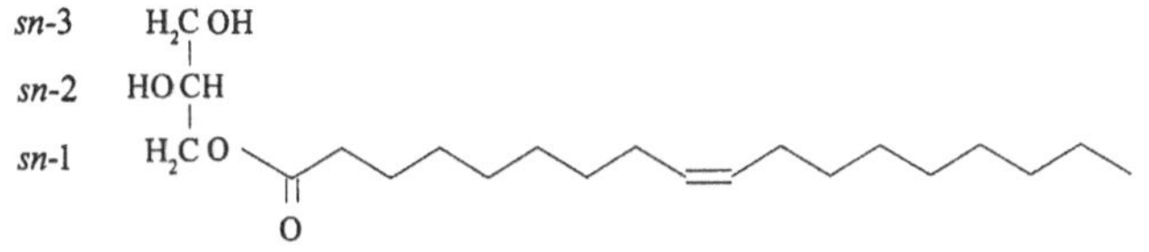

Considering the biological relevance and well-defined structure of the bicontinuous cubic phases, we report electrochemical investigations of UQ_{10} entrapped in a Pn3m cubic phase of aqueous monoolein (monoolein structure is shown in Fig. 1). A striking feature of the Pn3m monoolein/water phase is that it can coexist in equilibrium with excess of water (the boundary in the phase diagram at room temperature is observed at about 40 wt% of water) [2]. This is a crucial point when considering the potential application of cubic phases in electrochemical systems, since the cubic phase-based electrodes should function in excess of aqueous solution. In addition, as shown previously [2, 26], the Pn3m cubic phase can be obtained over a wide range of temperatures (5–92 °C), once the water content in the monoolein/water system is higher than 35 wt%.

We here present X-ray diffraction and FT-IR spectroscopy data on the structural features of the cubic monoolein/UQ_{10}/water phase.

Materials and methods

Ubiquinone-10 (UQ_{10}) from bovine heart (commercial title coenzyme Q_{10}, C-9629, Lot 57F9535, Sigma Chemical Co., St. Louis, MO) and monoolein (MO, glycerol 1-mono(9-octadecenoate), batch no. 1733-127, purity 98.1%, Danisco Ingredients, Denmark) were used as received. All other chemicals were of analytical grade. In all experiments double-distilled water was used.

Cubic phases were prepared in glass vials at 39 °C by slowly adding water on top of melted MO or UQ_{10} solution in melted MO. The vials were sealed and the samples were then equilibrated for 1 h at 39 °C and at 25 °C for a minimum of 24 h. Since UQ_{10} is colored, the equilibration could be followed visually. After equilibration, the cubic phase samples were completely homogeneous and transparent and showed no birefringence.

Small-angle X-ray diffractograms (SAXD) of the cubic phases were obtained using a Guinier camera after Luzzati et al. [27]. The sample-to-film distance was between 13 and 20 cm and copper K_α nickel-filtered radiation ($\lambda = 1.542$ Å) was used. The cubic phases were held in a thermostated sample holder constructed according to Hernqvist [28]. An exposure time of 24 h was used and measurements were performed at 25 °C.

FT-IR spectra were recorded with a Perkin-Elmer 16 PC spectrometer (Perkin-Elmer Ltd., England). Fifty scans were collected and Fourier-transformed to obtain a spectral resolution of 2 cm^{-1}. Samples of the cubic phases were assembled between ZnSe windows separated by 15 μm

Fig. 1 Structures of ubiquinone-10 (UQ_{10}) and monoolein (MO)

UBIQUINONE-10

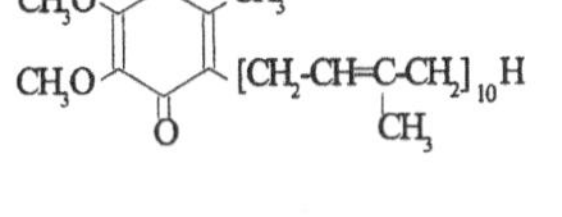

MONOOLEIN

Teflon spacers and a pair of ZnSe windows was used as a reference. Spectrum of the 6 wt% UQ_{10} solution in CCl_4 was recorded in a KBr cell of 30 μm pathlength, and the identical CCl_4-filled cell was used as a reference. Bands were simulated with mixed Gaussian-Lorentzian lineshape functions using Jandel Scientific Peakfit (Version 2.05) software (Jandel Scientific, Corte Madera, CA). Positions, bandwidths and amplitudes were varied until good agreement was achieved between actual and simulated spectra (correlation coefficients were no less than 0.9990).

Electrochemical measurements were performed with a VersaStat potentiostat-galvanostat (Princeton Applied Research, Princeton, NJ) in a thermostated glass cell (20 cm^3) at 25 °C using a three-electrode circuit with a platinum wire coil (geometric surface area of ca. 2 cm^2) as an auxiliary electrode and a saturated calomel electrode (SCE) as a reference. All potentials (E) in the text are referred to the SCE. The working electrode was a gold disk (geometric surface area 0.074 cm^2) soldered in a glass tube. The surface of the working electrode was polished using a polishing kit PK-4 (Bioanalytical Systems, Inc., West Lafayette, IN). After being thoroughly rinsed with pure water, the electrode was dried with a stream of nitrogen gas and finally plasma cleaned for 5 min in low-pressure air (0.3 Torr), using a radio frequency glow discharge apparatus Harrick PDC-3XG (Harrick Sci. Co., Ossining, NY). Immediately after the cleaning procedure a thin layer of the viscous cubic phase was applied to the surface of the electrode. The layer of the cubic phase was subsequently covered with a 30 μm thick dialysis membrane (MW cutoff 6000–8000, Spectrum Medical Ind., Inc., Los Angeles, CA), which was held in place by an O-ring. The thickness of the cubic phase layer under the dialysis membrane (ca. 300–350 μm) was determined using a measuring microscope (Marcel Aubert SA, Switzerland). After preparation and between experiments, the cubic phase-modified electrodes were stored in 0.1 M K-phosphate buffer (pH = 7.0) at room temperature. All electrochemical measurements were conducted in a nitrogen-purged (30 min) 0.1 M K-phosphate buffer solution as supporting electrolyte.

Results

Taking into account the phase diagram of the monoolein (MO)/water (W) system [2], in our study we have chosen a content of W of about 39 wt%. On the other hand, the ubiquinone-10 (UQ_{10}) content was limited to 0.5 wt%, since the samples involving higher concentrations of the coenzyme were partially birefringent. The sample of the cubic MO/UQ_{10}/W phase of 60.8 : 0.5 : 38.7 wt% composition stayed homogeneous and optically isotropic for a few months at room temperature. Therefore, the material pre-

sented is limited to the investigations of the ternary cubic phase specified.

Small-angle X-ray diffractograms of the cubic MO/W and MO/UQ_{10}/W phases

As the reference system, the cubic MO/W phase of 61 : 39 wt% composition was prepared and investigated by SAXD. Analysis of the diffractogram gave the following spacings d_{hkl} and Miller indices (hkl): 70.4 Å (1 1 0), 54.5 Å (1 1 1), 39.1 Å (2 1 1) and 32.5 Å (2 2 1) or (3 0 0), corresponding to the primitive cubic lattice of space group Pn3m with the unit cell axis $a = 96.8 \pm 2.2$ Å. Note that the determined value of a agrees closely with the values reported earlier [26, 29].

SAXD analysis of the MO/UQ_{10}/W phase of 60.8 : 0.5 : 38.7 wt% composition gave the spacings 69.6, 57.0, 49.0, 39.9, 34.8 and 32.5 Å. The indexings (1 1 0), (1 1 1), (2 0 0), (2 1 1), (2 2 0) and (2 2 1) or (3 0 0) correspond to the unit cell axis $a = 98.1 \pm 0.5$ Å of the Pn3m cubic phase.

FT-IR spectra of the cubic MO/W and MO/UQ_{10}/W phases

FT-IR spectroscopy was employed with the aim of elucidating the molecular features of the cubic phases under study. The obtained IR spectra were analyzed for changes in both the hydrophobic chain and the headgroup vibrational regions.

For the study of the hydrophobic chain conformation we have chosen the bands in the 1330–1390 cm^{-1} region, which are mainly due to localized wagging vibrations of CH_2 groups. It is well known [30–33] that the features observed over this wavelength interval are specific for different types of nonplanar (gauche) conformers in saturated alkanes and phospholipid bilayers: end-gauche (eg) conformer band at ~1340 cm^{-1}, double-gauche (gg) conformer band at ~1354 cm^{-1}, and both gauche-transgauche (gtg) and kink (gtg') conformers band at ~1368 cm^{-1}. In this spectral region, a feature at ~1378 cm^{-1} also appears due to the symmetric bending of the end CH_3 group (umbrella mode), whereas unsaturated lipids exhibit absorption band at ~1362 cm^{-1}, which is assigned to the wagging of CH_2 groups adjacent to the C=C bond [33]. An important point is that the band intensity of the umbrella mode is insensitive to chain length and conformation and may therefore be used as an internal standard with which to normalize the intensity of the conformation-sensitive wagging bands. According to the investigations performed by Casal and McElhaney [32], the total number of gauche bonds per chain can be calculated using

$$n = 2\left[(I_{\text{gtg}+\text{kink}}/I_{\text{u}}) \times 2.2 + (I_{\text{gg}}/I_{\text{u}}) \times 2.7\right] + (I_{\text{eg}}/I_{\text{u}}) \times 4.5 \, ,$$

(1)

where $I_{\text{gtg}+\text{kink}}$, I_{gg}, I_{eg} and I_{u} denote the intensities as integrated areas of the gtg + kink, gg, eg and umbrella modes, respectively.

Figure 2 shows the FT-IR spectra of the cubic phases MO/W (61:39 wt%) and MO/UQ$_{10}$/W (60.8:0.5: 38.7 wt%) together with the spectrum of the 6 wt% UQ$_{10}$ solution in CCl$_4$ over a region of the CH$_2$ wagging modes. As is seen from Fig. 2, the cubic phases and UQ$_{10}$ solution exhibit several overlapping bands. However, there is good reason to believe that the features of the cubic MO/UQ$_{10}$/W phase are basically determined by the vibrational modes of MO as the product of the UQ$_{10}$ concentration times the pathlength is increased by a factor of 24 in the CCl$_4$-based system compared to the UQ$_{10}$-containing phase.

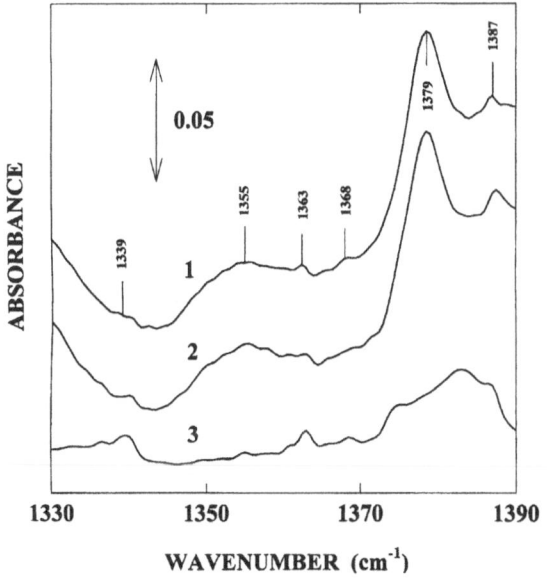

Fig. 2 Infrared spectra of the CH$_2$ wagging bands in the following samples: cubic MO/UQ$_{10}$/W phase of 60.8:0.5:38.7 wt% composition (1), cubic MO/W phase of 61:39 wt% composition (2) and 6 wt% UQ$_{10}$ solution in CCl$_4$ (3). The pathlength equals 15 μm for samples (1, 2), whereas it is increased to 30 μm for sample (3)

Table 1 presents the normalized areas of the bands due to eg, gg and gtg + kink conformers as a function of the cubic phase composition. The I values of the bands of interest were determined from curve-fitted spectra; the intensity of the umbrella mode was assigned a value of 1. To achieve a close agreement between the actual and simulated spectra (correlation coefficients ≥ 0.9993), the fitting procedure of the CH$_2$ wagging region was performed using six overlapping bands at 1339, 1355, 1363, 1368, 1379 and 1387 cm^{-1}. A linear baseline was selected in the process between 1335 and 1408 cm^{-1}. Additionally, the total numbers of gauche bonds per hydrocarbon chain (n) are also indicated in Table 1. These numbers were calculated using Eq. (1).

As illustrated earlier [15, 34–36], the headgroup interfacial arrangement in the liquid-crystalline phases of MO can be analyzed on a basis of the sn-3 C–OH, sn-2 C–OH, sn-1 CO–O and sn-1 C=O (see Fig. 1) stretching vibrations.

In the FT-IR spectra of the cubic phases MO/W (61:39 wt%) and MO/UQ$_{10}$/W (60.8:0.5:38.7 wt%), the headgroup-sensitive vibrational modes were observed at 1052 (sn-3 C–OH), 1122 (sn-2 C–OH), 1178 (sn-1 CO–O) and 1719 (hydrogen-bonded C=O groups) and 1744 cm^{-1} (free C=O groups) irrespective of the cubic phase composition. Moreover, both cubic phases exhibited practically identical bandwidths ($\Delta v_{1/2}$) of the corresponding bands.

It is worth noting that the broad bending mode of water in the cubic MO/UQ$_{10}$/W phase (peak-frequency at ca. 1647 cm^{-1}, $\Delta v_{1/2} = 125$ cm^{-1}) prevented the determination of the vibrational characteristics of the UQ$_{10}$ C=O groups. In common with other quinones [37] the C=O stretching modes of the coenzyme in CCl$_4$ were observed at 1650 and 1666 cm^{-1}.

Electrochemical behaviour of UQ$_{10}$ entrapped in the cubic MO/W phase

The cyclic voltammograms of the gold electrodes coated by the MO/W (61:39 wt%) and MO/UQ$_{10}$/W (60.8: 0.5:38.7 wt%) cubic phases at pH = 7.0 and potential scan rate 100 mV/s are shown in Fig. 3. As is evident from Fig. 3, cathodic and anodic peaks appear at -428 (E_{pc})

Table 1 Intensity (I) ratios of the end-gauche (1339 cm^{-1}), double-gauche (1355 cm^{-1}) and gauche-trans-gauche + kink (1368 cm^{-1}) bands relative to the umbrella band of the end CH$_3$ group (1379 cm^{-1}) and total number of gauche bonds per hydrocarbon chain (n)

Sample	$I_{\text{eg}}/I_{\text{u}}$	$I_{\text{gg}}/I_{\text{u}}$	$I_{\text{gtg}+\text{kink}}/I_{\text{u}}$	n
MO/W (61:39 wt%)	0.06 ± 0.02	0.76 ± 0.08	0.08 ± 0.02	4.7 ± 0.4
MO/UQ$_{10}$/W (60.8:0.5:38.7 wt%)	0.07 ± 0.02	0.69 ± 0.09	0.11 ± 0.01	4.5 ± 0.5

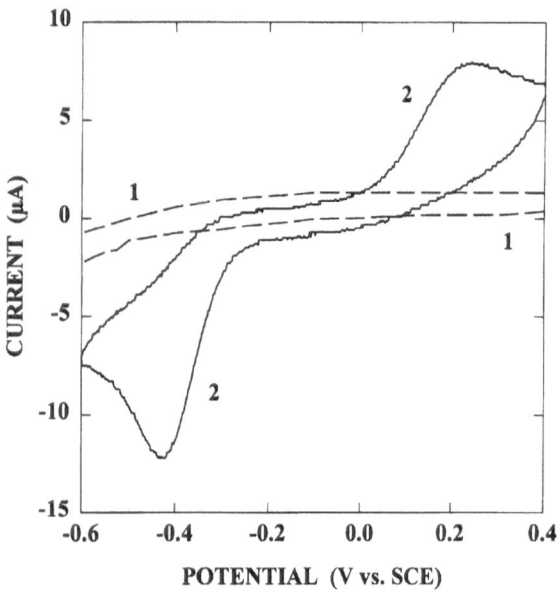

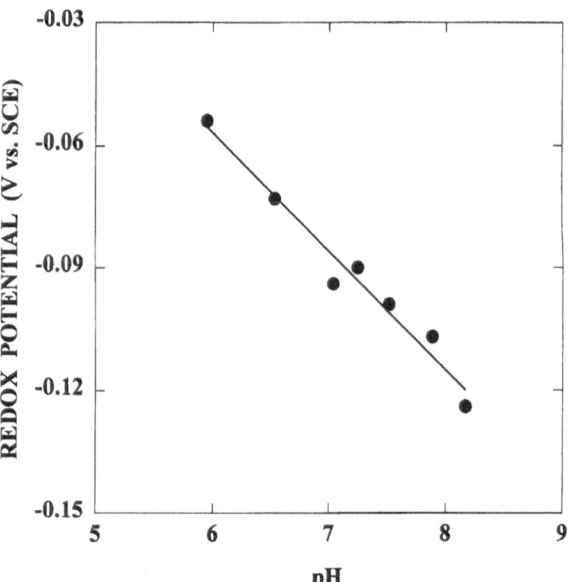

Fig. 3 Cyclic voltammograms of the gold electrodes coated by the cubic MO/W phase of 61:39 wt% composition (1) and the cubic MO/UQ$_{10}$/W phase of 60.8:0.5:38.7 wt% composition (2) in anaerobic 0.1 M K-phosphate buffer (pH = 7.0, 25 °C). Potential scan rate 100 mV/s, surface area of the gold electrode 0.074 cm^2

Fig. 4 The pH dependence of the formal redox potential of UQ$_{10}$ entrapped in the cubic MO/UQ$_{10}$/W phase (60.8:0.5:38.7 wt% in anaerobic 0.1 M K-phosphate buffer (25 °C)) is shown together with a linear fit of Eq. (2) to the experimental data

and 240 mV (E_{pa}) in the presence of UQ$_{10}$. The formal redox potential ($E^{0'}$) for the entrapped coenzyme, evaluated from the equation $E^{0'} = (E_{pa} + E_{pc})/2$, equals to -94 mV.

The pH dependence of $E^{0'}$ was examined to be able to determine the overall redox reaction of UQ$_{10}$. For this purpose the physiologically relevant pH range between 6 and 8 was selected. Figure 4 shows the results obtained, which can be expressed by the following equation:

$$E^{0'} (\text{mV vs. SCE}) = (117 + 17) - (29 \pm 2) \times \text{pH}, \quad r = 0.9902,$$

(2)

where r is the correlation coefficient.

The cathodic peak-current (I_{pc}) of the entrapped UQ$_{10}$ reduction is linearly dependent on $v^{1/2}$, with the slope of $28.4 \pm 0.2 \ \mu\text{A s}^{1/2}/\text{V}^{1/2}$ and zero intercept ($r = 0.9999$), at pH = 6.0 over the potential scan rate (v) range up to 200 mV/s as shown in Fig. 5. Similar to I_{pc}, the E_{pc} value is a function of v, and is shifted to more negative values by ca. 47 mV for each tenfold increase of v. It should be noted that the same results (within the experimental error) were obtained at pH = 8.0.

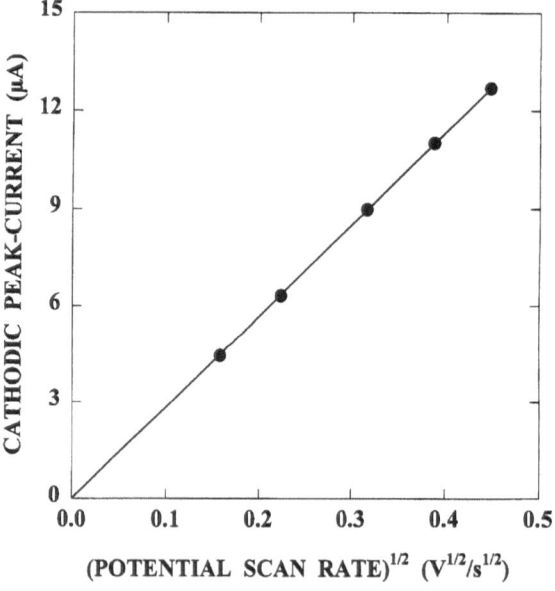

Fig. 5 The cathodic peak-current plotted as a function of the square-root of the potential scan rate for the gold electrode coated by the cubic MO/UQ$_{10}$/W phase of 60.8:0.5:38.7 wt% composition in anaerobic 0.1 M K-phosphate buffer (pH = 6.0, 25 °C). Surface area of the gold electrode was 0.074 cm^2

Discussion

According to the chemiosmotic mechanism of energy transduction in mitochondria, UQ$_{10}$ acts as a proton pump by undergoing sequential reduction and oxidation

reactions on the inner and outer surfaces of the inner mitochondrial membrane [38]. Therefore, several artificial [23–25, 39–43] and native membrane [39, 41] systems have been used with the aim to investigate the location and

Progr Colloid Polym Sci (1998) 108:76–82
© Steinkopff Verlag 1998

electrochemical characteristics of this diffusible redox carrier.

Conflicting models for the location of UQ_{10} have been reported. They range from the formation of UQ_{10} micelles spanning the entire lipid bilayer to different coexisting pools, either in the lipid bilayer midplane or closer to the lipid headgroups (see refs. [39–43]). It is believed that the lack of consensus between the proposed models stems in part from differences in the mole fractions of UQ_{10} (to lipid) incorporated into the membrane systems. A major part of these studies has used concentrations of the coenzyme far in excess of the value normally associated with mitochondrial membrane (1–2 mol% [44]).

To our knowledge, no investigations have been performed to evaluate the redox function of UQ_{10} within a well-defined matrix in which bilayer is formed solely by lipid monolayers. The lipid-containing matrices used for the electrochemical studies encompass a bilayer based on the stepwise formation of the aliphatic thiol (chemisorbed) and phospholipid (transferred by Langmuir–Blodgett technique) monolayers on gold [23], a poorly defined mixture of UQ_{10} and phosphatidylcholine on pyrolitic graphite [24] and the adsorbed monolayer of dioleoyl phosphatidylcholine on mercury [25].

In the present work, we have shown that UQ_{10} can be solubilized in the Pn3m cubic phase of MO at a 0.34 mol% concentration (with respect to MO). In addition, the SAXD data of the cubic MO/W (61:39 wt%) and MO/UQ_{10}/W (60.8:0.5:38.7 wt%) phases indicate that this low amount of UQ_{10} (ca. 3 molecules of the coenzyme per unit cell) leaves the crystallographic structure of the Pn3m cubic phase unaltered. Moreover, from the FT-IR spectra (cf. data in Table 1) it follows that these two cubic phases are characterized by practically identical number of gauche bonds per hydrophobic chain. Besides, the I_{eg}/I_u, I_{gg}/I_u and $I_{gtg+kink}/I_u$ ratios (Table 1) suggest that both systems contain similar concentrations of the eg, gg and gtg+kink conformers per chain. As in the case of the wagging vibrations of CH_2 groups, the headgroup-sensitive FT-IR features do not indicate any differences in the bilayer arrangement of the binary and ternary cubic phases.

Contrary to SAXD and FT-IR spectroscopy, electrochemical measurements have revealed a functional difference between the cubic MO/W and MO/UQ_{10}/W phases, which is due to the redox activity of UQ_{10} within the latter phase. Thus, the electrochemical data obtained for the cubic MO/UQ_{10}/W phase of 60.8:0.5:38.7 wt% composition provide means to evaluate certain functional characteristics of the entrapped coenzyme.

First, since the formal redox potential of UQ_{10} is pH-dependent (Fig. 4), it may be concluded that the headgroup of the entrapped coenzyme is accessible (or becomes accessible during the redox cycle) to the protons of aqueous phase. The slope of the $E^{0'}$–pH dependence is close to -30 mV (see Fig. 4 and Eq. (2)). Consequently, the overall redox reaction of UQ_{10} over the pH range 6–8 can be expressed by the following equation:

$$UQ_{10}\text{-}BQ + H^+ + 2e^- \leftrightarrow UQ_{10}\text{-}BQH^- , \qquad (3)$$

where UQ_{10}-BQ and UQ_{10}-BQH$^-$ stand for the benzoquinone and benzohydroquinone-anion forms of the UQ_{10} headgroup, respectively. Interestingly, similar redox behavior has been observed for a mixture of UQ_{10} with phosphatidylcholine [24].

Considering the difference between E_p values in Fig. 3 ($E_{pa} - E_{pc} = 668$ mV), the redox conversion of the entrapped coenzyme should be assigned to the type of electrochemically irreversible (slow) reactions [45]. On the other hand, Fig. 5 clearly indicates that the overall process (3) is diffusion controlled over the v range up to 200 mV/s.

For the electrochemically irreversible system, the E_{pc} shift observed (47 mV for each tenfold increase of v) provides a way of calculating the parameter $\alpha n_a = 0.64$, where α is the transfer coefficient and n_a is the number of electrons involved in the rate-determining step of electron transfer [45]. Further, knowing the $I_{pc} - v^{1/2}$ slope and αn_a, the following equation can be applied to calculate the apparent diffusion coefficient (D_{app}) [45]:

$$I_{pc} = 0.4958 n F A c_0 (D_{app} v)^{1/2} (\alpha n_a F/RT)^{1/2} , \qquad (4)$$

where $n = 2$ is the total number of electrons transferred, F is the Faraday constant, the electrode surface area $A = 0.074$ cm^2, c_0 is the bulk concentration of UQ_{10}, R is the molar gas constant and T is the absolute temperature.

The application of Eq. (4) to the data of Fig. 5 results in $c_0(D_{app})^{1/2} = 8 \times 10^{-10}$ mol/(cm^2 s$^{1/2}$). Under the assumption that the density of the cubic MO/UQ_{10}/W phase is close to unity, an average c_0 of the coenzyme in the phase of 60.8:0.5:38.7 wt% composition should be about 5.8×10^{-6} mol/cm^3. This gives $D_{app} = 1.9 \times 10^{-8}$ cm^2/s.

It is noteworthy that Rajarathnam et al. [41] have reported diffusion coefficient of $(1.1 \pm 0.2) \times 10^{-8}$ cm^2/s for a fluorescent derivative of UQ_{10} (the headgroup moiety modified by 4-nitro-2,1,3-benzooxadiazole; NBD) in unilamellar phospholipid vesicles of dimensions similar to those of the mitochondria. Thus, D_{app} determined by us for UQ_{10} fits the value for NBDUQ$_{10}$ sufficiently well if higher molecular weight of NBDUQ$_{10}$ is taken into account. This fact enables one to speculate that the location of UQ_{10} in the cubic phase is similar to that of NBDUQ$_{10}$ in the phospholipid vesicle, although the above SAXD and FT-IR spectroscopy experiments do not provide adequate data to either validate or disprove this hypothesis. Fluorescence emission and quenching studies carried out by

Rajarathnam et al. [41] indicate that the headgroup of $NBDUQ_{10}$ penetrates to some degree among the phospholipid acyl chains but not into the hydrophilic region while the major part of the isoprene chain (total length ca. 50 Å) is most likely located in the bilayer midplane. More recent NMR measurements of ^{13}C-labelled UQ_{10} in oriented model membranes [42] support this interpretation. In such a situation, our electrochemical results (see Eq. (3)) suggest that, since the charged species of UQ_{10} cannot be stabilized in the lipidic environment, the electronated headgroup of UQ_{10} (UQ_{10}-BQ^{2-}) reaches the lipid/aqueous interface where the protonation takes place giving the ultimate product UQ_{10}-BQH^-. As shown in ref. [24], the acid–base constant pK_a of the UQ_{10}-BQH_2 and UQ_{10}-BQH^- couple equals to 3.4 for the UQ_{10} adsorbed on pyrolitic graphite, whereas it shifts to 3.8 when the mixture of UQ_{10} with phosphatidylcholine is used.

In summary, the present work demonstrates that the bicontinuous lipid cubic phases look promising for the electrochemical investigations of membranous redox active biocompounds. The merits of these liquid-crystalline phases are their biological relevance and well-defined structure. As to the $MO/UQ_{10}/W$ system discussed above, it is evident that number of aspects of this ternary phase structure and redox activity are still to be determined. For example, to gain a better insight into the molecular organization of the phase components, the phase diagram of the ternary lipid–UQ_{10}–aqueous system has to be determined. Furthermore, the use of the spectroelectrochemical methods, e.g., Resonance Raman spectroscopy in a thin-layer electrochemical cell, may provide information about the rate-determining step and intermediate species of the entrapped UQ_{10} redox conversion. Studies along these lines are in progress.

Acknowledgements We would like to thank Prof. K. Larsson for many valuable discussions and for sharing his great knowledge on lipids with us.

References

1. Lindblom G, Larsson K, Johansson L, Fontell K, Forsen S (1979) J Am Chem Soc 101:5465–5470
2. Hyde ST, Andersson S, Ericsson B, Larsson K (1984) Z Kristallogr 168:213–219
3. Eriksson P-O, Lindblom G, Arvidson G (1987) J Phys Chem 91:846–853
4. Andersson S, Hyde ST, Larsson K, Lidin S (1988) Chem Rev 88:221–242
5. Larsson K (1989) J Phys Chem 93:7304–7314
6. Lindblom G, Rilfors L (1989) Biochim Biophys Acta 988:221–256
7. Luzzati V, Vargas R, Gulik A, Mariani P, Seddon JM, Rivas E (1992) Biochemistry 31:279–285
8. Delacroix H, Gulik-Krzywicki T, Mariani P, Luzzati V (1993) J Mol Biol 229:526–539
9. Delacroix H, Gulik-Krzywicki T, Seddon JM (1996) J Mol Biol 258:88–103
10. Larsson K (1994) Lipids – Molecular Organization, Physical Functions and Technical Applications. The Oily Press Ltd, Dundee, Scotland
11. Larsson K, Lindblom G (1982) J Dispersion Sci Technol 3:61–66
12. Ericsson B, Larsson K, Fontell K (1983) Biochim Biophys Acta 729:23–27
13. Buchheim W, Larsson K (1987) J Colloid Interface Sci 117:582–583
14. Razumas V, Kanapienienė J, Nylander T, Engstrom S, Larsson K (1994) Anal Chim Acta 289:155–162
15. Razumas V, Larsson K, Miezis Y, Nylander T (1996) J Phys Chem 100:11 766–11 774
16. Razumas V, Talaikytė Z, Barauskas J, Larsson K, Miezis Y, Nylander T (1996) Chem Phys Lipids 84:123–138
17. Nylander T, Mattisson C, Razumas V, Miezis Y, Hakansson B (1996) Colloids Surfaces A: Physicochem Eng Aspects 114:311–320
18. Landh T (1995) FEBS Lett 369:13–17
19. Prince RC, Dutton PL, Bruce JM (1983) FEBS Lett 160:273–276
20. Ksenzhek OS, Petrova SA, Kolodyazhny MV (1982) Bioelectrochem Bioenerg 9:167–174
21. Takamura K, Mori A, Kuso F (1982) Bioelectrochem Bioenerg 9:499–508
22. Schrebler RS, Arratia A, Sanchez S, Haun M, Duran N (1990) Bioelectrochem Bioenerg 23:81–91
23. Laval JM, Majda M (1994) Thin Solid Films 244:836–840
24. Sanchez S, Arratia A, Cordova R, Gomez H, Schrebler R (1995) Bioelectrochem Bioenerg 36:67–71
25. Moncelli MR, Becucci L, Nelson A, Guidelli R (1996) Biophys J 70:2716–2726
26. Czeslik C, Winter R, Rapp G, Bartels K (1995) Biophys J 68:1423–1429
27. Luzzati V, Mustacchi H, Skoulios A, Husson F (1960) Acta Crystallogr 13:660–667
28. Hernqvist L (1984) Polymorphism of Fats. PhD thesis, University of Lund, Sweden
29. Larsson K (1983) Nature 304:664
30. Snyder RG (1967) J Phys Chem 47:1316–1380
31. Maroncelli M, Qi SP, Strauss HL, Snyder RG (1982) J Am Chem Soc 104:6237–6247
32. Casal HL, McElhaney RN (1990) Biochemistry 29:5423–5427
33. Chia N-C, Mendelsohn R (1996) Biochim Biophys Acta 1283:141–150
34. Nilsson A, Holmgren A, Lindblom G (1991) Biochemistry 30:2126–2133
35. Nilsson A, Holmgren A, Lindblom G (1994) Chem Phys Lipids 71:119–131
36. Razumas V, Talaikytė Z, Barauskas J, Miezis Y, Nylander T (1997) Vibr Spectrosc 15:91–101
37. Bellamy LJ (1980) The Infrared Spectra of Complex Molecules. Chapman & Hall, London, pp 166–167
38. Mitchell P (1976) J Theor Biol 62:327– 367
39. Esposti MD, Bertoli E, Parenti-Castelli G, Fato R, Mascarello S, Lenaz G (1981) Arch Biochem Biophys 210:21–32
40. Stidham MA, McIntosh TJ, Siedow JN (1984) Biochim Biophys Acta 767:423–431
41. Rajarathnam K, Hochman J, Schindler M, Ferguson-Miller S (1989) Biochemistry 28:3168–3176
42. Metz G, Howard KP, van Liemt WBS, Prestegard JH, Lugtenburg J, Smith SO (1995) J Am Chem Soc 117:564–565
43. Jamiola-Rzeminska M, Kruk J, Skowronek M, Strzalka K (1996) Chem Phys Lipids 79:55–63
44. Capaldi R (1982) Biochim Biophys Acta 694:291–306
45. Bard AJ, Faulkner LR (1980) Electrochemical Methods. Fundamentals and Applications. Wiley, New York, pp 213–248

Progr Colloid Polym Sci (1998) 108:83–92
© Steinkopff Verlag 1998

N. Garti

New trends in double emulsions for controlled release

Prof. N. Garti
Applied Chemistry
School of Applied Science
The Hebrew University of Jerusalem
91904 Jerusalem
Israel

Abstract Double emulsions have significant potential in many applications since, at least in theory, they can serve as an entrapping reservoir for active ingredients that can be released by a controlled and sustained transport mechanism. Many of the potential applications are in pharmaceuticals, cosmetics and food.

In practice, double emulsions are thermodynamically unstable systems with a strong tendency for coalescence, flocculation and creaming.

During the last decade much work has been carried out in order to improve the stability and to control the release rates from double emulsions. The review will mention some of the more interesting studies making use of almost any possible combination and blend of monomeric emulsifiers, oils and stabilizers, polymerizable emulsifiers, macromolecular surfactants both natural occurring and synthetic, increase viscosity of each of the phases, microspheres and microemulsions in the internal emulsions, etc.

The presentation will stress also the most recent achievements in this area including: (1) the use of specially tailor-made polymeric emulsifiers to improve interface coverage and to better anchor into the dispersed phases; (2) droplets' size reduction by forming microemulsions or vesicles in the internal phase; (3) an improvement in the understanding of the release mechanisms; (4) the use of different filtration techniques in order to improve the monodispersibility of the droplets; and (5) use of various additives (carriers, complexing agents) to control the release via the reverse micellar mechanism.

Key words Double emulsions – slow release – sustained release – polymeric emulsifiers – steric stabilization

Introduction

Multiple emulsions are complex systems termed "emulsions of emulsions", in which the droplets of the dispersed phase themselves contain even smaller dispersed droplets. Each dispersed globule in the double emulsion is separated from the aqueous phase by a layer of oil phase compartments [1–10].

Several types of double emulsions have been documented. Some consist of a single, internal compartment while others have many internal droplets and are known as multiple compartment emulsions. A schematic presentation of some double emulsions is shown in Fig. 1. The most common double emulsions are of W/O/W but for some specific applications O/W/O emulsions can also be prepared.

Potential applications for double emulsions are well documented [1–10]. The most important applications are in pharmaceuticals, agriculture, cosmetics and foods. In most cases double emulsions are aimed for slow and sustained release of active matter from an internal reservoir

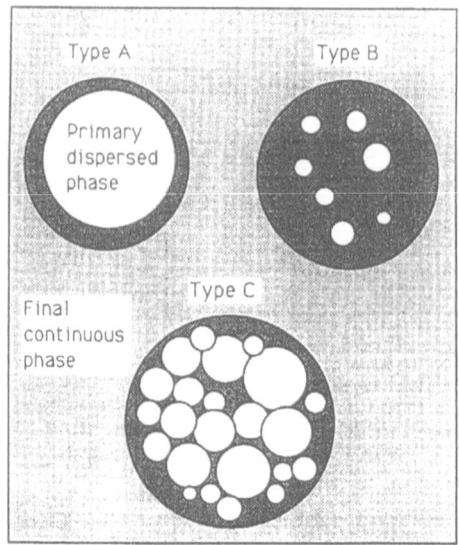

Fig. 1 Schematic presentation of the three common types of multiple emulsion droplets

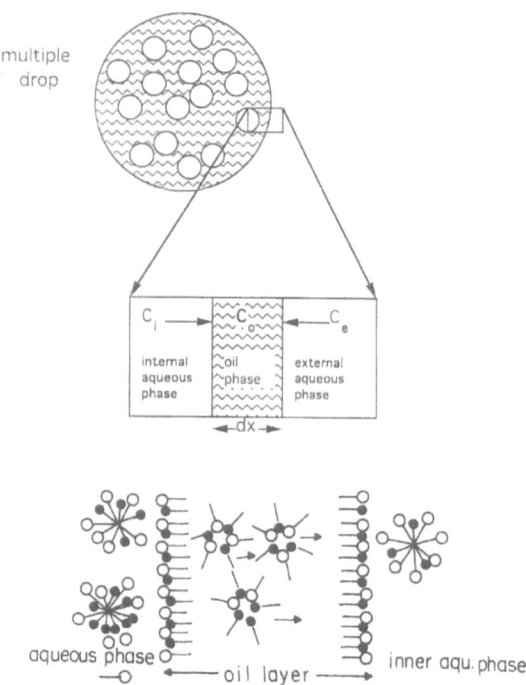

Fig. 2 Schematic illustration of the two possible transport mechanism. (a) Lamellar thinning and (b) reverse micellar transport of marker from the inner aqueous phase to the continuous aqueous phase

into the continuous phase (mostly water). In some applications the double emulsions can serve also as an internal reservoir to entrap matter from the outer diluted continuous phase into the inner confined space (removal of toxic matter).

Two types of emulsifiers are required to stabilize a W/O/W double emulsion; a hydrophobic one (in many cases a blend of two emulsifiers to obtain better results) for the internal interface and a hydrophilic one for the external interface. In most cases the emulsion is prepared in two steps. At first the W/O emulsion is prepared by high shear homogenization of the water and the emulsifier oil solution. In the second step the W/O emulsion is gently added with stirring (not homogenization) to the hydrophilic emulsifier water solution. The droplets' size distribution of a typical classical double emulsion ranges from 10 to 50 μm.

Double emulsions, in particular, are unstable thermodynamically due to the large droplets obtained when using monomeric emulsifiers. During the years of research attempts have been made to find proper and more suitable combinations of emulsifiers to improve the emulsion stability.

The double emulsions are usually not empty. A water-soluble active material is entrapped during the emulsification in the inner aqueous phase. It is well documented that because of the difference in osmotic pressure through a diffusion controlled mechanism the active matter tends to diffuse and migrate from the internal phase to the external interface, mostly through a mechanism known as reverse micellar transport (Fig. 2). The dilemma that researchers

were faced with was how to control the leak of water molecules as well as the emulsifier molecules, and mostly the active matter from the internal phase to the outer phase. It seemed almost impossible to retain the active material within the water phase upon prolonged storage. Attempts to increase the HLB of the external emulsifier or to increase its concentration in order to improve the stability of the emulsion worsens the situation and ended in faster release of the drug or the electrolytes (Fig. 3).

Stabilization by monomeric emulsifiers

Most of the research studies that were carried out in the years 1970–1985 searched for the proper monomeric emulsifier blend or combination (hydrophilic and hydrophobic), to be used at the two interfaces and the proper ratios between the two. Matsumoto [11–18] has established a "magic" weight ratio of minimum 10 of the internal hydrophobic to the external hydrophilic emulsifiers (see Fig. 4), and Garti [19–23] proved that there is a free exchange between the internal and the external emulsifiers, indicating the need to calculate an effective HLB number to optimize the stabilization of the emulsion. Many types of monomeric emulsifiers (mostly nonionic) were

Progr Colloid Polym Sci (1998) 108:83–92
© Steinkopff Verlag 1998

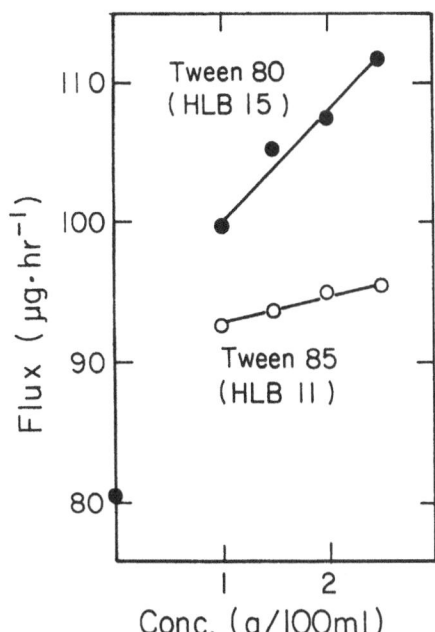

Fig. 3 Plot of the flux of a marker (electrolyte) as a function of the hydrophilic emulsifier concentration

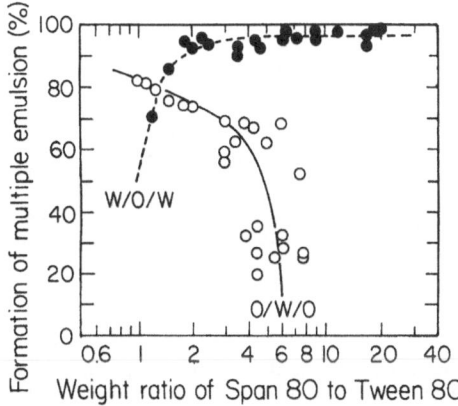

Fig. 4 Yield of formation of double emulsions of W/O/W and O/W/O as a function of the wt/wt ratio of the internal hydrophobic emulsifier, Span 80, and the external hydrophilic emulsifier, Tween 80

mentioned and used with various oils. Complete parametrization work was done on almost every possible variation in the ingredients and compositions [1–10].

Much work was devoted to establish the effects of osmotic pressure differences between the internal and the external phases on the stability of the emulsions, and on the release rates of the markers from the internal phase, and the engulfment of the internal droplets by the flow of water from the outer continuous phase to the inner droplets. Mono and multiple compartment emulsions were

prepared and evaluated in view of the enormous potential that these low viscosity liquid systems have in slow delivery of water soluble drugs.

Yet, only very limited improvements in the stability of the emulsions and in extending their shelf-life was recorded. There was practically no control of the release of the markers.

Stabilization by polymerizable emulsifiers

Florence [24–26] was the first to recognize the need to improve stability of the emulsion and to strengthen the "seal" of the external interface. He made some attempts to use polymerizable emulsifiers on the external interface, to polymerize them, in situ, and to obtain a cross-linked thick film of polymeric surfactant at the interface. The work was tedious and the results were somewhat disappointing. It is now clear that the polymerization process was not efficient due to the fact that the polymerizable emulsifiers were not properly aligned at the interface. No further attempts were made to explore this idea, mainly because the polymerizable surfactants are difficult to synthesize and to orient at the interface, and because there is only a slim chance that the polymerizable surfactant will be approved by the health authorities.

Stabilization by proteins and polysaccharides

Steric stabilization mechanisms by macromolecules (adsorbing onto the interface and forming full coverage of thick flexible and well-anchored moieties) proved to be good solutions for food colloids stability problems and for some food oil-in-water emulsions. The use of macromolecular amphiphiles and stabilizers such as proteins and polysaccharides, was adopted by scientists exploring stability of double emulsions. Gelatin [27], whey proteins BSA [28–32], HSA [33], casein [34, 35] and other proteins were mentioned and evaluated [32]. These proteins were used usually in combination with other monomeric emulsifiers [32]. A significant improvement in the stability of the emulsions was shown when these macromolecules were encapsulated onto the external interface. In most cases the macromolecule was used in low concentrations (max. 0.2 wt%) and in combination with a large excess of nonionic monomeric emulsifiers. In some cases casein alone served as the external emulsifier [35]. Furthermore, from the release curves it seems that the marker transport is more controlled (Figs. 5 and 6). Dickinson [31, 34, 35] also concluded that proteins or other macromolecular stabilizers are unlikely to completely replace lipophilic monomeric emulsifiers in double emulsions. However,

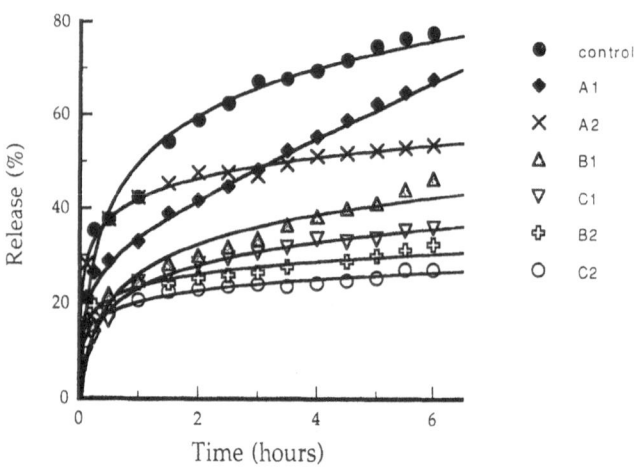

Fig. 5 Profile of chloroquine phosphate release from W/O/W multiple emulsions. A_1 – freshly prepared PVP emulsions, A_2 – PVP emulsions stored for 2 weeks, B_1 – freshly prepared gelatin emulsions, B_2 – gelatin emulsions stored for 2 weeks, C_1 – emulsion prepared with gum acacia, C_2 – acacia emulsion stored for two weeks

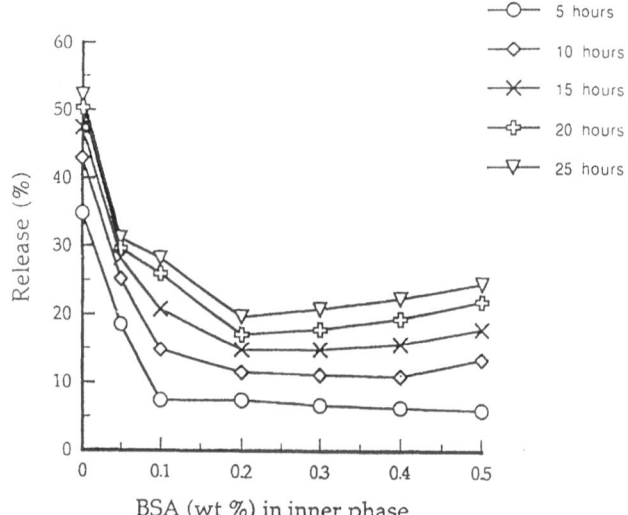

Fig. 7 Percent release of NaCl with time, from double emulsion prepared with 10 wt% Span 80 and various BSA concentrations in the inner phase and 5 wt% Span + Tween (1:9) in the outer aqueous phase

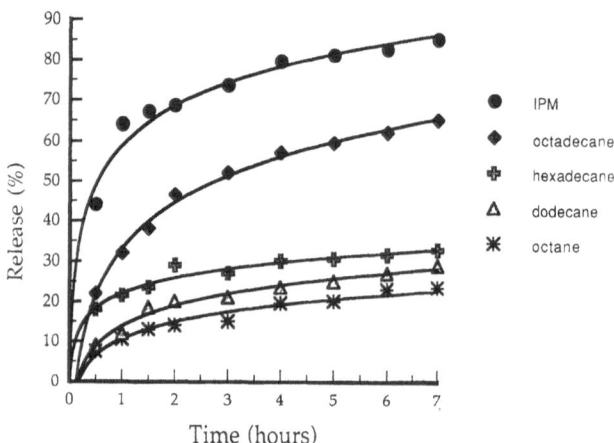

Fig. 6 Release of MTX from multiple W/O/W emulsions. The emulsions contained 2.5% Span 80 and 0.2% BSA as primary emulsifiers, with MTX (1 mg/ml) in the internal phase and the following oil phases: * – octane; △ – dodecane; + – hexadecane; □ – octadecane; and ● – isopropylmyristate

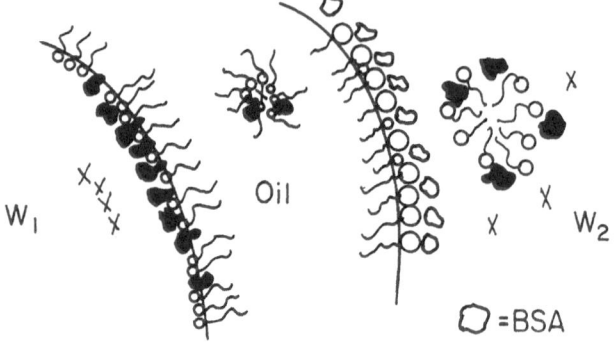

Fig. 8 Schematic illustration of possible organization and stabilization mechanism of BSA and monomeric emulsifiers (Span 80) at the two interfaces of double emulsion

proteins in combination with stabilizers do have the capacity to confer some enhanced degree of stability on a multiple emulsion system and therefore the lipophilic emulsifier concentration is substantially reduced.

The author of this review [32] has used BSA along with monomeric emulsifiers, both in the inner and the outer interfaces (in low concentrations of up to 0.2%), and found significant improvement both, in the stability and in the release of markers as compared to the use of the protein in the external phase only (Fig. 7). It was postu-

lated that while the BSA has no stability effect at the inner phase it has strong effect on the release of the markers (mechanical film barrier). On the other hand, BSA together with small amounts of monomeric emulsifiers (or hydrocolloids), serve as good steric stabilizers and improves stability, shelf-life, and slows down the release of the markers. The BSA plays, therefore, a double role in the emulsions: film former and barrier to the release of small molecules at the internal interface, and steric stabilizer at the external interface. The release mechanisms involving reverse micellar transport were also established (Fig. 8).

Table 1 Effect of colloidal microcrystalline cellulose on the release properties of lidocaine base incorporated in the inner water phase of W/O/W double emulsions (after Ref. [36])

Formulation	Percent of lidocaine base incorporated within formulation		Effective diffusion coefficient $D \times 10^6$ (cm^2/s)	Time for 30% release (h)	Release after 5 h (%)
2% CMC dispersion	1%	—	0.8	6.9	25.4
			0.9	6.2	27.0
W/O/W (4:6/1:1) 2% CMCC	1%	Innermost aqueous phase	0.03	186	4.9
			0.04	141	5.7

Stabilization by solid particles

Stabilization of certain food emulsions (margarine) was achieved by adsorption of solid fat particles, onto the water–oil interface, bridged by monomeric hydrophobic emulsifiers. In addition, it was clearly demonstrated that Colloidal Microcrystalline Cellulose (CMCC) can adsorb as solid particles onto water/oil emulsions interfaces and thus improve their stability by mechanical action. Oza and Frank [36] were the first to try the mechanical stabilization concept on double emulsions. The reports show some promise in improving stability and in slowing down the release of drug (Table 1). Recently (unpublished results), we have carried out simiar experiments with micronial particles of the α and β' polymorphs of tristearin mixed with polyglycerol–polyricinoleate (PGPR) as the internal emulsifiers. It was shown that the solid particles precipitate onto the water and bridge between them only if the lipophilic surfactant (PGPR) was first adsorbed onto the water–oil interface (the W/O emulsion) and served as an hydrophobic film for the solid particles to settle at the interface. The double emulsions prepared by this technique were more stable than those prepared by monomeric emulsifiers.

Stabilization by increasing viscosity

It is obvious that restricting the mobility of the active matter in the different compartments of the double emulsion will slow down coalescence and creaming, as well as slow down the transport of the drug or the marker from the water and through the oil membrane. Attempts to increase the viscosity of (1) the internal aqueous phase (adding gums/hydrocolloids), (2) the oil phase (fatty acids salts) and, (3) the external water (gums) was efficient only in applications that limited the use of these systems to cosmetic preparations, in which a semi-solid emulsions have potential applications (Table 2) (topical skin care, creams and body lotions uses) [37, 38].

Table 2 A detailed formulation of double emulsion stabilized by Hypermer A-60 in the inner phase and Synperonic PE/F127 in the outer phase (after Refs. [37, 38])

Ingredients	% w/w
Primary emulsion, W$_1$/O	
A. "Hypermer" A60	4.0
"Arlamol" HD or Paraffin oil perliquidum	24.0
B. MgSO$_4 \cdot$7H$_2$O	0.7
Water	71.3
Secondary emulsion, W$_1$/O/W$_2$	
A. Primary emulsion, W$_1$/O	80.0
B. "Synperonic" PE/F127	0.8
Water	19.2

Manufacture:
Primary emulsion, W$_1$/O
1. Heat A and B to 80 °C separately
2. Add B to A whilst stirring thoroughly ($\sim$700 rpm)
3. Homogenise for 1 minute with an Ultra Turrax homogeniser
4. Cool to room temperature whilst stirring ($\sim$500 rpm)
Secondary emulsion, W$_1$/O/W$_2$ (multiple)
1. Dissolve the "Synperonic" PE/F127 in the water at 5 °C, with stirring, to make B
2. Add the primary emulsion, W$_1$/O, to B whilst stirring intensively (850–1250 rpm)
3. Continue stirring for 30 minutes at 850 rpm at room temperature

Comments:
Viscosity at room temperature of W$_1$/O/W$_2$ (Brookfield LVT, spindle F, 1.5 rpm):
– "Arlamol" HD-based = 309 000 mPa s
– Paraffin oil-based = 165 000 mPa s

A preservative should be added for long-term stability.
"Hypermer" A60 is not recommended for cosmetic applications.

Formation of microcapsules in the internal phase

Improved sealing of the internal interface, or formation of a solid barrier on the interface will benefit the release properties of the double emulsions. Microspheres and nanoparticles using solid encapsulation techniques were tested to replace the classical W/O emulsion. The few experiments [39] that were carried out showed that release can be slowed down (Fig. 9), and that the stability of these systems is very limited. Such methods are applicable only for emulsions that can be freshly prepared prior to their

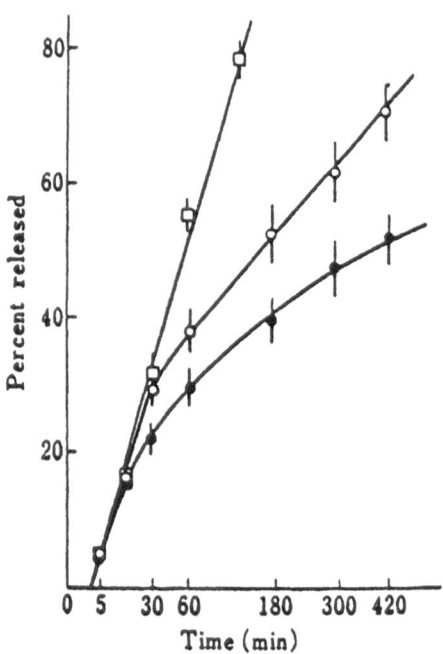

Fig. 9 In vitro release of 5-Fluorouracil from S/O/W emulsion and W/O/W emulsion in comparison to aqueous solution. ● – S/O/W emulsions; ○ – W/O/W emulsions; □ – aqueous solution

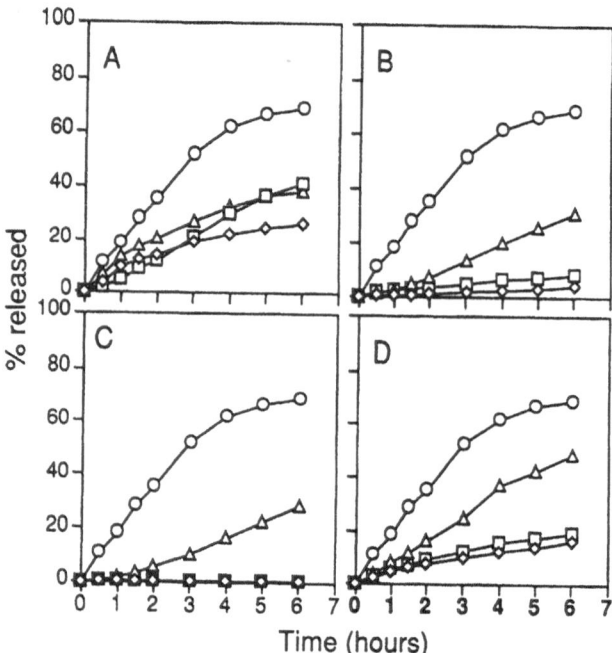

Fig. 10 In vitro release profile of CF from various formulations prepared with Span 20 (A), Span 40 (B), Span 60 (C) and Span 80 (D); ○ – CF solution; △ – vesicles; □ – W/O emulsion and ◇ – V/O/W emulsions

use. It is our hope that more efforts will be made in this direction.

Reducing the droplet sizes at the internal phase

Classical double emulsions are very polydispersed and the droplets are very large (10–50 μm). In order to reduce the sizes of the external droplets one must either have the possibility of shearing the external droplets without the fear that the marker will leak during the process and the emulsion will turn into simple O/W emulsion (see next paragraph), or try to reduce significantly the internal droplet size. Interesting work was conducted on O/W/O double emulsions in which high shear was applied during the single step of emulsification using a carefully selected single emulsifier. Extremely small and fine droplets of double O/W/O emulsions were observed. The authors have no logical explanation for their findings and no further attempts to analyze the results were made.

Vesicles in water-in-oil emulsions (V/W/O)

There are some potential applications in which the external phase is non-aqueous. Florence [40, 41] and others [42] described systems in which the aqueous suspension of vesicles are dispersed in the continuous oil phase. The

technique was exercised both for the study of its use as drug delivery systems and immunological adjuvants, as well as an intrinsically interesting colloidal system. The system is an emulsion prepared from a dispersion of niosomes in water, reemulsified in an oil using a surfactant mixture of low HLB to achieve stable W/O emulsion. The product, a V/W/O emulsion is a close analog to a O/W/O emulsion. The results are encouraging but no dramatic improvement was made on the release and transport phenomena (Fig. 10).

Emulsified microemulsions (L$_2$/W)

In an attempt to reduce the sizes of the internal droplets it was suggested, by Pilman et al. [43], to replace the internal emulsion by a microemulsion or a thermodynamically swollen stable reverse micelles (an L$_2$ phase). In a short report they conclude that such an option is feasible. One can emulsify L$_2$ reverse micelles or microemulsions to form very small, external droplets (dictated by the size of the internal droplets). The paper does not give sufficient experimental details nor does it have much stability and structural data, but it serves as seeded idea to retry the concept (Fig. 11). We [44] have prepared reverse microemulsions with AOT, in which small amounts of marker (2 wt% NaCl) were entrapped. The reverse micelles are

Progr Colloid Polym Sci (1998) 108:83–92
© Steinkopff Verlag 1998

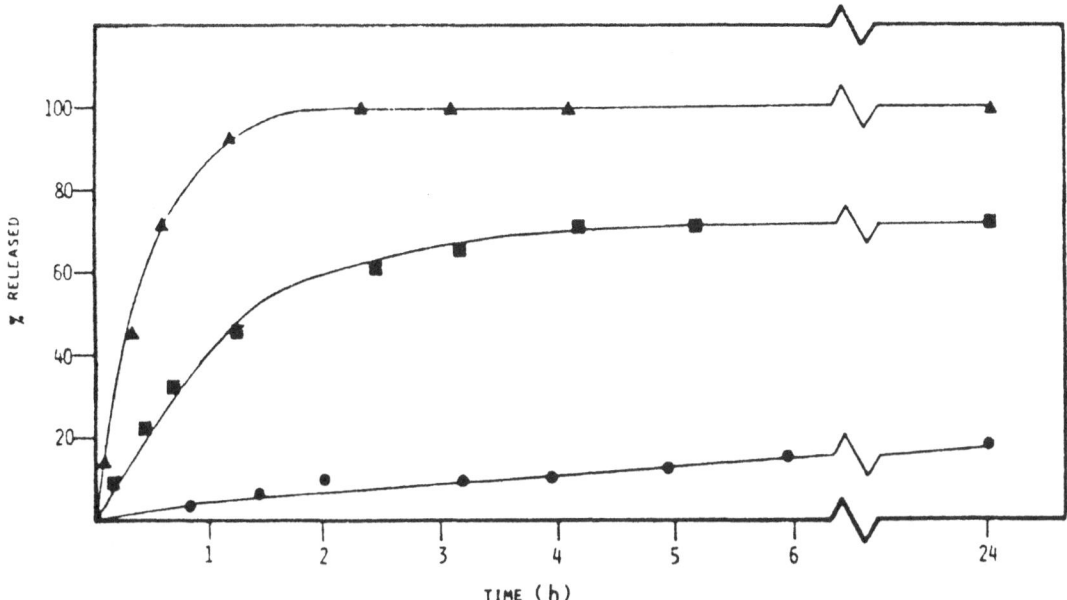

Fig. 11 Comparison of release patterns of methylene blue from different formulations of monolaurin-water-soybean oil systems at 20 °C. ● – protein stabilized emulsion of L_2-phase; ■ – L_2-phase; ▲ – water

thermodynamically stable. The L_2 phase was further added to a water solution containing various hydrophilic emulsifiers. The "emulsified microemulsion" has stability similar to the classical monomeric emulsion, with better (slower) release rates. The problem is that since the internal droplets are so small it is impossible to detect them by classical optical methods, and it requires the use of advanced techniques such as cryo-TEM, SAXS and SANS linked to self-diffusion NMR methods in order to evaluate the internal microstructure. The work is still in progress.

Stabilization by polymeric synthetic amphiphiles

The fact that proteins were excellent steric stabilizers for double emulsions, encouraged us to try and design an optimal synthetic polymeric emulsifier to be encapsulated both at the internal and the external interfaces. Only few commercially available polymeric surfactants exist on the market, many of which are designed for other applications. It was therefore, a difficult task for Garti and coworkers [45–47] to prepare a comb-like graft copolymer based on polyhydrogensiloxane grafted with polyethylene glycol side chains. A family of more than 16 products, shown schematically in Fig. 12, gave the option of preparing an entire scope of emulsifiers, lipophilic and hydrophilic, with (1) a range of grafting (various DGs), (2) a different side chain lengths, and (3) various lengths of the backbone (various DPs). The new polymeric emulsifiers (4 hydro-

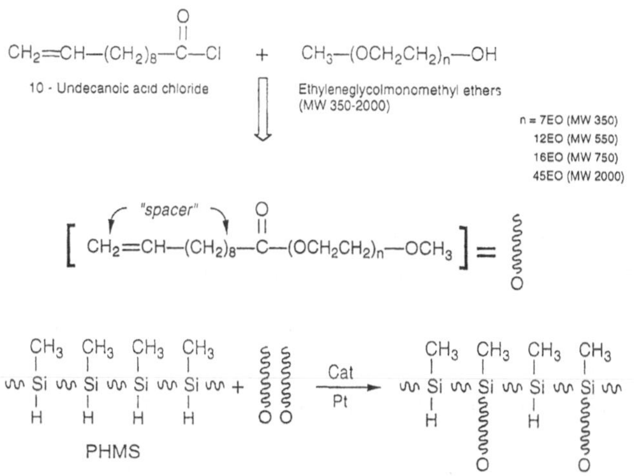

Fig. 12 Schematic presentation of the chemical pathway and structure of polymeric surfactants based on grafting an hydrophoilic side chained (with C_{11} spacer) onto a copolymer of polydimethyl siloxane and polyhydrogenmethylsiloxane

phobic and 12 hydrophilic) were prepared and their surface properties were evaluated. It was demonstrated that most emulsifiers can reduce interfacial tensions homogenize to very low values (below 10 mN/m) and therefore emulsification can be performed without the need of severe homogenization. The lipophilic emulsifiers were added to the oil to stabilize the internal interface and the hydrophilic emulsifiers were used at the external interface.

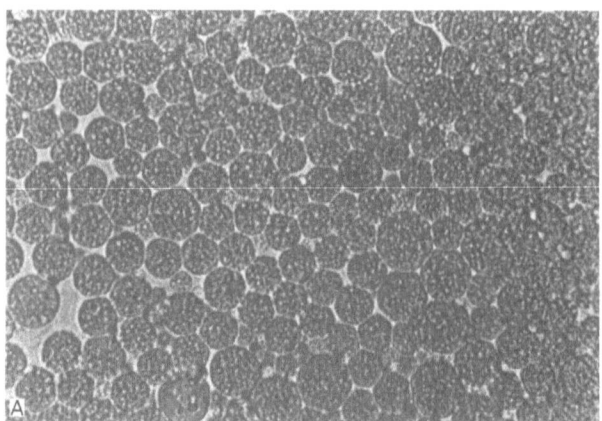

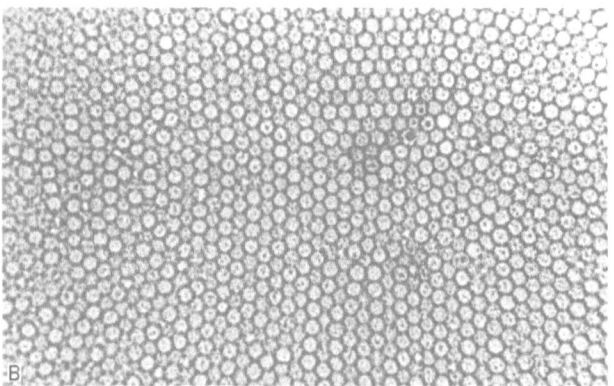

Fig. 13 Photomicrograph of typical w/o/w double emulsion stabilized with tailor-made grafted PEG side chains onto polydimethyl siloxane backbone (see text) after homogenization

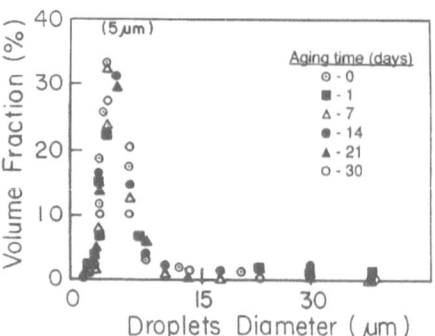

Fig. 14 Droplet size distribution of double emulsions stabilized with silicone Emulsifier I (Abil-EM-90) and siliconic Emulsifier II (PHMS-PDMS-52%-UPEG-45 EO) immediately after preparation and after 30 days of aging

Excellent double emulsions were obtained upon adding the W/O emulsion into the water phase without homogenization with small droplet and a narrow droplet size distribution, excellent stability (Figs. 13 and 14) and ability to keep the emulsions stable upon dilution. Thick films of polymeric emulsifier were coated on the interfaces. The release of markers from the emulsions was extremely slow (Fig. 15). When polymeric emulsifiers were used at both interfaces and the rates increased when other emulsifiers, such as PGPR (ETD-polyglycerol polyricinoleate) or Span 80 (sorbitan monooleate), were used in the internal interface.

However, the most surprising phenomenon was the fact that the double emulsions were easily homogenized without leaking the markers into the aqueous solution during the emulsification process. As a result of the homogenization process almost monodispersed double emulsion droplets could be obtained with very narrow size distribution ranging from 3 to 7 μm (Figs. 13 and 14).

A close examination of the release profiles of markers (NaCl or drug) from these double emulsions revealed that the release is done in three stages (Fig. 16). A long lag time (stage A) in which almost no release takes place is observed immediately after preparation (Fig. 16, slope A). The lag time depends on the nature of the internal lipophilic polymeric emulsifier and its concentration ratio to the monomeric lipophilic emulsifier. In the second stage (Fig. 16, slope B) the release slope is gradual, diffusion controlled, and complies with the Higuchi mechanism of release from a "slab into a sink" by reverse micellar mechanism [31]. In the third stage (Fig. 16, slope C) the release is very slow and actually stops. The fact that the release mechanism is via reverse micelles has a significant advantage since it allows us to control the transport through the liquid membrane film by controlling the number of micelles present in the oil membrane. The polymeric emulsifier that dissolves or aggregates in the oil phase does not have the characteristics of a reverse spherical micelle and is assumed to be a rather "open" aggregated random coil. Therefore, such structures cannot solubilize much water and cannot transport the water molecules and the marker that should dissolve in the water. Therefore, emulsions prepared with polymeric emulsifiers alone, practically do not release the marker upon aging (storage) the emulsion on the shelves. Very long lag times (Table 3) were detected and only minor quantities of marker were released at very slow rates. On the other hand, addition of variable concentrations of Span 80 (the lipophilic monomeric emulsifier) at the moment of use allowed to control both the lag time, the released amount of matter, and the rates. As one increases the amounts of Span 80 (already existing in the formulation, or dropwise added after the preparation and prior to using the double emulsion) more marker is released. The lag-time is shortened and the rates increase with the increase of added Span 80 (Fig. 17). In order to demonstrate that the reverse micelles are responsible for the release mechanism, the amounts of water, as a function of Span

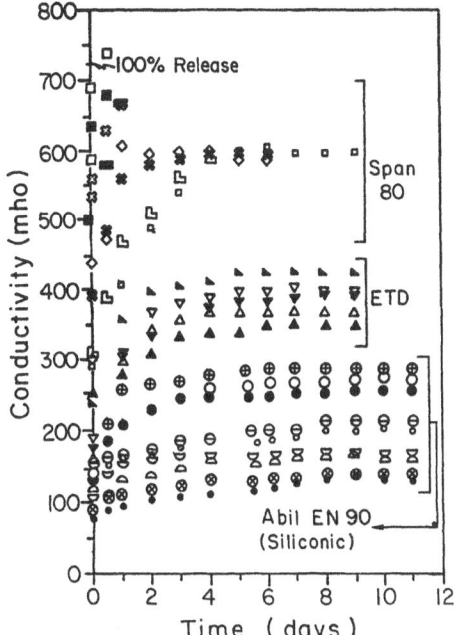

Fig. 15 Plot of conductivity of the outer aqueous phase (reflecting the concentration of NaCl in the outer phase, (% release) vs. time (days) in three sets of double emulsions. (1) All circles (○) indicate use of Abil EM-90 as hydrophobic Emulsifier I. The lower curves (●) indicate the most hydrophilic PHMS-PDMS-UPEG Emulsifier II and the upper curves (○) in each set indicate the most hydrophobic PHMS-PDMS-UPEG with 52% substitution and 45 EO units. Each circle symbol represents a different polymeric emulsifier. (2) All triangles represent use of polyglycerol polyricinoleate (ETD) as Emulsifier I and the curves are arranged again with increasing hydrophobicity of Emulsifier II. The lower curve (▲) represents the most hydrophilic emulsifier and the upper curve in the set represents the most hydrophobic one (▲). (3) All squares represent the use of Span 80 as Emulsifier I and the curves are arranged with increasing hydrophobocity of Emulsifier II

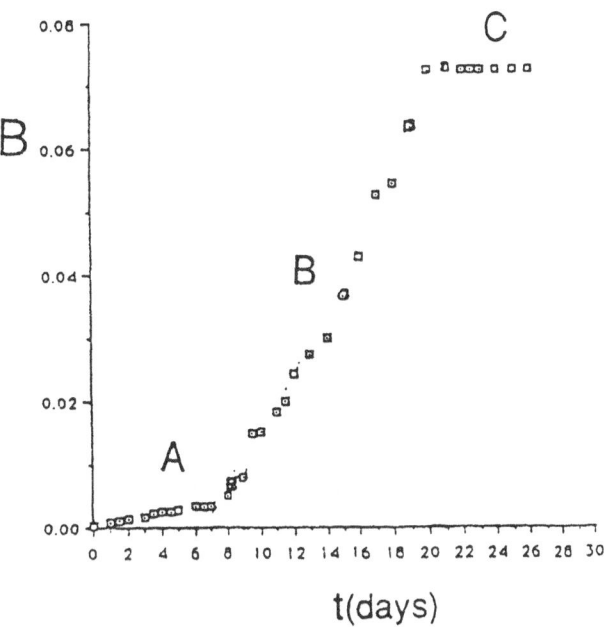

Fig. 16 Characteristic release profile of double emulsion in which a combination of Span 80 and siliconic emulsifier were used in the internal phase and an hydrophilic siliconic emulsifiers was used in the external phase

Table 3 The effect of the nature of the surfactant added to the internal phase of w/o/w double emulsion (Abil EM-90 or Abil EM-90 with different concentrations of Span 80) on the amount of water solubilized in the oil phase (after centrifugation) and the lag-time (in days, until some release of NaCl was observed) (after Ref. [47])

Surf.	Conc. (%)	H_2O conc. (%)	Lag time (days)
Abil-90	10	0.24	1
	15	0.46	4
	20	0.68	5
	30	0.78	8
(30% Abil) + Span 80	1	0.87	7
	5	1.12	5
	10	1.25	4
	20	1.4	2

concentrations in the oil phase, were measured (after centrifugation). It was found that water is present only in very minor quantities when the siliconic emulsifier is employed by itself. The water concentration increases as the amounts of Span 80 increases [47]. These findings are in good correlation with the release rates and the lag time.

It was therefore concluded that the polymeric internal emulsifier controls the release rates bot by improving the film formation on the interface and by restricting the formation of reverse micelles in the oil phase. It is assumed that the presence of the two emulsifiers (Span 80 and silicone lipophilic emulsifier) form a "reverse hemimicelle". These structures are capable of solubilizing less water and, therefore, less marker, a fact that leads to slower release and the ability to better control the release rates. New emulsions made with polymeric emulsifiers were prepared without the lipophilic monomeric emulsi-

fier. The emulsions remained stable on the shelf for over 6 months and did not show any leaking of the marker upon storage. Once the monomeric emulsifier was added dropwise, at various concentration levels, the release started and was completely controllable by both the rate of addition of the monomeric emulsifier and by its concentration (or the wt. ratio of the emulsifier to the polymer).

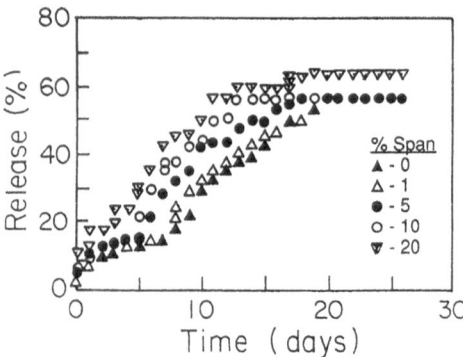

Fig. 17 Release profiles of double emulsions stabilized with polymeric hydrophobic emulsifier at the internal interface (Abil-EM-90) and a combination of hydrophilic siliconic polymeric emulsifier (PHMS-PDMS-52%-UPEG-45EO) with various concentrations of Span 80)

Conclusions

It seems that we have reached the stage where one can control now the size of the droplets, polydispersity, stability, shelf-life, dilution and shear, but most of all control both the start of the release (sustained release) and the rate (slow and controlled release). The main drawbacks that still exist are that we cannot make the droplets submicronial (for intravenous applications) and cannot find a proper, natural occurring polymer that will form such perfect films with excellent viscoelastic properties.

It is our belief that it is only a matter of time before natural or modified proteins will be capable of forming similar stable systems in which the release is properly controlled.

References

1. Davis SS, Hadgraft J, Palin KJ (1985) In Becher P (ed) Encyclopedia of Emulsion Technology. Marcel Dekker, New York, Vol 2, pp 159
2. Florence AT, Whitehill D (1981) J Colloid Interface Sci 79:243
3. Florence AT, Whitehill D (1982) Int J Pharm 11:277
4. Matsumuto S, Kang WW (1989) J Dispersion Sci Technol 10:455
5. Fox C (1986) Cosmetics Toiletries 10:1, 101–106, 109–112
6. Matsumoto S (1987) In Schick M (ed) Nonionic Surfactants: Physical Chemistry, Surfactant Science Series, Vol 23, Marcel Dekker, New York, pp 549–600
7. Whitehill W (1980) Chemist and Druggist 213:130, 132, 135
8. Garti N, Aserin A (1996) Adv Colloid Interface Sci 65:37
9. Garti N, Aserin A (1996) In Chattopadhyay AK, Mittal KL (eds) Surfactants in Solution. Marcel Dekker, New York, pp 297–332
10. Garti N (1996) In Berenholz Y, Lasic DD (eds) Handbook of Non-Medical Applications of Liposomes, Vol 3. CRC, New York, pp 143–199
11. Matsumoto S, Kida Y, Yonezawa D (1976) J Colloid Interface Sci 57:353
12. Matsumoto S, Khoda M, Murata S (1977) J Colloid Interface Sci 62:149
13. Matsumoto S, Ueda Y, Kita Y, Yonezawa D (1978) Agric Biol Chem 42:739
14. Kang WW, Matsumoto S (1988) 6th Int Conf Surface Colloid Sci, Hakone, Japan
15. Matsumoto S (1983) J Colloid Interface Sci 94:147
16. Matsumoto S, Inoue T, Khoda M, Ohta T (1980) J Colloid Interface Sci 77:564
17. Matsumoto S, Koh Y, Michiura A (1985) J Dispersion Sci Technol 6:507
18. Makino H, Fukui H, Matsumoto S (1984) The 37th Symp on Colloid Interface Chemistry, Morioka, Japan
19. Frenkel M, Schwartz R, Garti N (1983) J Colloid Interface Sci 94:174
20. Garti N, Frenkel M, Schwartz R (1983) J Dispersion Sci Technol 4:237
21. Magdassi S, Frenkel M, Garti N (1984) J Dispersion Sci Technol 5:49
22. Magdassi S, Frenkel M, Garti N (1985) Drug Ind Pharm 11:791
23. Magdassi S, Garti N (1986) J Controlled Release 3:273
24. Florence AT, Law TK, Whateley TL (1985) J Colloid Interface Sci 107:584
25. Omotosho JA, Whateley TL, Law TK, Florence AT (1986) J Pharm Pharmacol 38:865
26. Florence AT, Whitehill D (1981) J Colloid Interface Sci 79:243
27. Omotosho JA (1990) Int J Pharm 62:81
28. Omotosho JA, Law TK, Whateley TL, Florence AT (1986) Colloids Surfaces 20:133
29. Omotosho JA, Whateley TL, Florence AT (1989) J Microencapsulation 6:183
30. Omotosho JA, Whateley TL, Florence AT (1989) Biopharmaceutics & Drug Disposition 10:257
31. Dickinson E, Evison J, Owusu Apenten RK, Williams PA (1996) In Phillips GO, Wedlock DJ, Williams PA (ed) Gums and Stabilizers for the Food Industry, Vol 7. Oxford Univ Press, Oxford, pp 91
32. Garti N, Aserin A, Cohen Y (1994) J Controlled Release 29:41
33. Beissinger RL, Wasan DT, Sehgal LR, Rosen AL (1990) Multiple oil/water emulsion of hemoglobin as blood substitute UK Patent GB 2,221,912
34. Dickinson E, Evison J, Owusu Apenten RK (1991) Food Hydrocolloids 5:481
35. Evison J, Dickinson E, Owusu Apenten RK, Williams A (1990) In Hudson BJF (ed) Developments in Food Proteins. Elsevier, Amsterdam
36. Oza KP, Frank SG (1989) J Dispersion Sci Technol 10:163
37. Grossiord JL, De Luca M, Medard JM, Vantion C, Seiller M (1989) 5-eme Congress Intern Technol Pharm APGI, Paris, 1:172
38. De Luca M, Grossiord JL, Medard JM, Vantion C, Seiller M (1990) Cosmet Toilet 105:65
39. a) Hashida M, Liao MH, Muranishi S, Stzaki H (1980) Chem Pharm Bull 28:1659 b) Distefan FV, Shiffer OM, Elssama MS, Vandeview JW (1983) J Colloid Interface Sci 92:269
40. Florence AT, Omotosho JA, Whateley T (1989) In Rossof M (ed) Controlled Release from Drug Polymers and Aggregated Systems, VCH, New York
41. Florence AT, Yoshika T (1996) In Berenholtz Y, Lasic DD (eds) Handbook of non Medical Applications of Liposomes, Vol 3. CRC, New York, p 199
42. Albert EC, Matour R, Wallach DFH (1992) WP 09217179, US patent 4,911,928
43. Pilman E, Larsson K, Tornberg E (1980) J Dispersion Sci Technol 1:267
44. Garti N, Aserin A, Tiunova I, Binyamin H (1998) J Am Oil Chem Soc, submitted for publication
45. Sela Y, Magdassi S, Garti N (1994) Colloids Surfaces A 83:143
46. Sela Y, Magdassi S, Garti N (1994) Colloid Polym Sci 272:684
47. Sela Y, Magdassi S, Garti N (1996) J Controlled Release 33:1

Progr Colloid Polym Sci (1998) 108:93–98
© Steinkopff Verlag 1998

S. Engström
K. Alfons
M. Rasmusson
H. Ljusberg-Wahren

Solvent-induced sponge (L_3) phases in the solvent–monoolein–water system

This paper is dedicated to Professor Kåre
Larsson at the time of his retirement from
academic duties

S. Engström (✉) · K. Alfons
Food Technology
POB 124
S-221 00 Lund
Sweden

M. Rasmusson* · H. Ljusberg-Wahren
Camurus Lipid Research
Ideon Science Park
Sölvegatan 41
S-223 70 Lund
Sweden

* Present address
Chemical Physics
P.O. box 124
S-221 00 Lund
Sweden

Abstract The phase behavior of the monoolein–water system in the presence of polar solvents such as dimethyl sulfoxide, propylene glycol, polyethylene glycol ($M_w \approx 400$) and ethanol was investigated. These solvents share the common feature of being completely miscible with both monoolein and water. At water contents in the range 30–60% w/w, an isotropic liquid phase is found in all the four systems. The liquid phase shows many characteristic features of a sponge (L_3) phase, i.e. a long but narrow phase region, surrounded by two-phase regions which include a lamellar phase on the water-poor side and another liquid phase on the water-rich side. At the monoolein-rich end of the narrow phase region, a cubic liquid crystalline phase is in equilibrium with the isotropic liquid. At high solvent content, the sponge phase shows shear birefringence and scatters light in the case of ethanol. The formation of the sponge phase in the present systems is interpreted as a consequence of a subtle partitioning of the solvent between the monoolein and aqueous domains in the system, which in turn leads to the desired, slightly negative, interfacial mean curvature of the sponge phase.

Key words Sponge phase – L_3 phase – monoolein – phase diagram

Introduction

The monoolein–water system has been studied extensively over the past ten years. Its phase behavior is dominated by a large cubic liquid crystalline phase with a continuous membrane structure, which has been utilized for various purposes such as drug delivery [1, 2] and biosensors [3]. The cubic region in the monoolein–water system, in fact, consists of two cubic phases belonging to different space groups [4]. The structure of the cubic phases has been described by making use of the concept of minimal surfaces; its applicability has been confirmed with self-diffusion measurements by NMR.

The effect of adding amphiphilic molecules to the cubic phase in the monoolein–water system is nicely described by the so-called packing concept [5]. Hence, adding a bile salt transforms the cubic phase into normal structures such as hexagonal and micellar phases [6]. Adding a triglyceride, on the other hand, transforms the cubic phase to a reversed hexagonal and a reversed micellar phase, respectively [7]. Lecithin, which itself forms a lamellar liquid crystal in excess water, can be added at high amounts preserving the cubic structure, which transforms to a lamellar phase when lecithin exceeds the monoolein content [8].

In this study, the effect of adding a polar solvent to the monoolein–water system was investigated. The solvents used – dimethyl sulfoxide (DMSO), polyethylene glycol (PEG400), propylene glycol (PG), and ethanol (EtOH) – have one feature in common in that they are completely miscible with *both* monoolein and water. One relevant question is, therefore, where do these solvents partition in the monoolein–water system, and what effect the partitioning has on the phase behavior. It turns out from the present study that the phase behavior is qualitatively similar for all four solvents, with the formation of a sponge or L_3 phase as the most striking feature.

Materials and methods

DMSO (Sigma), propylene glycol (Sigma), polyethylene glycol (Hoechst), ethanol (95% aq) and monoglycerides

(GMO, Grindsted A/S, Brabrand, Denmark) were used as received. The monoglyceride was high in monoolein ($>90\%$ w/w) and its phase behavior in water is very close to that found for pure monoolein [4].

The samples were prepared in glass ampoules which were flame sealed and left standing until equilibrium was reached (hours to days). The samples were centrifuged in order to facilitate the separation of phases in equilibrium. The phase characterization was done visually by means of crossed polarizers and a polarizing microscope (Olympus BH-2). Some samples were investigated by means of small-angle X-ray diffraction in order to deduce the swelling behavior of the lamellar liquid crystalline phase in the presence of DMSO (Fig. 2). The phase maps drawn in Fig. 1 are based on the sample compositions shown by markers in each phase triangle.

Fig. 1 Phase maps showing the location of the isotropic liquid phase found in the solvent–monoolein–water systems, where the solvent is (a) dimethyl sulphoxide (DMSO), (b) polyethylene glycol $M_w \approx 400$ (PEG400), (c) propylene glycol (PG) and (d) ethanol (EtOH). The markers in the phase maps show the sample compositions from which the maps are drawn

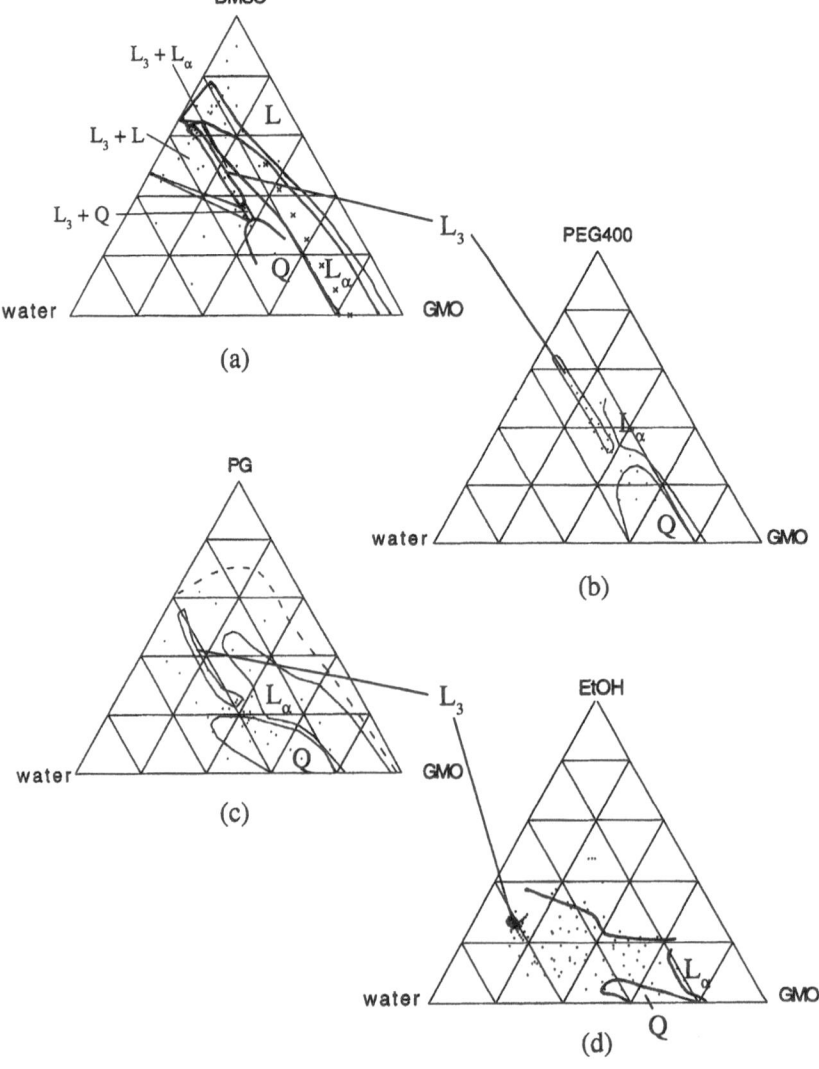

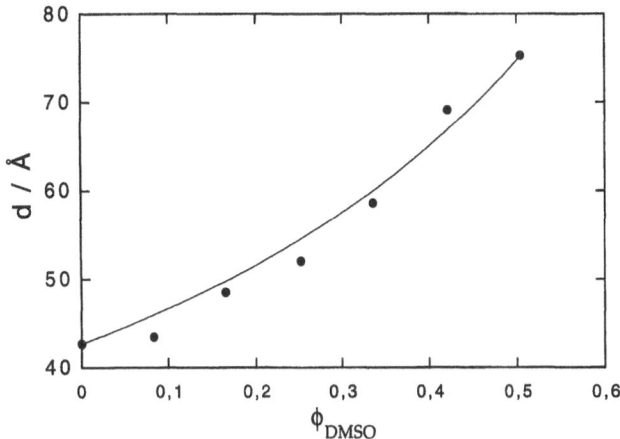

Fig. 2 The *d*-spacings obtained from X-ray diffraction at 20 °C on the lamellar liquid crystalline phase formed with various amounts of DMSO (marked with *x* in Fig. 1a) at constant water content (16% w/w) in the DMSO–monoolein–water system. The line shows a linear least-squares-fit of the data points to Eq. (1), where $d_0 = 42.7$ Å, $\phi_{GMO(0)} = 0.84$ and $\alpha = 0.72 \pm 0.02$ ($r = 0.99$)

Results and disucssion

The phase maps obtained for the four different solvents are given in Fig. 1a–d. In this section we first present and discuss the qualitative similarities found in this work between the different solvent–monoolein–water systems. The phase behavior and partitioning of DMSO is then discussed in some detail. Finally, a comparison is made between other sponge-phase containing systems based on ionic and nonionic amphiphiles.

The solvent–monoolein–water phase behavior

For the sake of simplicity we begin with a presentation of the qualitative phase behavior shown by all four solvents investigated in this study. It is obvious from Fig. 1 that all four systems give rise to a narrow one-phase region representing an isotropic liquid when a solvent is added to the monoolein–water system. The water content of the isotropic liquid phase is relatively constant within each system, but varies with solvent in the range 30–60 wt%, i.e. about 30% for DMSO and PEG400, 40% for PG, and 60% for EtOH. The extension of the narrow one-phase region also varies with solvent, being much less for ehtanol than for the other three solvents.

Several observations give strong indications that the phase formed by adding the polar solvents is a so-called L_3 phase, also referred to as a sponge phase due to its proposed internal structure. The isotropic liquid region is

(see Fig. 1), in all four cases, surrounded by two-phase areas which include, besides the isotropic liquid, a lamellar liquid crystalline phase (L_α) at the water-poor side and an isotropic liquid phase (L_1) at the water-rich side. A cubic liquid crystalline phase (Q) is in equilibrium with the liquid phase at high monoolein content. The isotropic liquid phase shows shear birefringence at high amount of solvent, and has a bluish appearance (especially in the ethanol system).

The DMSO–monoolein–water system

The most extensive investigation in this work was carried out on the DMSO–monoolein–water system. The phase map of this system is given in Fig. 1a. Four one-phase regions exist which represent a cubic liquid crystal (which in fact consists of at least two cubic phases), a lamellar liquid crystal (L_α) and two isotropic liquids. One of the isotropic liquid regions is an extension of the reversed micellar phase (L_2) found in the monoolein–water system at low water content. At high DMSO content, above about 80% w/w, this phase exists over the entire monoolein/water mixing ratio.

The other isotropic liquid region is narrow, centered around approximately 30% w/w water and a DMSO/monoolein weight ratio ranging from 1:1 to 19:1. Below 30% w/w water, the system enters into a two-phase region, isotropic liquid-L_α phase, and above 30% w/w water two isotropic liquids occur. At the low DMSO/monoolein ratio, the sponge phase is in equilibrium with a cubic liquid crystalline phase. At high DMSO content the phase is birefringent upon shearing. Thus, the narrow region shows, as stated above, all characteristics typical of a so-called L_3 or sponge phase.

It should be noted that the phase boundaries and tielines given in Fig. 1a are estimates based on the compositions shown in the phase map. The extension of the two- and three-phase regions should especially not be taken too seriously, but are given merely in order to indicate a conceivable appearance of the phase diagram. It is interesting to note that the large liquid region at low water content seems to phase separate into two liquid phases, one relatively rich in water but low in monoolein, and the other poor in water but rich in monoolein. However, we have not considered this area of the phase diagram any further.

Since DMSO (and the other solvents as well) is miscible with both monoolein and water, the partitioning of DMSO is crucial for the phase behavior. In order to obtain some information on the partitioning behavior in the system, X-ray diffraction measurements were performed on samples in the L_α phase at constant water

96

S. Engström et al.
Sponge phases in the solvent–monoolein–water system

content, varying the amount of DMSO. The d-spacings obtained are given in Fig. 2 as a function of the DMSO weight fraction. (Although it is more relevant to use volume fractions in the analysis below, we have assumed equal densities of the three components in order to facilitate the calculations. This simplification should have no influence on the qualitative conclusions drawn below). It is evident from Fig. 2 that the repeat distance increases with increasing DMSO content. In order to find out the location of DMSO in the lamellar phase, a simple model was applied in which the added DMSO behaves either as monoolein or water (see appendix).

If we assume that the monoolein bilayer thickness is preserved irrespective of the DMSO content, treating DMSO either as water or monoolein gives the following expression for the change in d-spacing as a function of DMSO fraction, ϕ_{DMSO}, at constant water fraction, $\phi_w = 1 - \phi_{GMO(0)} = 0.16$,

$$d = d_0/(1 - \alpha\phi_{DMSO}/\phi_{GMO(0)}) . \tag{1}$$

In Eq. (1) α is the fraction of DMSO behaving as water (i.e. $1 - \alpha$ is the fraction of DMSO behaving as monoolein).

This simple model has, of course, several drawbacks. The assumption of a constant bilayer thickness is most likely incorrect, since the inclusion of small DMSO molecules should make the bilayer thinner and less dense. Another problem is that the model does not describe the interfacial part of the system particularly well. Nevertheless, the analysis should at least give a rough estimate of the preferred location for DMSO. A linear least-squares-fit of the d-spacings to Eq. (1) gives $\alpha = 0.72$, i.e. 72% of the DMSO behaves as water.

The value obtained for α can also be used to estimate a monoolein bilayer/water partition coefficient for DMSO in the lamellar phase. If the partition coefficient, $P_{GMO/w}$, is defined as

$$P_{GMO1/w} = [(1 - \alpha)/\phi_{GMO}]/[\alpha/\phi_w] , \tag{2}$$

an expression for $P_{GMO/w}$ is easily obtained in terms of the fraction of added DMSO, ϕ_{DMSO},

$$P_{GMO1/w} = [(1 - \alpha)/\alpha][\phi_w/(\phi_{GMO(0)} - \phi_{DMSO})]$$
$$= 0.06/(0.84 - \phi_{DMSO}) , \tag{3}$$

which gives $P_{GMO/w} = 0.07$ ($\log P_{GMO/w} = -1.1$) when $\phi_{DMSO} \to 0$. If this value is compared to $\log P_{octanol/w} \approx -2$ [9], one may conclude that the tendency for DMSO to go to the monoolein domain in the lamellar phase is higher than the corresponding tendency to go to the octanol phase. The difference in partitioning behavior between the two systems probably reflects a difference in polarity between monoolein and octanol. At increasing DMSO content, $P_{GMO/w}$ increases slightly, being 0.09 ($\log P_{GMO/w} =$

-1.0) for $\phi_{DMSO} = 0.2$, i.e. DMSO partitions more into the monoolein domain of the lamellar phase.

Adding water to the lamellar phase in the DMSO system (and the other three systems as well) results in two different phases. At low DMSO content the lamellar phase transforms to a reversed bicontinuous cubic liquid crystalline phase, with a water swelling limit of about 35 wt%. This phase behavior is anomalous since the normal trend upon adding water to a polar lipid is an increase in interfacial curvature, i.e. reversed hexagonal → reversed bicontinuous cubic → lamellar. The monoolein–water system goes the other way, i.e. lamellar → reversed bicontinuous cubic, perhaps reflecting a subtle conformational change of the glycerol head group in the presence of more water.

At higher DMSO content, adding water to the lamellar phase results in the narrow sponge phase at a water content of about 30 wt%. The structure of the sponge phase is similar to the reversed cubic phase, consisting of a congruent lipid bilayer with a slightly negatively curved interface [10]. The difference between the cubic phase and the sponge phase is the strong bilayer–bilayer correlation in the former, which gives rise to a long-range order. In the sponge phase, this correlation has decreased because of a larger separation due to the added solvent.

Water alone cannot cause the formation of the sponge phase in the monoolein–water system; a solvent with sufficient tendency to go to the interfacial and bilayer domain has to be added. The role of the solvent in this case is therefore to make the interfacial region less negatively curved so that more water-solvent mixture can be incorporated. Adding water to the lamellar phase "drags" solvent from the monoolein and interfacial domains into the aqueous domain, thereby making the interfacial curvature slightly negative.

It is interesting to note that the amount of water needed to form the sponge phase for the solvents used in this work seems to correlate well with the octanol/water partitioning behavior of the solvents. The more lipophilic the solvent is (i.e. the higher the partition coefficient is $P_{oct/w}$ (EtOH) > $P_{oct/w}$ (PG) > $P_{oct/w}$ (DMSO) [9]), the more water is needed to form the sponge phase. The conclusion also implies that a solvent which is more prone to go to the lipid domain than to water should not give rise to a sponge phase in the monoolein–water system.

Comparisons with ionic and nonionic systems

L_3-phases are found in other systems as well, such as the AOT–NaCl–water system [11, 12] and the $C_{12}H_{25}$-$(OC_2H_4)_5OH$–water ($C_{12}E_5$–water) system [13]. In the AOT–water system a large lamellar phase dominates the

Progr Colloid Polym Sci (1998) 108:93–98
© Steinkopff Verlag 1998

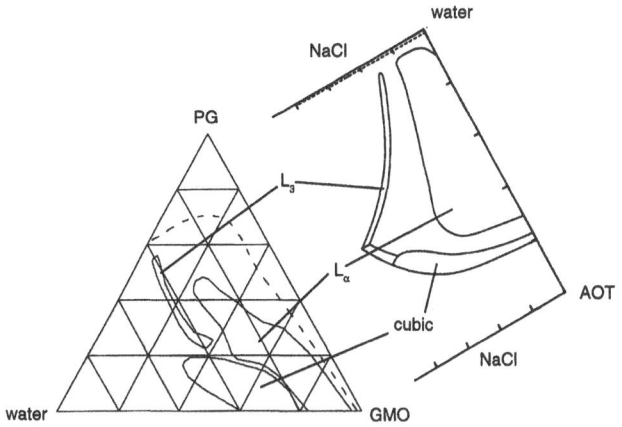

Fig. 3 A comparison of the phase behavior between the PG–monoolein–water system and the NaCl–AOT–water system (adapted from Ref. [12])

phase diagram, as shown in Fig. 3. In the presence of NaCl, a narrow L_3 phase is formed over the entire water range where the lamellar phase exists. At the low water side of the L_3 phase, it is in equilibrium with a cubic phase of reversed bicontinuous type. In the $C_{12}E_5$–water system, the L_3 phase occurs as a narrow region at elevated temperatures. At lower temperatures, the system forms a lamellar liquid crystal.

The formation of the L_3 phase in the AOT system is caused by NaCl screening the electrostatic repulsion between the negatively charged AOT head groups. This screening leads to the formation of a slight negatively curved interface, characteristic of the L_3 phase [10]. In the $C_{12}E_5$ system, a temperature increase makes the nonionic amphiphile more lipophilic, thereby promoting a negative interfacial curvature. It is thus evident that NaCl seems to play the same role for the formation of the L_3 phase in the ionic system as a temperature increase in the nonionic system.

An interesting comparison between the AOT system and the present systems can be made if we rotate the phase map of the AOT system in the way shown in Fig. 3. With some imagination it is seen that the water in our systems seems to play the same role as NaCl in the AOT system. The interpretation of the phase behavior concerning the formation of the sponge phase in our systems, and particularly the role of water, thus seems to be justified by the comparison with the AOT system. Water "drags" out solvent from the monoolein and interfacial domains in our systems, whereas NaCl screens the electrostatic repulsion between charged head groups. Both effects lead to a decrease in interfacial curvature.

Another polar molecule which has a profound effect on the monoolein–water phase behavior is sucrose [14]. If

sucrose, a highly water soluble substance, is added to the cubic phase, the phase transforms to a reversed hexagonal phase. The phase behavior, which implies that monoolein becomes more wedge-shaped, was interpreted in terms of a reduction of the monoolein's polar head group area caused by changes in hydrogen bonding [14]. It should be reviewed that a reduction of water in the monoolein–water system causes the transformation (reverse) cubic to (planar) lamellar, which in view of the sucrose results indicates that very subtle interactions in the head group region take place.

The temperature stability of the present sponge phases is several tens of degrees for the DMSO and PG systems. Moreover, at a given water content the sponge phase is favored over a lamellar phase upon a temperature increase. Increasing the temperature in the AOT system (at constant water content), on the other hand, leads to the transition $L_3 \rightarrow L_\alpha$, i.e. opposite to that found in the monoolein system and the $C_{12}E_5$ system. In the former case, the electrostatic repulsion between the charged head groups increases with increasing temperature, since $\varepsilon \propto 1/T$, leading to a planar interface. This effect dominates slightly over the increased hydrocarbon chain mobility, which would favor the opposite, i.e. $L_\alpha \rightarrow L_3$ [15].

In the solvent–monoolein–water case, the lack of ion–ion repulsion explains the difference. In a system with an ethoxylated fatty alcohol, the temperature dependence is qualitatively similar to the present system, but the temperature sensitivity is much higher due to the increased hydrophobicity of the ethoxy headgroup with increasing temperature [15], which in turn favors a negatively charged interface even more.

Work is now in progress to study the structure and dynamics of the sponge phases by means of X-ray diffraction and self-diffusion NMR. The effect of adding a fourth component to the sponge phase is also being investigated.

Appendix

The derivation of Eq. (1) is made under the following assumptions: (i) the structure is a lamellar liquid crystal, (ii) the lipid bilayer thickness, d_{lipid}, of the original lipid–water system is unaltered by the addition of the third

Fig. A1 Definition of symbols used in the derivation of Eqs. (A3) and (A4)

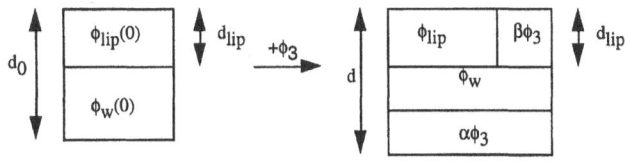

component, and (iii) a fraction α of the third component behaves like water, and $\beta = 1 - \alpha$ behaves like lipid. The situation is sketched in Fig. A1.

The following relation holds:

$$d(\phi_{\text{lipid}} + \beta\phi_3) = d_{\text{lipid}} = d_0\phi_{\text{lipid}}(\phi_3 = 0) , \qquad (A1)$$

where ϕ denotes a fraction and d is the measured spacing.

For the case when $\phi_w = \text{constant} (= 1 - \phi_{\text{lipid}}(\phi_3 = 0) = 1 - \phi_{\text{lip}}(0))$, Eq. (A1) may be rewritten as

$$d(\phi_{\text{lip}} + \phi_3 - \alpha\phi_3) = d_0\phi_{\text{lip}}(0) . \qquad (A2)$$

Since

$$\phi_{\text{lipid}} + \phi_3 = \phi_{\text{lip}}(0)$$

Eq. (A2) simplifies to

$$d = d_0/(1 - \alpha\phi_3/\phi_{\text{lip}}(0)) . \qquad (A3)$$

For the case where $\phi_w/\phi_{\text{lip}} = \text{constant}$, Eq. (A1) may be simplified to

$$d = d_0/(1 - (1 - \beta/\phi_{\text{lip}}(0))\phi_3) . \qquad (A4)$$

In this work $\phi_{\text{lip}} = \phi_{\text{GMO}}$ and $\phi_3 = \phi_{\text{DMSO}}$.

Acknowledgements We would like to thank Prof. Kåre Larsson, from whom we have learned most of the basics concerning polar lipids, in general and the monoolein system in particular. Johan Engblom and Håkan Wennerström are gratefully acknowledged as well. S.E. wishes to thank Prof. Barry Ninham and Dr. Stephen Hyde, Applied Math. Australian National University in Canberra for a stimulating visit which was the starting point for the present study. K.A. wishes to thank Camurus AB for generous support.

References

1. Ericcson B, Eriksson PO, Löfroth JE, Engström S (1991) ACS Symp Ser 469:251–265
2. Engström S, Ljusberg-Wahren H, Gustafsson A (1995) Pharm Tech 7: 14–17
3. Razumas V, Kanapieniené J, Nylander T, Engström S, Larsson K (1994) Anal Chim Acta 289:155–162
4. Hyde ST, Andersson S, Ericsson B, Larsson K (1984) Z Kristallogr 168: 213–219
5. Israelachvili JN, Marcelja S, Horn R (1980) Q Rev Biophys 13:121–200
6. Svärd M, Schurtenberger P, Fontell K, Jönsson B, Lindman B (1988) J Phys Chem 92:2261–2270
7. Lindström M, Ljusberg-Wahren H, Larsson K, Borgström B (1981) Lipids 16:749–754
8. Gutman H, Arvidsson G, Fontell K, Lindblom G (1984) In KL Mittal, B Lindman (eds) Surfactants in Solution. Plenum Press, New York, pp 143–152
9. Leo A, Hansch C, Elkins D (1971) Chem Rev 71:525–616
10. Anderson D, Wennerström H, Olsson U (1989) J Phys Chem 93:4243–53
11. Fontell K (1975) In Colloidal Dispersions and Micellar Behaviour, ACS Symp Ser No 9, American Chemical Society, Washington, pp 270–277
12. Balinov B, Olsson U, Söderman O (1991) J Phys Chem 95:5931–36
13. Harusawa F, Nakamura S, Mitsui T (1974) Colloid Polym Sci 252:613–619
14. Söderberg I, Ljusberg-Wahren H (1990) Chem Phys Lipids 55:97–101
15. Evans DF, Wennerström H (1994) In The Colloidal domain – where Physics, Chemistry, Biology and Technology Meet, Ch 11, VCH, New York

Progr Colloid Polym Sci (1998) 108:99–104
© Steinkopff Verlag 1998

H. Ljusberg-Wahren
J. Gustafsson
T. Gunnarsson
N. Krog
L. Wannerberger
M. Almgren

Micelles and liquid crystals in aqueous diglycerol monodecanoate systems

H. Ljusberg-Wahren (✉)
T. Gunnarsson · L. Wannerberger
Camurus Lipid Research
Ideon, Gamma 1
Sölvegatan 41
SE-223 70 Lund
Sweden

J. Gustafsson · M. Almgren
Department of Physical Chemistry
Uppsala University
Box 532
SE-751 21 Uppsala
Sweden

N. Krog
Danisco Ingredients
Edwin Rahrs Vej 38
DK-8220 Brabrand
Denmark

Abstract The ternary phase diagram of diglycerol monodecanoate–glycerol monodecanoate–water was studied with conventional methods. Aggregate structure in water-rich samples was examined by means of cryo-TEM (cryo-transmission electron microscopy). The phase properties of these commercially available lipids was compared with high-purity fractions of diglycerol monodecanoate and diglycerol didecanoate. The monoester of diglycerol was found to form elongated micelles with water which with increase of temperature or addition of diester condensed to a liquid phase that could coexist with water. Further addition of diglycerol didecanoate or glycerol monodeca-noate transforms this liquid phase to a highly swollen (>97% water) lamellar liquid crystalline phase. The phase behavior is viewed in relation to the behavior of common non-ionic surfactants and discussed with respect to aggregate structures observed by cryo-TEM.

Key words Diacyldiglycerol – monoacyldiglycerol – monoacyl glycerol – lamellar liquid crystal – phase behavior – vesicle

Introduction

In many industrial applications it is desirable to control the physico-chemical properties of aqueous dispersions. Surface active molecules like partly esterified polyalcohols have been used for this purpose for a long time by the food and pharmaceutical industry [1–3]. Well-known examples of such amphiphilic substances are fatty acid esters of glycerol, and polyglycerols. The industry has traditionally used unrefined reaction mixtures, containing molecules esterified at different positions and with varying numbers of ester groups. Better knowledge of the phase behavior of the different components that make up the commercially available additives is needed in order to explore the full potential of these types of compounds.

Physical properties of monoglycerides have been thoroughly studied [3–6] but much less attention has been directed to fatty acid esters of polyglycerols. Polyglycerols consist of polymerized glycerol with diglycerol as one component [7]. Esters of both medium and long-chain fatty acids with polyglycerol are commercially available, but work reported in the literature mainly concerns the physical properties of esters of long-chained fatty acids [3, 8, 9].

In this communication, the ternary phase diagram of the diglycerol monodecanoate–glycerol monodecanoate–water system is reported. The phase diagram is dominated by a lamellar liquid crystalline phase that within a narrow lipid composition swells to incorporate more than 97% w/w water. The phase behavior is viewed in relation to the behavior of common non-ionic surfactants and discussed with respect to aggregate structures observed by cryo-transmission electron microscopy (cryo-TEM).

Materials and methods

Materials

A reaction mixture of diglycerol and decanoic (capric) acid was purified by molecular distillation resulting in a diglycerol monodecanoate rich fraction (DGMD85), containing 85% monoester, 11% higher esters and 4% free diglycerol. High-purity (>99%) fractions of monoester (DGMD99) and diesters (DGDD99) were also obtained from the reaction mixture of diglycerol and decanoic acid by using preparative HPLC, see below. The glycerol monodecanoate (GMD), was also purified by molecular distillation to a monoester content of more than 97%.

"Purified water" of USP (United States Pharmacopoeia) quality was used in all studies of phase behavior.

Liquid chromatography

The HPLC system consisted of Gilson 305 and 306 pumps, a Gilson 805S manometric module, a Gilson 811C dynamic mixer, a Rheodyne 7125 injector equipped with a 200 or a 20 μl loop for preparative and analytical purposes respectively, a Sedex 45 evaporative light scattering detector operating at 30 °C and a Shimadzu C-R5A Chromatopac reporting integrator. During preparative sampling a stainless-steel split (15:85) was attached between the column and the detector.

The columns used were stainless-steel columns, 250 × 10 and 250 × 2 mm i.d. for preparative and analytical purposes, respectively. They were packed with Lichrospher 100 diol, 5 μ, (Merck, Germany) at 800 and 950 bar, respectively. The flow rates were 2.8 ml/min (10 mm column) and 0.3 ml/min (2 mm column) and the column temperature was 55 °C. A gradient was run according to the following scheme: time 0 min (t_0) – 0% B, t_3 – 0% B, t_{34} – 80% B, t_{35} – 80% B, t_{40} – 0% B, t_{60} – 0% B.

Solvent mixture A consisted of hexane:1-propanol:formic acid (93:5:0.5) and solvent mixture B of hexane:1-propanol:water:formic acid (16:77:5:0.5). Hexane, 1-propanol and water (all HPLC-quality) were purchased from Fisher Scientific (UK) and formic acid (p.a.) were purchased from Riedel de Haën (Germany).

500 mg of the reaction mixture was dissolved in 2 ml of solvent A and 175 μl aliquots were injected on the preparative column each run. The eluates corresponding to the peaks of di- and mono-esters, respectively, were collected

in Teflon-lined screw-cap test tubes. The monoester as well as diester eluates from eight different fractionations were mixed and evaporated under nitrogen flow. The two ester fractions were further freeze dried until constant weight and checked for purity on the analytical column.

Physical characteristics of samples

The phase diagram in Fig. 2 is based on approximately 150 samples (1 g) prepared in the following manner. Mixtures of lipids were put in 10 ml injection vials and heated to 40 °C to melt the lipids. Water was added and the vials sealed. The samples were first put in the freezer (−18 °C) for at least 6 h and then kept at 25 °C to equilibrate for at least two days. The samples were centrifuged at 3000 rpm for 0.5 h before examination in polarized light. Birefringent samples were further examined under light microscope with polarized light to determine the presence of crystals.

Hydration of lipids in light microscopy

Only small quantities of the pure (>99%) mono- and diesters were available for phase studies which were performed in the following manner. About 3 mg lipid was placed on a glass plate and a cover glass was attached. The sample was first held at 40 °C for 5 min and then an additional 5 min at room temperature in the microscope before examination. The interface between lipid and air was focused at 200 times magnification before a small drop of water was added to the side of the cover glass. The water was sucked into the slit between the two glass plates and when the water reached the lipid interface, the interaction between the two components was recorded with a video camera.

Cryo-TEM

Samples studied by cryo-TEM were diluted to a water content of more than 99% w/w. Electron microscopy investigations were performed with a Zeiss 902 A instrument, operating at 80 kV. Substrates were prepared by a blotting procedure performed in a chamber with controlled temperature and humidity [10]. A drop of the sample solution was placed onto a copper EM-grid coated with a perforated polymer film. Excess solution was thereafter removed with a filter paper, leaving a thin film (<μm) of the solution on the EM-grid. Vitrification of the thin film was achieved by rapid plunging of the grid into liquid ethane held just above its freezing point. The vitrified specimens were then transferred in the cold state to the microscope and examined at about 100 K.

Results

This study concerns the phase properties of different esters of two polyalcohols, glycerol and diglycerol with decanoic (capric) acid, and mixtures of these esters. The esters of glycerol and diglycerol consist of different isomers depending on which hydroxyl group(s) the fatty acid(s) are attached to and on how the two glycerol units of the diglycerol are connected. All monoesters of diglycerol have three free hydroxylgroups while the diesters of diglycerol and monoglycerides have only two. The linear isomers of diglycerol monoester (DGMD), and diester (DGDD), are shown in Fig. 1, as well as HPLC chromatogram for the purified esters. It is obvious from the chromatogram that there is a greater resemblance in polarity between the different isomers of the monoester than the diesters.

The phase diagram of the diglycerol monodecanoate (DGMD85)–glycerol monodecanoate (GMD)–water system at 25 °C is dominated by a large lamellar phase (Fig. 2). A narrow band of lamellar liquid crystalline phase with a ratio of GMD to DGMD85 around 15:85 extends towards the water corner. More than 97% w/w water can be incorporated into the lamellar phase at this lipid ratio. At higher or lower ratios the lamellar phase exhibits a more limited swelling with water (40–50% w/w). The

lamellar phase is easily dispersed in water with formation of vesicles. Figure 3 shows cryo-TEM micrographs of the vesicles at two lipid compositions, weight ratios GMD:DGMD85 of 1:1 and 1:9, respectively. The micrograph of vesicles with lipid composition 1:9 shows membrane disruptions that are absent in vesicles with a 1:1 ratio of

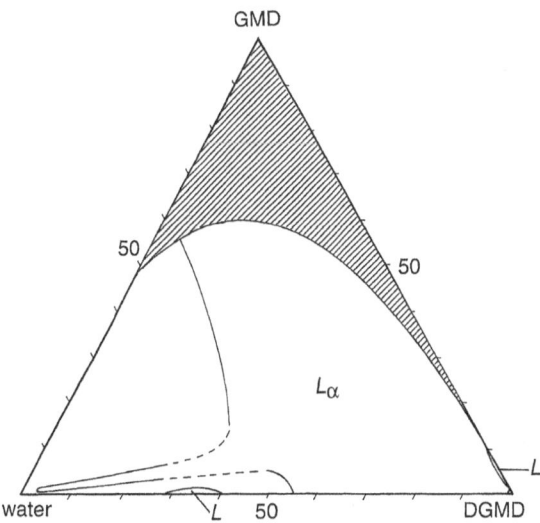

Fig. 2 Ternary phase diagram of the diglycerol monodecanoate (DGMD85)–glycerol monodecanoate (GMD)–water system at 25 °C. Only one-phase regions are shown. The shaded area indicates presence of crystals. L_α stands for a lamellar liquid crystalline phase and L stands for an isotropic liquid phase

Fig. 1 HPLC chromatogram showing isomers of >99% pure (a) diglycerol monodecanoate, DGMD and (b) diglycerol didecanoate, DGDD. Inset structures of the linear isomers

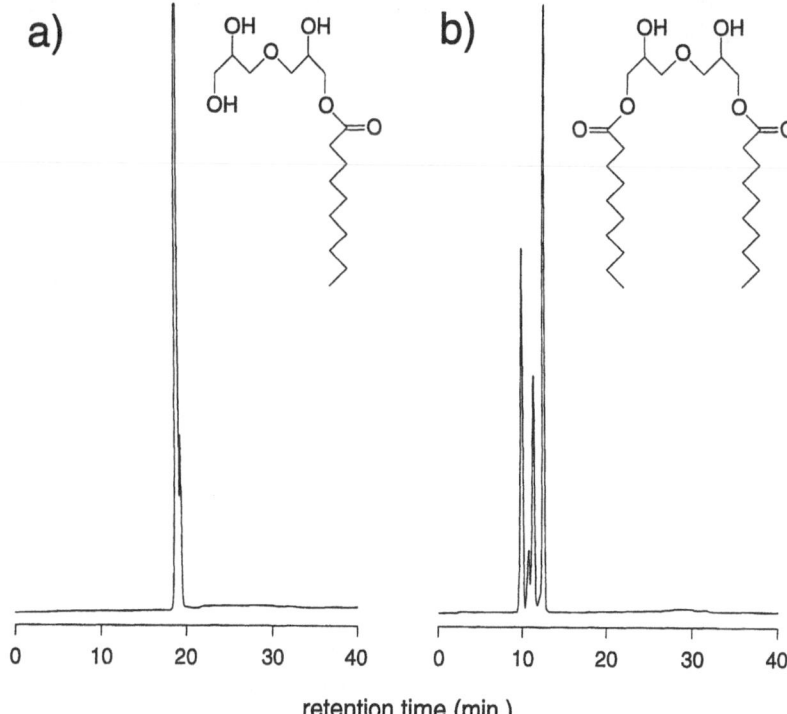

retention time (min.)

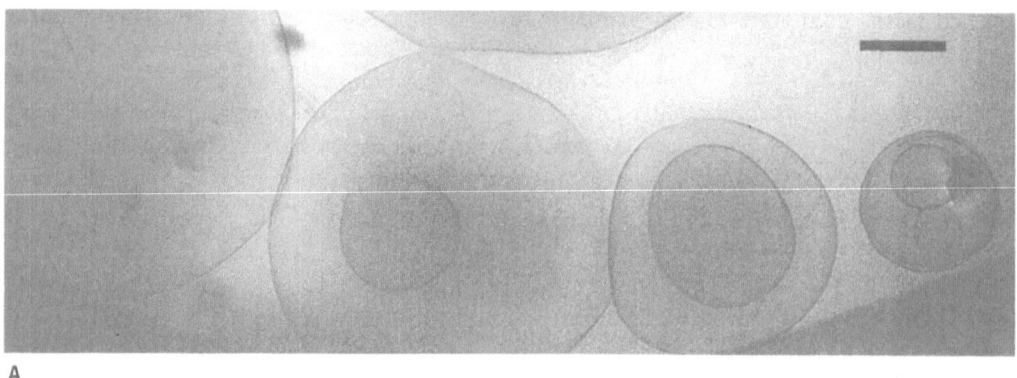

A

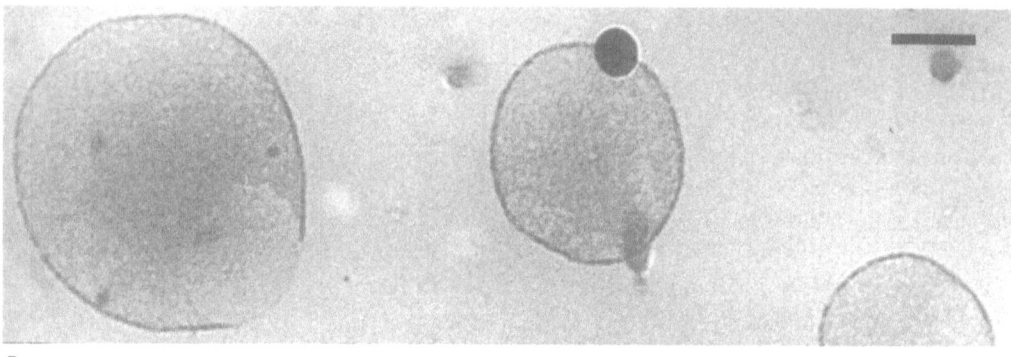

B

Fig. 3 Cryo-TEM micrographs of vesicles with different lipid composition from samples with 99% w/w water. Bar equals 100 nm: (A) weight ratio, glycerol monodecanoate (GMD):diglycerol monodecanoate (DGMD85), 1:1. Somewhat deformed vesicles of more or less intact membranes. (B) weight ratio, glycerol monodecanoate (GMD):diglycerol monodecanoate (DGMD85), 1:9. Vesicles of perforated membranes

the lipids. The membrane defects are probably pores piercing the bilayer structure [11, 12].

Samples with a composition corresponding to the area enclosed by dotted lines exhibit an inhomogenous texture but do not separate into two phases (Fig. 2). Insufficient time for equilibration (1–3 month) could be the reason for this inhomogeneity. The ^{2}H NMR studies on some samples enriched in heavy water were made in order to ascertain the single-phase character of the highly swollen lamellar phase (>70%). Spectra showing a doublet without interference of an isotropic peak was observed from the samples in the narrow band, suggesting that there is a single lamellar phase [13].

Two regions with isotropic liquid exist in the phase diagram (Fig. 2). One is melted DGMD85, which can solubilize 10% w/w monoglyceride but only a limited amount of water. The other is found in the binary system DGMD85 and water at intermediate water concentration (60–70% w/w). This phase can incorporate less than 5% w/w of the monoglyceride. It was observed that this phase had a limited time of existence. After 3 months the one-phase region was considerably reduced. This shift in equilibrium is probably due to hydrolysis of the ester bonds.

Since only small quantities of the pure (> 99%) mono- and diesters of diglycerol were available, their interactions with water were mainly studied by light and electron microscopes. Two solution phases were formed at 25 °C in dilute aqueous samples of the DGMD99(< 10% w/w), as previously observed with DGMD85. However, below 18 °C DGMD99 formed only a single isotropic phase in the dilute regime. The cryo-TEM micrograph of Fig. 4 shows that the micellar phase of DGMD99 at 17 °C contains thread-like micelles. In diluted mixtures of DGMD99 and DGDD99 (1:1) vesicles were present (see Fig. 5). These vesicles show membrane defects similar to the ones shown by the vesicles from GMD and DGMD85 at compositions corresponding to the narrow band of lamellar liquid crystalline phase (see Fig. 3B).

The diglycerol didecanoate (DGDD99) dispersed spontaneously in water, but did not exhibit any birefringence when examined in the light microscope. When water was added to the HPLC purified monoesters (DGMD99) a birefringent phase appeared initially. This birefringent phase was rapidly replaced by an isotropic low viscous phase. The monoester purified by distillation (DGMD85) had the same appearance when exposed to water under

Progr Colloid Polym Sci (1998) 108:99–104
© Steinkopff Verlag 1998

Fig. 4 Cryo-TEM micrograph from the micellar phase of diglycerol monodecanoate (99% w/w water) at 17 °C. Bar equals 100 nm

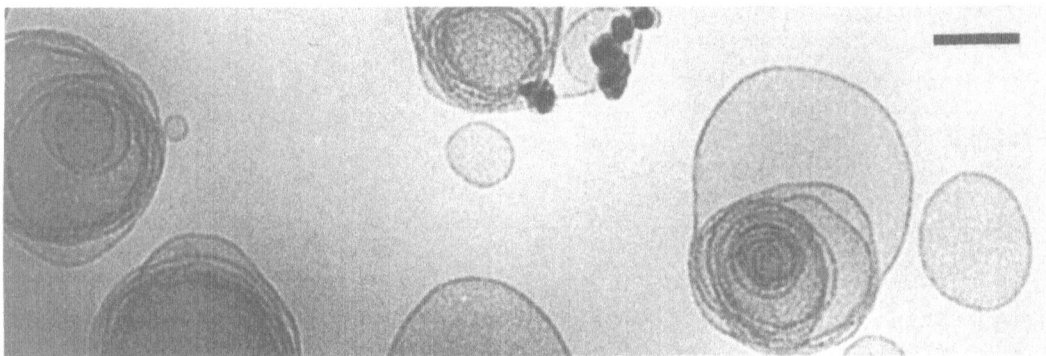

Fig. 5 Cryo-TEM micrographs of vesicles with perforated membranes from a diluted mixture of diglycerol monodecanoate and diglycerol didecanoate (99% w/w water). Bar equals 100 nm

similar conditions, but the extinction of the birefringent phase took much longer time. These observations are in agreement with the phase diagram reported here and show that the two qualities of DGMD used both form an isotropic low viscous phase when diluted in water at room temperature, and that a liquid crystalline phase exists at low water concentrations.

Discussion

Two observations of the phase behavior of diglycerol monodecanoate (DGMD)–glycerol monodecanoate (GMD)–water system are of primary interest: (a) Diglycerol monoesters can form a micellar solution with water. (b) There is an uncharged lamellar liquid crystalline phase that swells to more than 97% w/w water at a well-defined ratio between diglycerol monoester and monoglyceride.

Monohexanoin is the only monoglyceride reported to form a micellar solution at room temperature [4]. Other monoglycerides form L_2 phases or lyotropic liquid crystalline phases above their Krafft point with excess water.

Esters of diglycerol and fatty acids have been studied much less extensively than glycerol esters, but earlier work indicates similarities in properties between monoglycerides and esters of polyglycerol and fatty acids [3, 8, 9]. The present studied diglycerol ester (DGMD85) was, at water contents above 70% w/w, found in a liquid–liquid coexistence. Such equilibria are commonly observed with non-ionic surfactant–water mixtures, where they usually show a strong temperature dependence [14, 15]. Note that a single isotropic phase is formed below room temperature in dilute aqueous solutions of diglycerol monodecanoate (DGMD99). According to the cryo-TEM micrographs it contained thread-like micelles.

It is tempting to draw parallels between our system and those from other non-ionic surfactants, such as polyethylenglycol derivatives and polyglycosides [14–17]. Increased temperature induces in common binary non-ionic surfactant–water systems a transition into less curved aggregates. In the present mixed amphiphilic system the addition of the monoglyceride (GMD) has an effect corresponding to a temperature increase, i.e., it brings about a transition from a solution phase to a lamellar phase.

With respect to both the coexistence of two isotropic phases and the micellar–lamellar transition, there are thus obvious similarities in the temperature dependence of binary non-ionic surfactant systems and the phase behavior of the present mixed systems.

The highly swollen lamellar phase of the present diagram, located in a narrow band near the isotropic phases, is in the absence of charged amphiphiles presumably stabilized by undulation forces from fluctuations of flexible bilayers [18]. The thickness of the bilayers is an important parameter for its flexibility [18, 19]. We expect the present bilayers to be rather thin, since the fatty acid in the amphiphilic molecules have only 10 carbon atoms, which may be a contributing reason to the flexibility. Furthermore, the presence of solvent-filled defects (curvature defects) in the bilayers has been suggested to influence the bilayers in favor of a high flexibility [20]. The defective vesicles found in our system were observed at compositions corresponding to the location of the most swollen lamellar phase and in close vicinity to the micellar phase. This composition region is also where one would expect to find curvature defects in the bilayers, since their function is to allow for an aggregate structure with curvatures intermediate to intact bilayers and cylindrical micelles [11]. It is worth noting that also the addition of the diglycerol diester (DGDD99) induce the formation of defective membrane structures. It seems as if DGDD and GMD influence the phase behavior of diglycerol monoester DGMD in similar ways.

In conclusion, our results show that diglycerol fatty acid esters in addition to liquid crystalline phases also form micellar phases. Furthermore, diglycerols partly esterified with medium chain length fatty acids have features in their phase properties that are akin to amphiphilic polyethelenglycols derivatives rather than to monoglycerides of comparable alkyl chain length.

Acknowledgements We thank Susanne Nilsson and Pia Nilsson for skilful technical assistance and Allan Johansson for drawing the phase diagram. Greger Orädd is acknowledged for performing the NMR measurements.

References

1. Larsson K (1997) In Friberg S, Larsson K (eds) Food Emulsions. Marcel Dekker, New York, pp 39–65
2. Attwood D, Florence AT (1983) Surfactant Systems, their Chemistry Pharmacy and Biology. Chapman and Hall, London
3. Krog NJ (1997) In Friberg S, Larsson K (eds) Food Emulsions, Marcel Dekker, New York, pp 67–135
4. Larsson K (1967) Z Physik Chem Neue Folge 56:173–198
5. Larsson K (1994) Lipids – Molecular Organization, Physical Functions and Technical Applications. The Oily Press, Dundee
6. Lutton ES (1965) J Amer Oil Chemists' Soc 42:1068–1070
7. Aitzetmüller K, Böhrs M, Arzberger E (1979) Fette Seifen Anstrichm 81: 436–441
8. Hemker W (1981) J Amer Oil Chemists' Soc 58:114–119
9. Kumar NT, Sastry YSR, Lakshminarayana G (1989) J Amer Oil Chemists' Soc 66:153–157
10. Bellare JR, Davis HT, Scriven LE, Talmon Y (1988) J Electron Microsc Tech 10:87–111
11. Gustafsson J, Orädd G. Lindblom G, Olsson U, Almgren M (1997) Langmuir 13:852–860
12. Gustafsson J, Orädd G, Almgren M Langmuir (1997) 13:6956–6963
13. Persson N-O, Lindman BJ (1975) Phys Chem 14:1410–1418
14. Kratzat K, Schmidt C, Finkelman H (1994) J Coll Int Sci 163:190–198
15. Strey R, Schomäcker R, Roux D, Nallet F, Olsson U (1990) J Chem Soc Faraday Trans 86:2253–2261
16. Ekwall P, Mandell L, Fontell K (1969) Molec Cryst Liq Cryst 8:157–213
17. Rybinski W (1996) Curr Opp Colloid Int Sci 1:587–597
18. Safynia CR, Sirota EB, Roux D, Smith GS (1989) Phys Rev Lett 62:1134–1137
19. Szleifert I, Kramer D, Ben-Shaul A, Gelbart W M, Safran SAJ (1989) Chem Phys 92:6800–6817
20. Gustafsson J (1997) PhD Thesis, Uppsala

Progr Colloid Polym Sci (1998) 108:105–110
© Steinkopff Verlag 1998

O. Söderman
B. Balinov

Microstructures in solution and ordered phases of surfactants. Self-diffusion in the AOT/octanol/water system

O. Söderman (✉)
Physical Chemistry 1
Center for Chemistry
and Chemical Engineering
P.O. Box 124
S-22100 Lund
Sweden

B. Balinov
Nycomed Imaging
Nycoveien 1-2
Postboks 4220 Torshov
N-0401 Oslo
Norway

Abstract The observation of the presence of bicontinuous structures in both ordered and disordered phases in surfactant systems has proved to be of great importance in understanding the phase behavior of such systems. In this communication we show how the presence of such structures can be conveniently monitored by the NMR self-diffusion method. The system chosen for this study, AOT/octanol/water has a reversed cubic phase that melts into a liquid solution (L_2) phase upon dilution with water. The NMR data give evidence in favor of a situation where the microstructure in the two phases is similar.

From the values of the water diffusion coefficients, it appears that the microstructure is bicontinuous, albeit with a somewhat too large obstruction factor than would be predicted from the theoretical description of water diffusion in bicontinuous structures. We suggest that this behavior can be rationalized by the presence of "necks" or dead-ends in the structures.

Key words Microstructure – surfactant – AOT/octanol/water – NMR self-diffusion method

Introduction

One of Kåre Larsson's major scientific contributions is the recognition of the importance of bicontinuous structures in surfactant (or lipid) cubic liquid crystalline phases [1–3]. These phases are very important in a number of situations, the most striking case being probably the role they play in biological systems [4].

Of equal importance is the microstructure of surfactant solution phases. One of the major recent advancements in this field is the realization that the closed micellar aggregate is but one of many structures found in such systems. Also here one finds bicontinuous structures. Examples are L_3 (or sponge) phases [5] and bicontinuous micellar solutions [6]. Such phases often undergo a disorder–order transition into bicontinuous cubic phases. In fact, this process conveys important information about the micro-

structure in the disordered phase. The topology of the surfactant film is often retained during this process. As a consequence, the microstructures of the ordered and disordered phases are often similar. One way to prove this to be the case is to perform self-diffusion studies across the phase boundary [5]. If the microstructure in the two phases is similar, we expect the value of the self-diffusion coefficient to make no sudden jumps across the phase boundary.

Self-diffusion coefficients can conveniently be measured with the NMR Pulsed Field Gradient (PFG) method [7, 8]. This technique yields component-resolved and accurate values of the diffusion coefficients of isotropic phases. The measuring time is often rapid (on the order of 10 min or so per sample) and the technique requires no isotopic labeling or addition of probe molecules.

In this contribution we will describe such a NMR PFG study of the solution and cubic liquid crystalline phases found in the system AOT/octanol/water. This system has

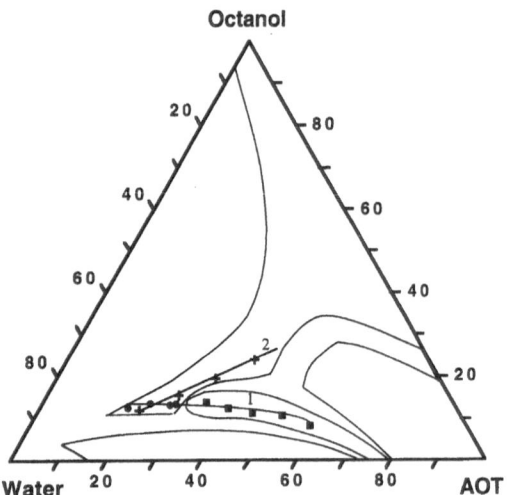

Fig. 1 The isothermal ternary phase diagram for AOT/octanol/ water adapted from [9]. Only one-phase areas are shown. Also indicated are the sample compositions for which the water and AOT diffusion coefficients were determined (the actual compositions based on heavy water have been converted to compositions based on ordinary water, assuming that the molar ratios between the components are kept constant). The lines denoted "1" and "2" are the paths referred to in the text

been used as a model system in drug delivery studies [9, 10]. The system has a rather rich ternary isothermal phase diagram which includes both an extended solution (L_2) phase and a cubic liquid crystalline phase (cf. Fig. 1). On the basis of the NMR PFG data, we discuss the microstructure in the two phases, with particular emphasis on the role octanol plays in the different structures formed. One further reason for performing this study is the recent report by Hahn and Wokaun, who studied diffusion by means of forced Rayleigh scattering in the AOT/octanol/ water system [11].

Experimental

Materials

AOT (sodium bis(2-ethylhexyl)sulfosuccinate) was obtained from Sigma and used without further purification. Octanol was from BDH, while heavy water was from Dr A.G. Glaser.

Sample preparation

Samples were prepared by weighing appropriate amounts into glass tubes that were flame-sealed. After equilibration, the tubes were opened and appropriate amounts were transferred into 5 mm NMR tubes. The viscous cubic phases were transferred by means of a syringe with a thick

needle. Samples were made with heavy water. The following densities were used in converting weight fractions to volume fractions: $\rho_{AOT} = 1.139$ g/cm^3, $\rho_{D_2O} = 1.104$ g/cm^3 and $\rho_{Oct} = 0.827$ g/cm^3.

Experimental procedures

The PFG NMR experiments were performed on a JEOL FX 60 spectrometer, monitoring the ^{1}H signal at 60 MHz. A standard Hahn echo was used, following procedures recommended in [7]. The gradients were generated by means of a home-built gradient driver. The strength of the gradients were calibrated using substances with known diffusion coefficients. AOT diffusion coefficients were measured by following the combined peak of the octanol and AOT methylene groups. Since octanol diffusion is considerably more rapid, its contribution to the echo attenuation is negligible at higher field gradient strengths, and at these strengths the AOT diffusion coefficient can be determined. The procedure was checked by monitoring the signal from the isolated AOT signal from the succinate head group. The signal-to-noise of this peak is poor, and hence the methylene peak was used to obtain data with higher precision.

Small angle X-ray scattering data were performed on a Kratky small angle system equipped with a position-sensitive detector.

All experiments were carried out at 25 °C.

Results and discussion

Review of phase behavior

AOT is a double-chained surfactant, where each chain is branched. As such, it has a bulky hydrophobic part, and its binary phase behavior with water is typical for surfactants of this type. Thus in the binary AOT/water system, a lamellar phase extends from roughly 15 wt% AOT up to roughly 75 wt% AOT. Upon increasing the AOT concentration further, a narrow cubic region follows, and finally a reversed hexagonal phase extends all the way out to pure surfactant. Thus in terms of curvature, the surfactant film goes from zero curvature in the lamellar phase to reversed curvature as the AOT concentration increases. This can be rationalized from the increase of counter-ion concentration, that screens the electrostatic interactions among the head groups, leading to a decreased curvature.

In Fig. 1 we present the ternary isothermal AOT/ octanol/water phase diagram (please note that only one-phase regions have been included). Upon addition of octanol, the cubic (and reversed hexagonal) phase swells with water. By weight, the amount of water that can be incorporated into the cubic phase is roughly 60%, at which point the molar ratio of octanol to AOT is around 1.5. Upon increasing the water amount even further, the cubic phase undergoes a transition into a L_2 region. The L_2 region is separated from the cubic phase by a narrow two-phase region. This L_2 phase dominates the phase triangle, and it swells up to a maximum amount of water of ≈ 75 wt%.

The evolution in phase behavior caused by the addition of octanol can be rationalized from curvature arguments. Adding a co-surfactant with a small head group creates an effective surfactant with an even larger hydrophobic volume. Thus the curvature is driven towards reversed phases. At high enough water content, the phase melts into a disordered liquid solution on account of the lowered inter-aggregate interactions.

Small angle X-ray scattering

Four samples in the cubic phase were investigated by small angle X-ray scattering. Only one Bragg peak was obtained from each sample. Based on 6–8 observed Bragg peaks, Fontell reported that the cubic phase in the binary AOT/water system as well as the corresponding phase in the ternary AOT/decanol/water system, belong to the Ia3d space group [12]. In particular, he detected no changes in the space group up to rather high water contents (roughly 50 wt%). We shall therefore assume that the cubic phase in the octanol system indeed belongs to the Ia3d space group. Given in Table 1 are the compositions of the samples studied by X-ray as well as the unit cell size. The unit cell size varies smoothly with the water content over the interval studied, indicating that there is no change in space group taking place.

There are two ways of describing the arrangements of the surfactant and water in the unit cell of cubic phases. In one approach, the underlying microstructure is described in terms of a minimal surface which, for a reversed structure, is draped with a surfactant film [2]. The minimal surface corresponding to the Ia3d phase is the gyroid, which has a coordination number of 3.

In the other approach, originally introduced by Luzatti and co-workers [13], the structure is described in terms of two subvolumes of water (for a reversed structure) where each forms a labyrinth of connected rods. This structural model has been termed the Interconnected Rod (ICR) model.

As a general rule, the former model is more applicable to dilute phases, while the latter may be more useful in describing more concentrated phases (for a detailed discussion of this topic, see [14]). In this work, we have chosen to use the ICR approach. Using results from Luzatti et al., we have calculated the area of the rods making up the unit cell, and the result is given in Table 1. On the assumption that all the octanol is located at the hydrophobic/hydrophilic interface, and on the further assumption that the area of AOT is 68 Å^2 and octanol 25 Å^2, the calculated area from the sample compositions is also given in Table 1. As can be seen, the two sets of areas agree reasonably well, except perhaps at the highest water content, where the ICR model may be less applicable. This indicates that in the cubic phase, a substantial fraction, if not all of the octanol resides at the interface.

NMR self-diffusion data

In order to obtain information about the microstructure in the cubic and disordered L_2 phase at high water content, we have performed a self-diffusion study of samples in the cubic and L_2 phases. Given in Fig. 2 are the reduced water diffusion coefficients (the actual value of the self-diffusion coefficient divided by the value for bulk water at the same temperature), while in Fig. 3 we present the AOT diffusion coefficient (the data correspond to samples along path 1 in

Table 1 Small angle X-ray scattering data for the AOT/octanol/water system

Φ_{AOT}[1]	Φ_{oct}[1]	Φ_{wat}[1]	d [Å][2]	a [Å][3]	A [10^4 Å^2][4]	A_{calc} [10^4 Å^2][5]
0.281	0.159	0.559	48.9	119.8	7.67	8.61
0.375	0.131	0.495	42.2	103.3	5.70	5.96
0.431	0.126	0.443	37.2	91.1	4.32	4.35
0.494	0.096	0.409	34.0	83.2	3.52	3.47

[1] Volume fractions of AOT, octanol and water.
[2] Bragg distances.
[3] Unit cell size, calculated for the Ia3d space group, from the relation $a = d\sqrt{6}$.
[4] Area of the surfactant film per unit cell, calculated assuming that all AOT and octanol form a surfactant film and using as the area for AOT and octanol, 68 and 25 Å^2, respectively.
[5] Area calculated for the labyrinths in the ICR model of the Ia3d structure, using relations from [13].

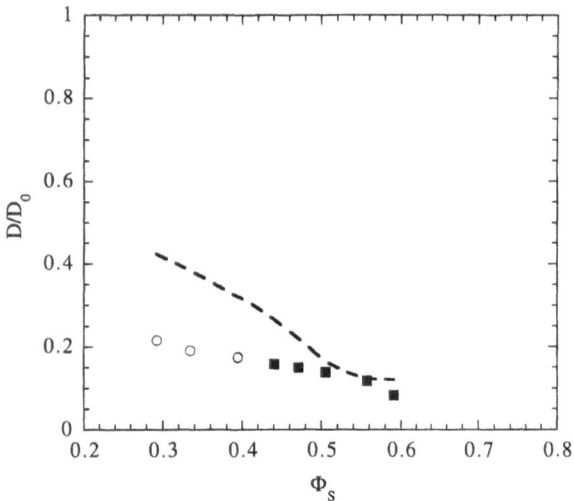

Fig. 2 The reduced water diffusion coefficients along path 1 (see Fig. 1) vs. the volume fraction of AOT plus octanol, Φ_S. The solid squares refer to samples in the cubic phase, while the circles refer to samples in the L_2 phase. The dashed line is the prediction of a model for the diffusion in the ICR structure (see text for details)

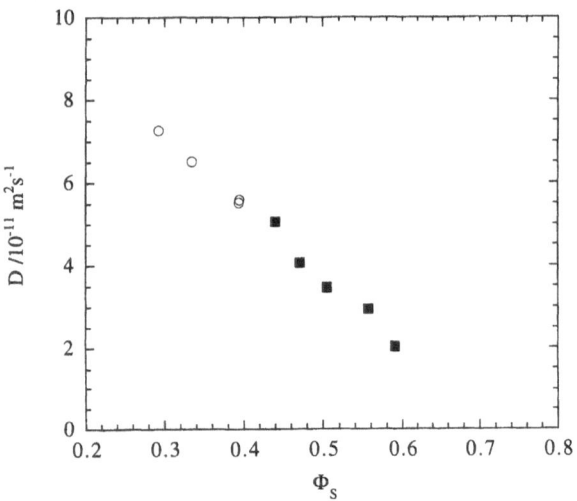

Fig. 3 The AOT self diffusion coefficients along path 1 (see Fig. 1) vs. the volume fraction of AOT plus octanol, Φ_S. The solid squares refer to samples in the cubic phase, while the circles refer to samples in the L_2 phase

the phase diagram in Fig. 1). Both sets of data are presented vs. the volume fraction of AOT and octanol, Φ_S. A striking observation that can be made in both sets of data is the continuous change in the diffusion behavior over the studied composition range. In particular, there is no discontinuity at the phase transition from the cubic to the L_2 phase. This fact suggests that the microstructure in the L_2 phase is topologically related to the microstructure in the cubic phase. This follows since the self-diffusion behavior

of the water and surfactant depends on the microstructure [5, 15], and the absence of a discontinuity in the diffusion behavior thus suggests that the microstructure evolves smoothly without any major changes in topology along path 1 in the phase diagram.

Having said this, the next logical question is: What is the microstructure? From the phase diagram we know that the structure has to be reversed. Two extreme structures are then constituted by closed (reversed) micelles and a bicontinuous structure, visualized in term of the ICR or minimal surface models, as described above. The former can be excluded for the following two reasons. First, for a structure of closed water droplets, the self-diffusion of the water and surfactant is expected to be similar, since they both form the diffusing unit. This is clearly not the case (we recall that the diffusion of HDO in D_2O is 1.9×10^{-9} m^2 s^{-1} at 25 °C). Secondly, for a spherical aggregate, the hydrodynamic radius can be estimated from the value of the diffusion coefficient and the relation $R_H = k_B T / (6\pi\eta D)$, where η is the viscosity of the medium. Taking η to be the value of octanol and the value for the reduced diffusion coefficient as 0.2, one obtains a value of R_H equal to 0.5 Å, clearly an unphysical result. Thus we conclude that the structure is bicontinuous along path 1 as indicated in Fig. 1.

In order to rationalize the data further, we will make use of the obstruction effects for surfactant and water diffusion calculated by Andersson and Wennerström [15]. We have chosen to use their results for the ICR structures. The arguments below could also be made on the basis of minimal surface structures. Starting with the water diffusion, they obtained for the volume diffusion in the labyrinths of the ICR structures:

$$D_V/D_0 \approx 0.636 - 0.264\Phi_S \ (\text{for } \Phi_S < 0.8) , \tag{1}$$

where D_0 is the bulk diffusion coefficient of water.

For surface diffusion of a particle confined to diffuse along the surface (as would be the case for the surfactant in the present system) of the interconnected network, Andersson and Wennerström's results can be written as an expansion to second order in Φ_S:

$$D_S/D_0^S \approx 0.607 - 0.133\Phi_S - 0.131\Phi_S^2 , \tag{2}$$

where D_0^S is the lateral diffusion along the surface of the channels forming the labyrinth.

Now, one problem in interpreting the water data is the need to account for an ensemble of water molecules that is "bound" to the surfactant film. Since the water exchange between the surface and "free" (free in the sense that they experience no perturbation in their translational diffusion properties due to the presence of the surfactant film) states is rapid on the experimental time scale (on the order of

100 ms), one detects a population weighed average of bound and free water.

Hence we write

$$D_{obs} = P_B D_B + (1 - P_B)D_F . \tag{3}$$

We shall take the number of bound water per AOT molecule as 15 and per octanol as 5. This corresponds to a layer of bound water which is roughly two water diameters thick. This is in keeping with the generally accepted notion that the surface-induced perturbation of water is rather short-ranged.

The bound water will be described with Eq. (2) above, i.e. we consider those water molecules as diffusing along the surface of the water channels, while the remaining water is described by Eq. (1) above. Furthermore, we shall take as the value for the bound water lateral diffusion $D_0^S = D_0/4$. As discussed in [5], this appears to be a reasonable number.

We are now in a position to predict the water diffusion along path 1 in the phase diagram, and the prediction is included as a dashed line in Fig. 2. We note that the observed diffusion coefficients are considerably lower than the predicted ones.

This is an interesting observation. We note that in a related study on the AOT/brine system, which has a similar phase behavior, albeit with the concentration of salt in the water as a tuning parameter, this was not observed. In an analysis similar to the one performed above, it was indeed possible to rationalize the water diffusion data in terms of a bicontinuous minimal surface structure. In particular, at high water contents (in the brine system, a L_3 phase is formed which can take up roughly 90 wt% brine) the limiting reduced water content was 2/3, to which our predicted values extrapolate to.

This would suggest the presence of additional obstruction effects, not accounted for in the analysis above. We envision two mechanisms that may cause such additional obstruction effects. One is the formation of "necks" formed in the cylindrical part of the labyrinths. Such necks could be stabilized by a non-ideal mixing of AOT and octanol, such that the octanol content is higher at the cylindrical parts, which would reduce the penalty in curvature energy caused by the unfavorable low curvature of such necks. Another conceivable candidate for reducing the water diffusion would be the presence of dead-ends in the structure. Again such dead-ends could be stabilized by the enrichment of octanol at the dead-ends. We note that the molar ratio of octanol to AOT increases along path 1 (cf. Fig. 1). We also note that from path 1 in the phase diagram, the L_2 phase can also evolve along path 2. Along this path the water diffusion decreases, as indicated in Fig. 4, indicating that eventually a situation with closed water droplets is at

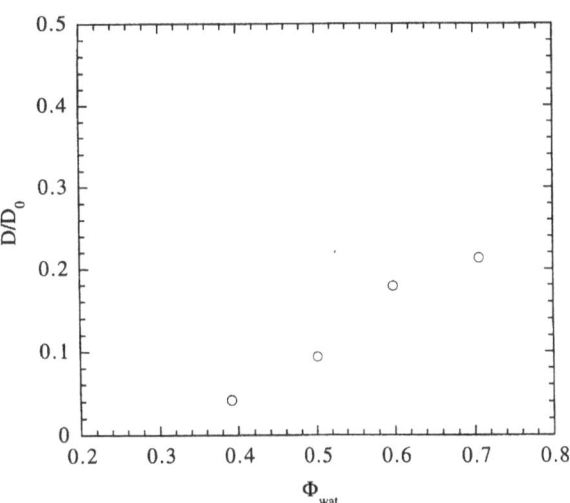

Fig. 4 Reduced water diffusion coefficient along path 2 in the phase diagram (see Fig. 1) vs. the volume fraction of water

hand. At that point all connections between the water droplets have been cut off.

We now turn to the AOT diffusion coefficients. As already noted above we see no discontinuity at the phase transition from cubic to L_2, again indicating a situation with similar microstructures in the two phases. Also striking is the (almost) linear increase in the AOT diffusion with addition of water. This increase is larger than what was found in the AOT/brine study referred above. In addition, it is not in accordance with the predictions of Eq. (2). One reason for this increase is the presence of increasing amounts of octanol in the surfactant film upon increasing the water content along path 1. Since the octanol molecules diffuse considerably faster along the surfactant film than does AOT, we expect the AOT diffusion to become more rapid. With regard to the extra obstruction effects needed to account for the water diffusion, the formation of necks is not expected to influence the surfactant diffusion to any considerable extent. The presence of dead-ends is slightly different in this regard. At first, one would think that such dead-ends would also decrease the AOT diffusion. However, one should note that AOT/octanol forms an extended solution phase, indicating that AOT may dissolve as monomers in the octanol phase as the amount of octanol increases. Such an effect would lead to an increase in the AOT diffusion coefficient.

Finally, it is interesting to compare the results of this study with the results of a diffusion study on the AOT/octanol/water system performed by means of forced Rayleigh scattering [11]. In this study the diffusion of a photochromic highly water soluble probe molecule was measured. The diffusion behavior of the probe molecule along path 2 in the phase diagram in Fig. 1 is opposite to what we find for water: upon addition of water the diffusion of the probe molecule *decreases* (Fig. 4 in [11]). At high water contents, the ratio of the probe molecule diffusion coefficient in the microemulsion to that found in the water solution used for making the microemulsion of the dye is ≈ 0.01, while for water we find roughly 0.2. This is a considerable difference.

Hahn and Wokaoun interpret this finding in terms of a microstructure with no open pathways for the dye molecules over the distance range probed (which in their experiment is on the order of 20 μm). Such a structure is clearly not compatible with the water data presented in Fig. 2. The only conceivable way to reconcile the two data sets would be in terms of the presence of narrow necks, as described above. The probe molecule is a fairly large molecule and it seems plausible that for certain radii of such necks, only water could pass through them, while the large and bulky probe molecules would be effectively hindered.

easily applied to isotropic phases. Both ordered, such as cubic liquid crystalline phases and solution phases can be investigated. Of particular interest are cases where an ordered phase is in equilibrium with a disordered phase. This follows since the microstructure in the ordered phase is often known, and if the diffusion behavior of the components evolves smoothly upon going from the ordered phase to the disordered, it is conceivable that the microstructure in the two phases is similar.

We have shown this to be the case for the cubic and L_2 phases in the AOT/octanol/water system. An interesting observation in this system is that the observed obstruction effects due to the presence of the surfactant aggregates determined from the water diffusion coefficients are somewhat larger than what would be expected from assumed pictures of the structures. This would indicate that there is an extra effect causing the water diffusion to slow down somewhat. We suggest that this is caused by the presence of "necks" and/or dead-ends and that it is the presence of octanol in the surfactant film that provides the system with an extra degree of freedom to encompass such structures.

Conclusion

In this communication we have attempted to show the usefulness of the NMR PFG diffusion method in studying microstructures in surfactant systems. The method can be

Acknowledgments We are grateful to Björn Håkansson for performing the SAXS experiments, the data from which are shown in Table 1. This work was financially supported by the Swedish Natural Science Research Council (NFR).

References

1. Lindblom G, Larsson K, Johansson L, Fontell K, Forsén S (1979) J Am Chem Soc 101:5465
2. Hyde S, Andersson S, Ericsson B, Larsson K (1984) Z Kristallogr 168:213
3. Larsson K (1989) J Phys Chem 93:7304
4. Landh T (1996) Thesis, Lund University, Lund
5. Balinov B, Olsson U, Söderman O (1991) J Phys Chem 95:5931
6. Monduzzi M, Olsson U, Söderman O (1993) Langmuir 9:2914
7. Stilbs P (1987) Prog Nucl Magn Reson Spectrosc 19:1
8. Söderman O, Stilbs P (1994) Prog Nucl Magn Reson Spectrosc 26:445
9. Osborne DW, Ward AJI, O'Neill KJ (1988) Drug Develop Ind Pharm 14:1203
10. Osborne DW, Ward AJI, O'Neill KJ (1990) Drugs Pharm Sci 42:349
11. Hahn C, Wokaun A (1997) Langmuir 13:391
12. Fontell K (1973) J Colloid Interface Sci 43:156
13. Gulik A, Luzzati V, Rosa MD, Gambacorta A (1985) J Mol Biol 182:121
14. Ström P, Anderson DM (1992) Langmuir 8:691
15. Anderson DM, Wennerström H (1990) J Phys Chem 94:8683

Progr Colloid Polym Sci (1998) 108:111–118
© Steinkopff Verlag 1998

K. Lindell
J. Engblom
M. Jonströmer
A. Carlsson
S. Engström

Influence of a charged phospholipid on the release pattern of timolol maleate from cubic liquid crystalline phases

This paper is dedicated to Prof. Kåre Larsson at the time of his retirement from academic duties

K. Lindell (✉) · J. Engblom · S. Engström
Food Technology
Center for Chemistry
and Chemical Engineering
University of Lund
P.O. Box 124
S-221 00 Lund
Sweden

M. Jonströmer
Astra Draco AB
P.O. Box 34
S-221 00 Lund
Sweden

A. Carlsson
Scotia LipidTeknik AB
P.O. Box 6686
S-113 84 Stockholm
Sweden

Abstract The release of timolol maleate, a drug for treatment of glaucoma, from cubic liquid crystals consisting of charged and uncharged lipids was investigated in vitro. It was shown that the release of drug was dependent on the ionic strength of the receptor medium and the relative amount of a charged phospholipid in the cubic phase. In the absence of charged lipids, the release of timolol was complete, irrespective of the medium used, i.e. distilled water or saline solution. In the presence of a negatively charged phospholipid, diacylphosphatidylglycerol, the release of drug from the cubic phase was incomplete when distilled water was the receptor medium. The residual drug was dependent on the amount of charged lipid in the cubic phase. In saline solution, on the other hand, all of the drug was released. The results are interpreted as an effect of electrostatic attraction between the positively charged drug with the negatively charged bilayer surface, which is screened in the presence of saline. The possible influence of the cubic geometry is discussed as well. This work shows that lipids offer many possibilities in the formulation of sustained drug release systems.

Key words In vitro release – timolol maleate – cubic liquid crystal – X-ray crystallography – lipid-drug interaction – glyceryl monooleate – distearoylphosphatidylglycerol – electrostatic screening

Introduction

Glyceryl monooleate (GMO, monoolein), a metabolite in fat digestion, has interesting phase properties in water which make it an interesting excipient for drug delivery. In excess water, GMO forms a reversed bicontinuous cubic liquid crystal, whose structure was thoroughly investigated by Larsson and co-workers [1, 2]. The labyrinth structure of the cubic phase, with its relatively narrow water and lipid channels (ca. 1–5 nm wide), gives rise to a sustained release of incorporated drug [3, 4] due to increased diffusion path length and interaction between the large lipid/water interface and the diffusing drug. The

objective of this work was to investigate the latter mechanism further.

Timolol maleate (TM) is a drug used for the treatment of glaucoma. It is usually administered as a low-concentrated (0.34 wt% TM) aqueous solution into the cul-de-sac (inside the lower eyelid), with a desired local effect. Although an eyedrop formulation is generally accepted by patients, it suffers from a disadvantage from a delivery point of view, since the administered drop of drug solution is rapidly cleared from the eye-pocket by blinking and efficient drainage. To compensate for that loss of drug, the eyedrops are taken frequently and the drug is overdosed with a risk of systemic effects, i.e. the drug enters into the circulation. Various strategies to

112

K. Lindell et al.
Release of timolol from cubic liquid crystals of GMO/DSPG/water

overcome this problem has been suggested, ranging from lenses to in situ gelling systems [5, 6]. The latter formulation has the advantage in that it combines the eyedrop concept, simple to use, with the lens concept, better release characteristics.

Most in situ gelling systems suggested in this context are based on polymers which in aqueous solution respond to external *stimuli*. One example is the Gelrite® system which consists of a 1% aqueous solution of low acetyl gellan gum which undergoes a sol–gel transition in the presence of electrolytes, such as those present in tear fluid [7]. A product based on this concept is currently on the market. Another system makes use of the amphiphilic cellulose derivative EHEC, which, in the presence of ionic surfactants, undergoes a sol–gel transition when the system is heated from room to body temperature [8]. A common feature of both systems is that they contain about 98–99 wt% water, and that the drug release, although delayed, will probably not be plausible for once-a-day therapy of most of the currently used ophthalmic drugs.

In this work we have investigated an alternative system based on polar lipids, which, when carefully composed, may be able to undergo a sol–gel type of transition as an effect of a temperature and/or an electrolyte effect. We have given earlier several examples of such systems based on lipids [9]. We have used timolol maleate as a model drug and some comparisons are made with the polymer systems mentioned above. However, it should be stressed that the present system is not applicable for administration into the eye due to the irritation caused by the monoglyceride, but the system is chosen merely for being well-characterized. There are other potentially interesting systems based on phospholipids alone, which may be more suitable for ophthalmic drug delivery [10].

Of particular interest in this work was the interaction between the timolol maleate salt and internal lipid/water interface of the cubic phase. The uncharged monoolein was partially replaced by a charged phospholipid and the in vitro release pattern studied. The obtained results are discussed in terms of interactions between charged ions of various polarity with charged lipid surfaces. A general finding is that the release of the drug is substantially lower than from a corresponding polymer system, and that the incorporation of charged lipids may provide an opportunity to tune the drug release pattern further.

Experimental section

Materials and sample preparation

Glyceryl monooleate (monoolein, GMO) from lot TS-ED 173, with ≥90% oleic acid in the 98.1% monoglycerides,

was purchased from Grinsted A/S, Denmark. The (mono)sodium salt of distearoylphosphatidylglycerol (glycerol 1,2-dioctadecanoate 3-phospho-1′-glycerol, DSPG) from Lot B12142/2 was kindly provided by Genzyme Pharmaceuticals and Fine Chemicals, UK. Timolol (hydrogen)maleate (TM), sodium chloride and sodium azide were obtained from Sigma, USA. Double-distilled water was used in all experiments.

The samples were prepared in glass ampoules, which after sealing in a nitrogen atmosphere were centrifuged 6 times at $1000g$ for 5 min, inverting each time. The samples were then stored in a dark place at room temperature until phase equilibrium was reached (normally after about 2–3 weeks). All samples contained 0.01 wt% sodium azide to avoid microbial growth and degradation of the lipids. Equilibrium was confirmed by visual inspection. Crossed polars and polarizing microscopy were used to detect anisotropic regions.

In vitro drug release

The release of timolol from the formulations was studied in an USP rotating paddle apparatus (Prolabo Dissolutest, France). Diffusion cells to be put at the bottom of the flasks with a cylindrical 2.0 mm counterbore (diffusion surface 4.91 cm^2) were used. The cells were filled with the formulation and weighted (to enable the calculation of the exact amount of drug). The average weight of a fill was 0.90 g. No cover net or membrane was needed due to the rigidity of the cubic liquid crystalline formulations. All samples were preheated at 37 °C for 45 min, and the experiment was started as each cell was immersed in a flask filled with 200 ml of either distilled water (with 0.01% sodium azide), or 0.90 wt% NaCl (aq, with 0.01% sodium azide) which had been equilibrated at 37 °C. The experimental set up was regarded as to allow for so-called perfect sink conditions, since the theoretic end-point concentration of timolol would be more than 200 times lower than the initial concentration in the test formulations. The paddle stirring rate was 20 rpm. 1.5 ml samples of the receptor medium were taken for HPLC analysis of timolol at certain time intervals throughout the release experiments. The volume fractions removed were replaced by fresh medium. This dilution effect as well as any possible concentration effects due to evaporation were corrected for in the subsequent data treatment. Analysis of the timolol concentration was performed by reversed phase liquid chromatography using a Kromasil KR100 column and a mobile phase of 60% phosphate buffer (50 mM, pH = 2.8) and 40% methanol. The flow of the mobile phase was 0.75 ml/min and the injection volume 20 μl. The timolol peak, which appeared after about 3 min, was detected using a variable UV-detector operating at 295 nm.

Crystallography

SWAXD studies were performed on a Gunier camera modified after Luzzati et al. [11]. The radiation used was Ni-filtered Cu K_α ($l = 1.542$ Å). The structure of the formulations was determined by X-ray diffraction before and after the in vitro release experiments.

Results and discussion

In vitro drug release

General considerations

Table 1 gives the sample compositions. The aim of this study was to investigate the influence of distearoylphosphatidylglycerol (DSPG) on the release pattern of the water-soluble drug timolol maleate (TM) from a cubic liquid crystalline phase, which mainly consisted of pure monoolein (GMO) and water. Three different formulation compositions were chosen for this purpose. The first (reference) system contained no phospholipid, the second contained DSPG corresponding to twice the molar amount of TM and the third contained DSPG of four times the amount of TM. The formulations were therefore named according to the relative TM:DSPG ratios, i.e. 1:0, 1:2 and 1:4. The concentration of TM, sodium azide and water was kept at the same level in all formulations. A concentration of 0.34 wt% TM corresponded to 8 mM in the initial formulations.

The release profiles of timolol from each of the formulations to either a *low ionic strength receptor medium* consisting of distilled water with 0.01 wt% (1.5 mM) sodium azide (open symbols), or to a *high ionic strength physiological saline medium* with 0.90 wt% (150 mM) sodium chloride and 0.01 wt% sodium azide (closed symbols) are shown in Figs. 1–3 (TM:DSPG = 1:0, Fig. 1), (TM:DSPG = 1:2, Fig. 2) and (TM:DSPG = 1:4, Fig. 3). The figures show the fraction of timolol released, given in percent of the total formulation drug content, as a function of time.

No significant difference was seen between the release profiles of timolol to the two different receptor media when only the nonionic lipid GMO, without phospholipid, was included in the formulation (Fig. 1). The *apparent* release of timolol to the low electrolyte medium was obviously retarded in the presence of DSPG (Figs. 2 and 3, open symbols). This effect was most pronounced for the formulation with the highest amount of DSPG (Fig. 3). The release of timolol to the high electrolyte medium, on the other hand, seemed much less affected by the presence of DSPG in the formulations, as only small differences in the release profile could be detected (Figs. 1–3, closed symbols). The trend was yet similar to that of the low electrolyte receptor medium, as the steepness of the profiles seemed to decrease slightly with increasing DSPG content.

The release of timolol to the low ionic strength medium from the two formulations containing phospholipid appeared to level off and come to an end before completion. This statement is supported by a prolongation of the sampling time in some release experiments. The fraction of

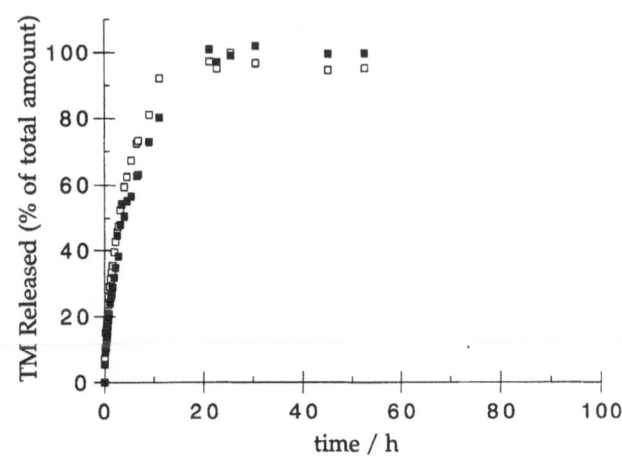

TM:PG/1:0

Fig. 1 In vitro release profile of timolol from a cubic liquid crystal formulation consisting of nonionic GMO as the only lipid component and denoted TM:PG/1:0 (composition given in Table 1). Open symbols indicate the release into a low electrolyte receptor medium (1.5 mM NaN₃) and closed symbols show the release to a high electrolyte medium (150 mM NaCl + 1.5 mM NaN₃)

Table 1 Name and composition of samples

Sample name	GMO [wt%]	TM [wt%]	DSPG [wt%]	H₂O [wt%]	Na Azide [wt%]	TM:DSPG [mol:mol]
1:0	69.66	0.34	—	29.99	0.01	1:0
1:2	68.40	0.34	1.26	29.99	0.01	1:2
1:4	67.14	0.34	2.52	29.99	0.01	1:4

TM:PG/1:2

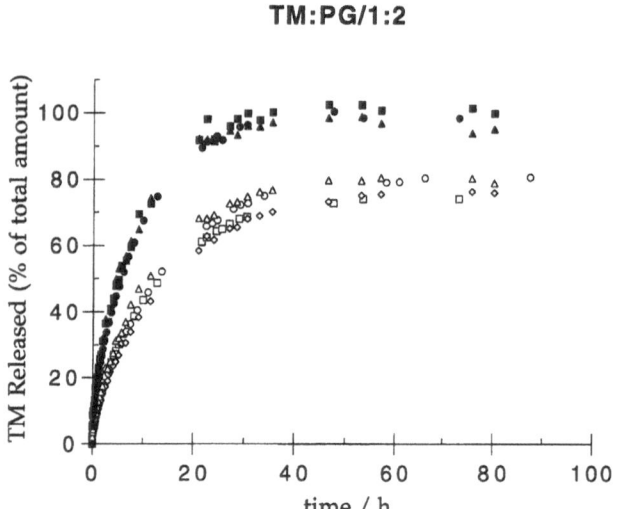

Fig. 2 In vitro release profile of timolol from a cubic liquid crystal formulation, denoted TM : PG/1 : 2, with GMO and a minor amount of the anionic phospholipid DSPG as the lipid components. The composition is given in Table 1. The symbols have the same meaning as in Fig. 1

TM:PG/1:4

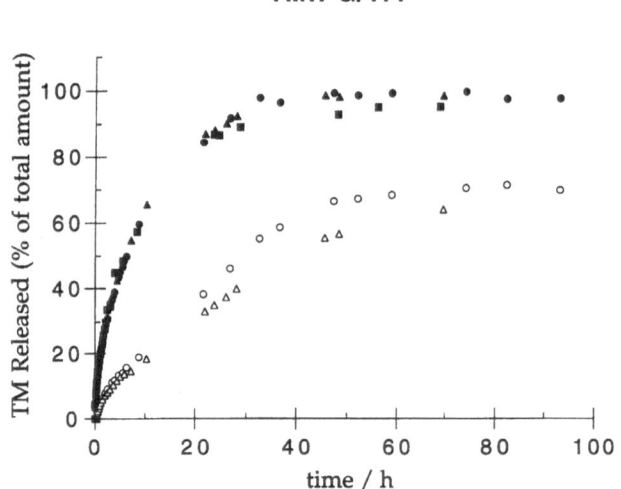

Fig. 3 In vitro release profile of timolol from a cubic liquid crystal formulation, denoted TM : PG/1 : 4, with GMO and some DSPG as the lipid components. The composition is given in Table 1. The symbols have the same meaning as in Fig. 1

timolol released from the formulation with the highest DSPG amount, for example, was still in the range of 60–65% after more than one week's dissolution testing, which corresponds to 180 h (cf. the open symbol end-point in Fig. 3). This observation may be explained by a strong interaction between the cationic organic timolol molecules and the lipid bilayers, negatively charged from DSPG, in the liquid crystalline formulation which could lead to the

TM:PG/1:0

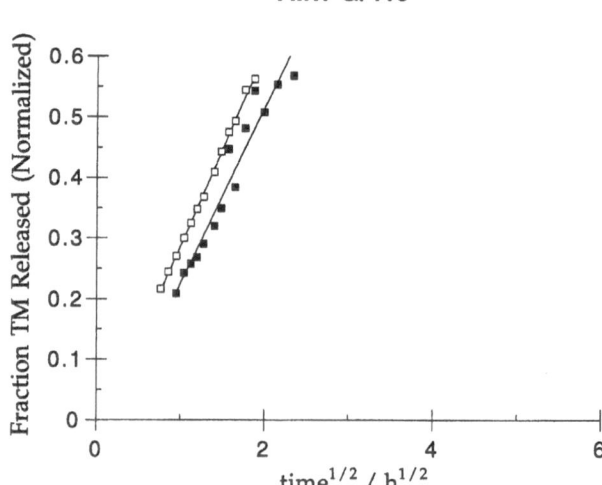

Fig. 4 The fraction of timolol released from the formulation called TM : PG/1 : 0, normalized to the actual end-point level as a function of the square root of time. The symbols have the same meaning as in Fig. 1

"binding" or complexation of a certain amount of the drug. The electrostatic contribution to this binding seems to be essential, since it appeared to be counteracted by screening at a high ionic strength, whereas approximately 100% of the drug was released to the physiological saline receptor medium.

Release mechanisms and kinetics

In order to enable a more adequate comparison of the release mechanisms and kinetics, the release data was normalized against the plateau end-points, which were regarded as the completed release of the "unbound" timolol. Figures 4–6 show plots of the normalized (unbound) drug fractions released, plotted as a function of the square root of time for the three different formulations investigated in this work. The interval 0.2–0.6 was chosen to reduce any possible contributions from differences in osmotic pressure and variations in swelling characteristics or depletions that may affect the initial and later parts of the release curves. Except from an apparent change of the intercepts, the slope of the plots seemed rather unaffected by the type of receptor medium. The curves are fairly well-fitted to a linear relationship and quite parallel. The slopes are all in the same range, indicating that the (dominating) release mechanism and rate may be the same for the "unbound" fraction of timolol in all three formulations, independent of the type of receptor medium. A linear square root of time dependence is in accordance with the expected drug release from a matrix system [12]. Figure 7

Progr Colloid Polym Sci (1998) 108:111–118
© Steinkopff Verlag 1998

TM:PG/1:2

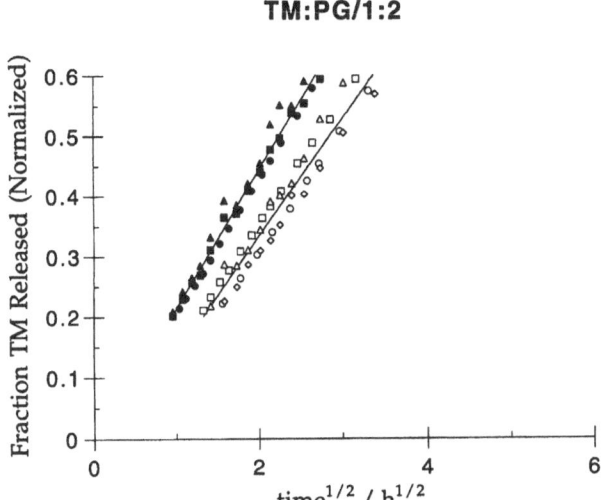

Fig. 5 The fraction of timolol released from the formulation called TM:PG/1:2, normalized to the actual end-point level as a function of the square root of time. The symbols have the same meaning as in Fig. 1

TM:PG/1:4

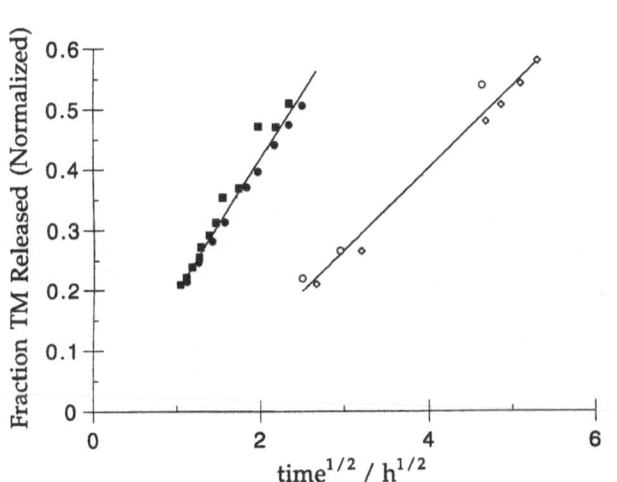

Fig. 6 The fraction of timolol released from the formulation called TM:PG/1:4, normalized to the actual end-point level as a function of the square root of time. The symbols have the same meaning as in Fig. 1

Release to physiol. saline

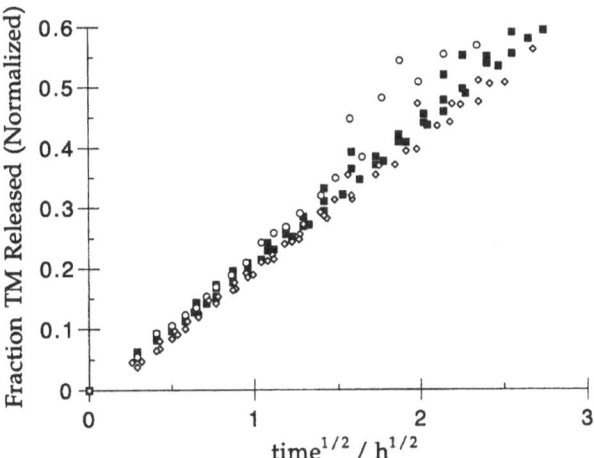

Fig. 7 The fraction of timolol released into the high electrolyte medium from the three different formulations used in this work, normalized to the actual end-point level as a function of the square root of time. Open circles represent the TM:PG/1:0 formulation, closed squares correspond to the TM:PG/1:2, and open diamonds are the TM:PG/1:4 case

experiments, while the physiologic saline (high electrolyte) medium was unchanged (clear). We believe that these observations have their origin in the different swelling characteristics of this formulation in the two media. The completely swollen formulation resulted in three phases at equilibrium, i.e. a cubic phase (P-surface), water (low electrolyte) and a small amount of lamellar phase (cf. Razumas et al. [13]). This resulted in a slight budding from the lamellar phase and the partial change of the liquid crystalline structure may have affected the release pattern for that particular system to some extent. It should be pointed out that the erosion of the formulation was only marginal, as no visible loss of material from the sample holders could be detected at the end of those experiments. Interestingly, the high electrolyte saline medium seemed to counteract the phase transition induced by DSPG, leaving a different cubic phase (D-surface) in equilibrium with the excess saline medium.

Other related systems

Before the present work was initiated, we made some studies on a similar system including GMO, water, TM and another negatively charged phospholipid – a diacyl phosphatidyl inositol (PI). The inositol phospholipid is similar to DSPG but has a sugar (inositol) unit attached to the hydrophilic head group in place of the smaller glycerol group in DSPG. Unfortunately, this phosphatidyl inositol

shows the plots of all three formulations in the high electrolyte medium to illustrate the similarities of the (normalized) data in the first half of the in vitro release experiments.

The formulation with the largest difference between the "unbound" drug release to the different receptor media was the formulation with the highest amount of DSPG, cf. Fig. 6. The low electrolyte medium in this case turned slightly hazy and opalescent with time during the release

turned out to be chemically unstable. The preliminary in vitro release data, however, indicated that the effective interaction with timolol was stronger than for the current DSPG formulations. Incorporation of twice the timolol molar amount of PI (TM : PI = 1 : 2) lead to a final plateau level of released timolol into a low electrolyte medium (distilled water) corresponding to only 20% of the total amount of drug.

We have previously studied the in vitro release of timolol from similar cubic liquid crystalline formulations based on lipid mixtures of GMO and soybean phosphatidyl choline (SPC), which is zwitterionic at neutral pH [14]. The results indicated a small, but still significant, retardation of the timolol release, as compared to the reference system with only the nonionic GMO in the formulation.

Drug-formulation interactions

We therefore conclude that the "strength" of the timolol interaction with the lipid bilayers, in these types of (fully water swollen) cubic liquid crystalline formulations with mixtures of GMO and *small amounts of phospholipids*, is decreasing in the order PI > DSPG > SPC. Electrostatic interactions seem to be one important factor for this effect. Both PI and DSPG carry one negative charge per molecule while SPC is zwitterionic at neutral pH, indicating that other types of interactions also contribute, e.g. hydrogen bonding and hydrophobic interactions. Note that we then assume that the very smallest amounts of phospholipids will not significantly change the type of liquid crystalline cubic symmetry, the width of the water channels or the effective diffusive path length. This assumption will fail at higher degrees of phospholipid incorporation, e.g. already at the TM : DSPG ratio of 1 : 2, as described below. The "binding" of timolol to the lipid bilayers seems, neverthe-

less, to be the dominating event in these types of phospholipid-doped cubic GMO/water systems.

The in vitro release of other model drug substances from GMO-based formulations (without phospholipids) has been studied by others [15, 16]. Incomplete release of a number of water-soluble drugs, e.g. chlorpheniramine maleate, diltiazem HCl and propranolol HCl, into a high ionic strength medium (100 mM phosphate buffer) was reported. A correlation between drug surface activity and absorption to the cubic formulation was seen. The GMO qualities were not the same as in this work. It is possible that impurities like negatively charged fatty acids may have contributed to the binding of the cationic drugs. In our opinion, it is more likely that the difference in affinity of the drugs used in our and Chang et al. work is due to a more pronounced partitioning into the lipid bilayer in the latter case. The formulations were, in most cases, prepared without water, which might have affected the release pattern as well.

Crystallography

General considerations

All formulations studied were stiff, visually transparent and isotropic when viewed between crossed polars at 37 °C. However, X-ray spectroscopy (SAXD) revealed differences in cubic symmetry (Table 2). The simple GMO–water system, thoroughly studied elsewhere [1], shows two different bicontinuous cubic phases of space group, Ia3d and, at higher water content, Pn3m. The second phase can exist in equilibrium with excess water. The formulations of the present investigation were prepared at 30 wt% water and even with the highest amount of DSPG (2.5 wt%) the Ia3d-phase was obtained at 37 °C. When placed in excess water for the drug release studies, the

Table 2 Crystallographic data on the formulations before (initial) and after drug release in water (aq.) and physiological salt solution (phys.), respectively (lattice parameter, a; radius of curvature of an IPMS, $\langle R \rangle$; water channel length, L_w; volume fraction of water, ϕ_w). The drug concentration (TM) in the initial formulation was 0.34 wt%

	TM : PG mol : mol	Space group	IPMS	a [Å]	$\langle R \rangle$ [Å]	L_w [Å]	$L_w a^{-1}$	ϕ_w [v/v]
Initial	1:0	Ia3d	G	137.5	34.1	272.8	2.0	0.30
Phys.	1:0	Pn3m	D	89.1	34.8	69.6	0.8	0.37**
Aq.	1:0	Pn3m	D	88.9	34.7	69.5	0.8	0.37*
Initial	1:2	Ia3d	G	137.5	34.1	272.8	2.0	0.30
Phys.	1:2	Pn3m	D	92.5	36.1	72.3	0.8	0.39**
Aq.	1:2	2-phase	D	91.1	35.6	71.2	0.8	0.38**
			P	133.6	40.8	163.2	1.2	0.45**
Initial	1:4	Ia3d	G	137.5	34.1	272.8	2.0	0.30
Phys.	1:4	Pn3m	D	92.3	36.1	72.1	0.8	0.39**
Aq.	1:4	Im3m	P	138.5	42.3	169.2	1.2	0.47**

* Estimated from the phase diagram of Hyde et al. [1].
** From Eq. (3), based on *.

formulations reached their swelling limit. In the simple GMO–water system the expected phase transition into Pn3m occurred, however, with DSPG a third cubic phase appeared which we interpret as having a symmetry of Im3m. The Im3m phase was also observed by others in the GMO–water system in presence of for example DSPG or lysozyme [13, 17]. Furthermore, when the drug release studies where undertaken in physiological salt solution (150 mM NaCl), formation of the Im3m phase was suppressed in favor of the Pn3m, probably due to the screening of DSPG-headgroup electrostatic repulsion. No crystallographic effect was seen when including the water soluble Timolol maleate (0.34 wt%) in the formulations.

Geometrical considerations

In order to resolve the observed differences in drug release rate from a morphological point of view, the structural aspects of cubic phases have to be considered in some more detail. Bicontinuous cubic phases can be described by so-called "infinite periodic minimal surfaces" (IPMS), i.e. surfaces composed of saddle points of zero mean curvature. Each spacegroup may fit a number of different IPMS, however, the most common ones found in lipid/surfactant systems are the G- (Ia3d), D-(Pn3m) and P-surface (Im3m) [18, 19]. These are also the ones found in the present system.

The average magnitude radii of curvature of an IPMS, $\langle R \rangle$, can be derived from

$$\langle R \rangle = (-Ha^3/(2\pi\chi))^{1/3} \tag{1}$$

where H is the homogeneity index for various IPMS (G: 0.7665, D: 0.7498, P: 0.7163); a is the lattice parameter for the cubic unit cell; and $\chi = 2 - 2g$ is the Euler–Poincaré index per unit cell (G: -8, D: -2, P: -4). g is the genus of the IPMS, cf. the number of water channel outlets per unit cell divided by two.

The water channel length, L_w, can be estimated independently of hydration level, ϕ, and lipid bilayer half-thickness, l, from $A_{UC} = -2\pi\chi\langle R \rangle^2$ and $L_w = A_{UC}/(2\pi\langle R \rangle)$. A_{UC} is the IPMS area per unit cell,

$$L_w = -\chi\langle R \rangle. \tag{2}$$

Equation (3) allows for an estimate of the hydration level of the cubic phase. The lipid bilayer half-thickness can easily be derived from the lattice parameter of the lamellar phase, $l = 17.1$ Å at 37 °C. However, here it is possible, and perhaps more accurate, to use the fully swol-

len GMO–water Pn3m phase ($\phi_w = 0.37$, $l = 15.6$ Å) [1]:

$$\phi_s = 1 - \phi_w = (3(\langle R \rangle/l)^2 - 1)/(2(\langle R \rangle/l)^3). \tag{3}$$

Reasons for minor deviations in the release pattern

From our results, it appears that the DSPG-containing formulations exhibit a slightly slower drug release rate, even when the more direct "binding" or complexation of a fraction of the drug has been accounted for by normalization to the plateau (end-point) fraction that was actually released. One explanation may be found in the morphological differences between the various cubic phases. The included DSPG allows for a cubic phase of higher hydration level (P) than the original one (D), which should favor the release of hydrophilic substances. However, the normalized water channel length, $L_w a^{-1}$, is longer for the P-surface (1.2) compared to the D-surface (0.8), i.e. the diffusion path length in the water region is apparently longer which should retard the release rate. In addition, the differences in genus (see above) may also affect the release rate to some extent.

Concluding remarks

This work shows that lipids can offer many possibilities for the formulation of sustained drug release systems. The release of timolol from the currently investigated formulations was significantly slower than reported for polymeric in situ gelling systems [7, 8]. There may, moreover, be some opportunities for the fine tuning of the release profiles by introducing lipids that are oppositely charged to the drug.

The choice of receptor medium can strongly influence the in vitro release pattern, as illustrated in this study. Physiologically relevant solutions are preferred in order to enable correlations with in vivo conditions.

Strong interactions between organic ions (drug) and oppositely charged lipid or amphiphilic excipients has to be taken into account in the formulation development. Such interactions may otherwise result in unexpected effects. For example, if a surfactant is added in order to enhance the release characteristics of an ionic drug with low solubility by increasing the wettability of the drug particles; this may instead lead to the precipitation of a sparingly soluble complex with even worse dissolution properties.

Acknowledgments We thank Kåre Larsson, Tommy Nylander and Ulf Olsson for valuable discussions. Jenny Kedström is acknowledged for technical assistance on some in vitro release experiments.

References

1. Hyde ST, Andersson S, Ericsson B, Larsson K (1984) Z Kristallogr 168: 213–219
2. Larsson K (1994) Lipids – Molecular Organization, Physical Functions and Technical Applications. The Oily Press, Dundee
3. Engström S (1990) Lipid Technol 2: 42–45
4. Ericsson B, Eriksson PO, Löfroth JE, Engström S (1991) In: Dunn R, Ottenbritte RM (eds) Polymeric Drugs and Drug Delivery Systems. ACS Symp Ser, Vol 469, pp 251–265
5. Lee VHL (1990) J Ocular Pharmacol 6:157–164
6. Mitra AK (ed) (1993) Ophthalmic Drug Delivery. Marcel Dekker, New York
7. Rozier A, Mazuel C, Grove J, Plazonnet B (1989) Int J Pharm 57:163–168
8. Lindell K, Engström S (1993) Int J Pharm 95:219–228
9. Engström S, Lindahl L, Wallin R, Engblom J (1992) Int J Pharm 86:137–145
10. Bergenståhl BA, Claesson PM (1997) In: Friberg SE, Larsson K (eds) Food Emulsions. Marcel Dekker, New York, pp 87–91
11. Luzzati V, Mustacchi H, Skoulios A, Husson F (1960) Acta Crystallogr 13: 660–667
12. Cardinal JR (1984) In: Langer RS, Wise DL (eds) Medical Applications of Controlled Release, Vol I, CRC Press, Boca Raton, FL, pp 41–67
13. Razumas V, Talaikyte Z, Barauskas J, Larsson K, Miezis Y, Nylander T (1996) Chem Phys Lipids 84:123–138
14. Engblom J (1991) MSc Thesis, Food Technology, University of Lund, Sweden
15. Chang CM, Bodmeier R (1997) Int J Pharm 147:135–142
16. Chang CM, Bodmeier R (1997) J Pharm Sci 86:747–752
17. Buchheim W, Larsson K (1987) J Colloid Interface Sci 117:582–587
18. Engblom J, Hyde ST (1995) J Phys II (France) 5:171–190
19. Hyde S, Andersson S, Larsson K, Blum Z, Landh T, Lidin S, Ninham B (1997) The Language of Shape, Elsevier, Amsterdam, pp 199–235

Progr Colloid Polym Sci (1998) 108:119–128
© Steinkopff Verlag 1998

L. Nyberg
R.D. Duan
Å. Nilsson

Sphingomyelin – a dietary component with structural and biological function

L. Nyberg (✉)
Swedish Dairies Association
S-223 70 Lund
Sweden

R.D. Duan
Department of Cell Biology 1
Lund University Hospital
S-221 85 Lund
Sweden

Å. Nilsson
Department of Medicine
Lund University Hospital
S-221 85 Lund
Sweden

Abstract Sphingomyelin (SM) is a constituent of most eucaryotic cells, particularly in the plasma membrane and related cell membranes. Due to the widespread occurrence, SM is a significant component in foods, present primarily in milk, meat, fish and egg. SM is comprised of three components; a long chain sphingoid base backbone, a fatty acid and a phosphorylcholine polar head group. The fatty acids in SM are mostly very long and saturated and this influences the physical properties. In the cell membrane SM was earlier considered to be only a structure element, which interacts with cholesterol and forms a system for bilayer stabilization. In the last decade there has been an increasing interest in sphingolipid metabolism, since their hydrolysis products are found to have important signalling effects on cellular functions, such as cell growth, cell differentiation and also programmed cell death – apoptosis. The potential biological effects of SM have recently increased the interest in the metabolism of dietary SM. In 1994 a collaborative project on digestion and absorption of milk SM was initiated in Lund. The results so far show that SM digestion is extended all over the small intestine and occurs mainly in the middle and lower parts. This coincides with the intestinal distribution of an alkaline SMase, which may be important for digestion. The capacity of SM digestion is limited. Also after administration of small amounts, all of the small intestine and colon are exposed to SM and its bioactive metabolites. When rats were fed a mixture of SM and cholesterol, the uptake of SM in the rat intestine was further reduced. A novel alkaline SMase has been identified in human bile. In human milk, the bile salt stimulated lipase was shown to have ceramidase activity. The potential biological effects of SM metabolites, in intestinal tumour development in adults and in regulation of optimal development of the gut mucosa in breast-fed infants, are interesting areas for further studies.

Key words Sphingomyelin – sphingomyelinase – ceramide formation – intestine – milk

Introduction

Sphingolipids are a group of components, which have, until now, received little attention as dietary ingredients. Based on recent findings, this will probably or certainly change in the future. They were first described more than 100 years ago by Thudichum [1], in a study on the chemical constitution of the brain. This explains the naming in accordance with the neuronal system, although this leaves the false impression that they are unique to neuronal tissues and obscures the fact that they are present in all eucaryotic cells. Thudicum called the compounds he had discovered an "enigmatic group of new compounds". The prefix *sphingo* is said to originate from the Greek myth of the sphinx and indicates the mystery that is associated with these molecules. For a long time studies of the sphingolipids were focused on their role as structural components in biological membranes and on the elucidation of pathways for their biosynthesis and degradation [2]. In the last decade there has been an increasing interest in sphingolipid metabolism, since hydrolysis products originating from both endogenous and dietary sphingolipids may have important signalling effects on cellular functions, such as cell growth, cell differentiation and also programmed cell death – apoptosis [3, 4]. Naturally, this has also increased the interest in the digestion and absorption of dietary sphingolipids, and in the potential biological effects that exogenous sphingolipids may have.

Structure and composition of sphingomyelin and other sphingolipids

Sphingomyelin (SM) is both a sphingolipid and a phospholipid. Unlike other phospholipids SM does not have a glycerol backbone. In 1927 Pick and Bielschowsky [5] established the structure of SM to be *N*-acylsphingosine-1-phosphorylcholine. 50 years later Shapiro and Flowers [6] showed that all SMs of biological origin has a D-*erythro*-configuration. SM is comprised of three components; a long chain sphingoid base back-bone, a fatty acid and a phosphorylcholine polar head group (Fig. 1). A fatty acid in amide linkage in the second position of the sphingoid base constitutes ceramide, the central building block of all sphingolipids. In SM a phosphodiester bond exists between ceramide and phosphorylcholine at the C1 position. Naturally occurring SM differs in the nature of the

sphingoid base and fatty acid, although the prevalent long-chain base of most mammalian tissues is sphingosine, an 18-carbon amine diol (1,3-dihydroxy-2-amino-4-octadecene), with a trans double bond between carbons 4 and 5 [7]. There are reports indicating that the trans double bond is crucial for some biological activities [8]. The fatty acids in SM are mostly very long and saturated, typically with an alkyl chain length of 16–24 carbon atoms. The principal acyl groups found in SMs derived from most tissues are palmitic (C16:0), nervonic (C24:1), lignoceric (C24:0) and behenic acid (C22:0) [7]. In the central nervous system, stearic acid (C18:0) is most abundant.

The composition of SM isolated from bovine butter milk has been obtained by plasma spray tandem mass spectrometry and GLC analysis [9]. The results show (Table 1) that the composition of sphingoid bases and fatty acids in milk SM is quite complex. Like other SMs it contains primarily saturated fatty acids, with palmitic acid (C16:0) as the dominating acyl group, besides a large fraction of very long, saturated fatty acids with 22–24 carbons. The dominating sphingoid bases in milk are sphingosine and a 16-carbon amine diol with one double bond.

In most tissues more than half of the fatty acids in SM have 20 carbons or more, while the paraffinic residue of the sphingosine base has only 13–15 carbon atoms. Thus the two hydrocarbon chains, comprising the hydrophobic part of SM, differ in length by more than 7 methylene residues [2]. In comparison phosphatidylcholine (PC), the other phosphorylcholine containing phospholipid, has two acyl chains of almost equal length. Hence, ceramide confers a more asymmetric structure onto SM than the corresponding diacylglycerol moiety of PC [2]. In SM the average number of cis double bonds per molecule is 0.1–0.35, while for PC it is 1.1–1.5. The interfacial region of SM is more polar than in PC, containing the amide bond between the primary amino group on carbon 2 and the acyl chain, the hydroxyl group attached to carbon 3 and the trans double bond between carbons 4 and 5, whereas in

Fig. 1

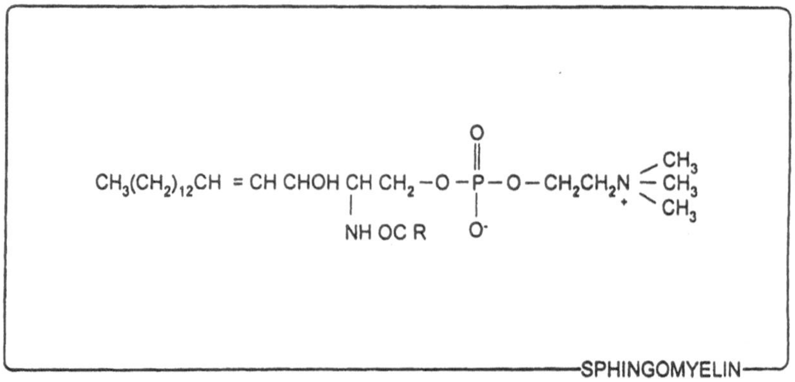

$$CH_3(CH_2)_{12}CH = CH\ CHOH\ CH\ CH_2 - O - \overset{\overset{\displaystyle O}{\|}}{\underset{\underset{\displaystyle O^-}{|}}{P}} - O - CH_2CH_2\overset{+}{N} \overset{\diagup CH_3}{\underset{\diagdown CH_3}{\langle}} CH_3$$

NH OC R

SPHINGOMYELIN

Table 1 Fatty acid and sphingoid base composition of sphingomyelin from bovine milk

Fatty acid	wt%	Sphingoid base	wt%
C14:0[a]	3	C16:0	3
C16:0	34	C16:1	23
C18:0	4	C17:1	3
C18:1	3	C18:0	3
C18:2	<1	C18:1	64
C20:0	<1		
C20:1	<1		
C22:0	17		
C23:0	21		
C24:0	14		
C24:1	1		

[a] Number of carbon atoms:number of double bonds.

PC it is composed of carbon 1–2 of the glycerol backbone and the associated fatty ester linkages. The higher polarity of SM compared to PC allows for stronger interaction with water, and intra- and intermolecular hydrogen bonding within the lipid bilayer [10].

There are many different kinds of sphingolipids; more than 300 structures have been reported to occur in nature [11]. Free long-chain bases occur in small amounts [12, 13]; but most exist as ceramides or as SM and other complex sphingolipids. Simple or more complex carbohydrates at position 1 form cerebrosides, gangliosides, sulfatides, etc. Neutral glycosphingolipids contain neutral sugars such as glucose, galactose, N-acetylglucoseamine, N-acetylgalactoseamine and fucose. Acidic glycosphingolipids contain charged functional groups such as phosphate (phosphoglucosphingolipids) or sulfate (sulfatoglycosphingolipids) as well as charged sugar residues such as glucoronic acid (in some plant glycosphingolipids) or sialic acid in all gangliosides [14].

Physical properties of SM

The large fraction of long and saturated acyl groups in the SM structure influences its physical properties. Naturally occurring SM, in contrast to PC, is practically insoluble in chloroform, in the absence of a polar solvent [7]. In X-ray diffraction and differential scanning calorimetry (DSC) studies of bovine brain SM [15], the bilayer gel → liquid crystal transition occurred in the physiological temperature range at 30–40 °C, which is a high temperature for a naturally occurring lipid. Similar behavior was observed in a study of SM isolated from bovine butter milk [9]. There is a growing interest concerning the properties of SM in membrane sciences, in particular because membranes are seen to undergo thermotropic phase transition within the physiological temperature range [16]. Biological membranes may have bilayer domains of sphingolipids in which some of the hydrocarbon chains are able to penetrate completely across the hydrocarbon width of the bilayer [17]. Such packing arrangements may have profound effects on the properties. Membranes containing high concentrations of SM have increased microviscosity [18]. The high degree of hydrogen bonding in SM compared to PC also results in enhanced membrane stability and reduced permeability to various electrolytes and non-electrolytes [19]. Increase in SM concentration in eucaryotic cell plasma membranes has been shown to decrease the fluidity of the lipid bilayer [20, 2] in a similar way to that observed for cholesterol [10].

The cubic phase is a liquid crystalline phase, consisting of a lipid bilayer with water on each side and curved in space so that the average curvature is zero. This phase can be dispersed into particles, Cubosomes, with interesting functional properties [21, 22]. Milk SM has been shown to be efficient in dispersing the cubic phase of monoolein [23]. As expected, due to the crystalline character of its acyl chains at room temperature, SM was useful in surface bilayer stabilization of the dispersion.

Cellular and tissue distribution of SM

Sphingolipids are constituents of most eucaryotic cell membranes in animals, plants and some lower forms of life [3]. Membranes from animal cells of different sources exhibit a wide range of SM content and composition, and may also be influenced by environmental factors such as diet [24]. Due to the widespread occurrence, SM is present in several common food constituents, primarily in milk, meat, fish and egg (see Table 2). The SM content of most mammalian tissues ranges from 2 to 15% of total phospholipids, depending on the species and tissues examined [25]. Erythrocytes, peripheral nerve and brain tissues usually have high levels of SM, ranging between 20 and 30% of total phospholipids. SM accounts for 4–6% of the phospholipids in liver and 25% in pancreas [26, 27], and is also an important component of the serum lipoproteins, constituting 23% of phospholipids in VLDL, 25% in LDL and 13% in HDL [27].

Table 2 Sphingomyelin content in food (mg/kg)

Milk	100	[87]
Butter	350–1000	[92, 105]
Cheese	300–700	[105]
Beef	300	[88]
Egg	115	[106]
Salmon	125	[88]

In analysis of fractionated subcellular membranes from rat liver [25, 28], the plasma membrane, Golgi apparatus and lysosomes were shown to have the highest SM content, whereas nuclear and mitochondrial membranes had the lowest. SM is thus mainly distributed throughout the outer cell membrane, and the secretory and endocytic compartments. There appears to be an inverse correlation between the amounts of SM and PC in many membranes [29]. One consequence of this is that the total choline containing lipids can be kept constant while varying the membrane fluidity [30]. The distribution of lipids between the outer and inner faces of plasma membranes is asymmetric, with most of the SM and a large part of the PC located in the external leaflet [2]. SM in the outer layer of the plasma membrane should enhance the rigidity of this membrane and may present an appropriate surface for interaction with proteins [7].

During aging, there is a striking increase of SM in the cell membranes of the arteries and aorta. In human aorta the SM to PC ratio changes from a value of 0.4 at birth to 2.4 at age 90 [31], due to decreased activity of the first enzyme in the SM-degrading pathway. A more pronounced increase of SM occurs during the development of atherosclerosis [29], then it may reach 75% of the total phospholipids. A substantial portion of the SM found in arteries and atherosclerotic lesions appears to arise from synthesis in the arterial tissue, accompanied by a decreased turnover [31]. Serum lipoproteins may also be an important source for aortic wall SM [32].

The lipid composition in the nervous system continually changes throughout the lifetime. In human brain, SM and cerebrosides gradually replaces PC [33]. A similar age-dependent change in the ratio has also been noted for the lens of the eye; for humans the SM content rises to 70% of the total phospholipid and the PC content falls to 5% [2].

Synthesis of SM

In mammalian cells, de novo sphingolipid biosynthesis begins in the microsomes of the endoplasmatic reticulum. The initial precursors palmitoyl-CoA and serine [34] are irreversibly condensed, with loss of the carboxyl group of serine, by the action of serine palmitoyl-transferase, a pyridoxal 5′-phosphate-dependent enzyme [35]. This is probably a rate limiting step in the synthesis of sphingolipids, as the enzyme has lower activity than the other enzymes of ceramide synthesis and utilizes substrates that are shared by other pathways [36]. The enzyme has a high degree of specificity for palmitoyl-CoA [37], corresponding to the high prevalence of the 18-carbon sphingoid bases [38]. The 3-ketodihydrosphingosine thus formed is

reduced to dihydrosphingosine (sphinganine) by a microsomal NADPH-dependant reductase [39].

In the next step a fatty acid is amide-linked to dihydrosphingosine by ceramide synthase to form dihydroceramide, which apparently occurs very fast in vivo, since free sphingoid bases are not detected as intermediates in sphingolipid biosynthesis [40]. It has been suggested that introduction of the 4,5-trans-double bond occurs after addition of the amide-linked fatty acid [35]. Therefore, free sphingosine is not an intermediate of sphingolipid biosynthesis de novo [3]. Different mechanisms have been proposed for synthesis of dihydroceramide, both acyl-CoA requiring and acyl-CoA independent pathways [41, 42]. The principal pathway is likely to be transfer of an acyl group from acyl CoA to a sphingoid base [14], reversal of the ceramide hydrolysis by a ceramidase being less important [43]. Fumonisins, a family of mycotoxins which are toxic and carcinogenic for animals, have been shown to inhibit ceramide synthase and thereby block the synthesis of complex sphingolipids [3, 44]. This inhibition may involve widespread disruption of cellular functions and has been suggested to account for the toxicity and carcinogenicity of these mycotoxins.

Introduction of various head groups to ceramide results in the formation of more complex sphingolipids. Formation of SM by addition of phosphorylcholine to ceramide involves the transfer of choline phosphate from PC, liberating diacylglycerol through the action of phosphatidylcholine: ceramide choline phosphotransferase [35].

SM catabolism

Sphingomyelinases (sphingomyelin phosphodiesterase) catalyze the cleavage of SM to phosphorylcholine and ceramide [45]. Acid, neutral and alkaline forms of sphingomyelinase (SMase) have been described. The acid SMase is a lysosomal hydrolase with optimal activity at 4.5 [46, 47]. Sphingolipid activator proteins (saposins) are a family of four small, lysosomal proteins, which are required under certain conditions for substrate breakdown and enhanced enzyme activity [7, 14, 48]. The acid lysosomal SMase mainly functions in the breakdown of substrate arising from the endocytosis of extracellular lipid and cell membrane turnover [14] but is also thought to be involved in signal transduction [49]. Selective absence of lysosomal SMase activity results in accumulation of SM in cells scattered throughout the spleen, bone marrow, lymph nodes, liver and lungs in patients with the autosomal recessive disorder Niemann-Pick Disease. Other forms of this disease may have normal lysosomal SMase activity but still elevated levels of SM [35]. Evidence has been provided for a defect in cellular cholesterol esterification in these cases [50], due to the strong interaction between SM and cholesterol, accumulation of SM also occurs.

At least two different neutral SMases have been identified [31, 51], a plasma membrane associated enzyme with requirement for divalent ions (Mg^{2+} or Mn^{2+}), which appears to be located on the outer, extracellular side of the plasma membrane [52] and a cytosolic, Mg^{2+}-independent enzyme. These SMases, which can be activated by several cytokines and hormones, act on endogenous SM [53] and are involved in the turnover of SM to produce ceramide and other sphingolipid metabolites for cell signalling (see below).

A bile salt dependent SMase activity with alkaline pH optimum was first identified by Nilsson in human and pig intestinal contents and rat intestinal brush border [54, 55]. A similar enzyme was recently shown to occur in human bile [56] and was purified to homogenity [57].

The next step in SM catabolism is cleavage of the amide-linked fatty acid. This is catalyzed by ceramidases (*N*-acylsphingosine deacylase). At least three distinct forms of ceramidase have been described, differing with respect to pH-optima, substrate specificity, subcellular location, tissue distribution and metabolic function [58]. An acid ceramidase has been described, with optimum at pH 4.8 and requirement for bile salts [59]. Genetic deficiency of acid ceramidase results in Farber's disease, characterized by a marked accumulation of ceramide with a high percentage of hydroxy fatty acids [60].

Less is known concerning the tissue ceramidases with neutral or alkaline pH optimum, although such an enzyme was recently purified from skin [61]. This enzyme activity, which is not deficient in Farber's disease, has been demonstrated in both microsomal, plasma membrane and cytosolic fractions [62, 63]. Ceramidases with activity at neutral to alkaline pH have been proposed to function in signal transduction pathways [58].

Sphingoid bases released from ceramide are further degraded in a two-step process. First, they are phosphorylated at the first position to sphingosine 1-phosphate, by a sphingosine kinase. The next step is cleavage between the second and third positions by a sphingosine 1-phosphate aldolase, which yields ethanolamine phosphate and a 16-carbon aldehyde [35]. The products are rapidly utilized [7]; palmitaldehyde is oxidized to palmitic acid and some ethanolamine phosphate is incorporated into the polar headgroup of phosphatidylethanolamine (or PC and SM after methylation).

SM cycle – signal transduction

Membrane lipids were classically considered as biologically inert molecules, whose major function was to maintain the structural integrity of the cell and to provide a permeability barrier. Today lipids are known to be active participants in cellular life and in the pathways by which extracellular signals elicit an intracellular response [64]. Living organisms have evolved a multitude of signalling pathways to ensure efficient inter- and intracellular communication. Transduction of extracellular signals is believed to be mediated by transient molecules, known as second messengers [65], the primary messenger being the signal itself. Hydrolysis of plasma membrane lipids in response to extracellular stimulation generates a variety of lipid signal transduction molecules. Diacylglycerol (DAG) and inositol triphosphate (IP_3) were the first recognized lipid second messengers, in the early 1980s [66]. Hydrolysis of phosphatidylinositol (PI) by a PI-specific phospholipase C, led to the formation of IP_3 and DAG, which were identified as potent stimulators of Ca^{2+} mobilization and protein kinase C (PKC) activity, respectively.

More recently the sphingolipid metabolites ceramide, sphingosine and sphingosine 1-phosphate were identified as important lipid mediators [49]. SM hydrolysis by SMase initiates an evolutionary conserved and ubiquitous intracellular signalling pathway [67]. Activation of a SM cycle in a responsive cell turns on antiproliferative pathways [68], such as inhibition of cell growth, induction of differentiation, initiation of programmed cell death (apoptosis), activation of tumor suppressors or induction of specific cell cycle arrest.

Two key observations suggested that SM-metabolites could be novel signalling molecules [49]. First, sphingosine was found to be a potent inhibitor of PKC activity [69]. Second, DAG was found to stimulate the hydrolysis of SM to ceramide [70]. There seems to be several similarities between the SM and PI signalling pathways [49]. The phospholipid substrates are hydrolyzed by specific phospholipases to yield structurally similar lipid intermediates, ceramide and DAG. Both exists in mammalian cells at the level of 1–2% of cellular lipids. They serve not only as second messengers, but also as backbones of all phospholipids. As structural lipids they are found in bound and free form, distributed through all cellular membranes. In response to activation of cell surface receptors at signalling pools the levels increase 1.4–2 times [71].

The effects of sphingosine on PKC activity led to a postulate [72] of the existence of a "SM cycle" in cells, analogous to the "PI cycle". The SM cycle is composed of at least four components: (1) SM as substrate, (2) SMase which hydrolyzes SM, (3) ceramide and phosphorylcholine resulting from SM hydrolysis and (4) enzymatic pathways for resynthesis of SM.

The existence of an SM-cycle was confirmed with $1\alpha,25$-dihydroxy vitamin D_3 treatment of HL-60 leukemia cells. Vitamin D_3, which is known as an inducer of monocytic differentiation in this cell line, caused a rapid hydrolysis of SM, and thus generation of ceramide and phosphorylcholine [73]. A cytosolic, magnesium-independent neutral SMase was shown to be activated by vitamin D_3. After four hours the amounts of SM and ceramide had returned to basal levels. Cell-permeable ceramides, such as N-acetyl-sphingosine were found to mimic the effects of vitamin D_3 in inducing monocytic differentiation [73].

Activation of SMases in the SM cycle has been linked to several cytokines and hormones, such as tumour necrosis factor (TNF), interleukin-1, interferon and dexamethasone [74, 75]. Cytokines are agents, which inhibit cell growth and turn on anti-proliferative programs or induce differentiation [76]. The SM cycle may thus provide a signalling function nearly opposite to the PI mediated pathway, which leads to growth activation [77].

Although the highest concentration of SM is found in the outer leaflet of the plasma membrane, recent studies suggest that there may be functionally distinct pools of SM involved in signal transduction, located in the inner leaflet of the plasma membrane and in specific lysosomal/endosomal compartments [78]. SM pools would more easily facilitate ceramide generation in closer proximity to key targets, which are modulated by ceramide during signalling [49].

Biologic effects of ceramide

Ceramide appears to be the second messenger for the SM cycle [79]. It has been shown to play important roles in a variety of signal transduction systems. A number of direct targets for ceramide action have been identified and it appears to act by stimulating a ceramide activated protein kinase (CAPK) [80], and a ceramide activated protein phosphatase (CAPP) [81], which are involved in a complex signalling cascade of reversible phosphorylation/dephosphorylation of protein substrates [49] and resulting in modulation of a number of intracellular targets. These effects appear to result in profound changes in cell growth behaviour [68].

In contrast to ceramide, dihydroceramide is inactive in antiproliferation, differentiation and apoptosis [82]. The lack of activity is not due to decreased uptake or increased metabolism but probably arises from the lack of activity on CAPP [83]. These considerations raise the intriguing possibility that the sphingolipid double bond serves to impart critical biological activities to ceramide, thus disso-

ciating the early metabolic steps in sphingolipid biosynthesis from biologic activity [82].

Although the generation of ceramide from SM hydrolysis mainly plays an antiproliferative role, it is worthwhile noting that a cross communication between SM pathway and MAP kinase pathway has been identified, mediating proliferative effects on many growth factors and mitogens. Activation of MAP kinase by exogenous SMase, ceramide or sphingosine-1-phosphate has been reported in either HL-60 cells [84] or Swiss 3T3 cells [85]. Such a communication is probably mediated by sphingosine-1-phosphate, which has been found to activate phospholipase D and generate phosphatidic acid, a strong activator of MAP kinase [86].

Digestion and absorption of dietary SM

Although there have been few systematic analyses, SM is a significant component in food [87, 88], occurring mainly in milk, meat, fish and egg (Table 2). In milk, SM is a constituent of the complex biological membrane that surrounds the milk fat globule. One litre of bovine milk contains about 100 mg SM, representing about one third of total milk phospholipids.

Rather few studies concern the digestion, uptake and subsequent metabolism of dietary SM. After some early studies in Lund [54, 55], the field has received minor attention. Recently, the potential biological effects of dietary SM have increased the interest in this area. In 1994 a collaboration between the Gastroenterology Division of the Medicine Department at Lund University Hospital and the Research Department of the Swedish Dairies began, in part due to an initiative by Professor Kåre Larsson, on studies of the digestion and absorption of milk SM. The Swedish Dairies had developed a method for large scale purification of milk SM [89] and necessary amounts of SM thus became available for nutritional studies.

Enzymes in digestion of SM

The early studies [54, 55] indicated that the digestion and absorption of sphingolipids follow a pattern, which is quite different from that of the glycerolipids. The studies identified an SMase activitiy in human and pig intestinal contents and rat intestinal brush border with alkaline pH optimum. The presence of a bile salt dependent mucosal enzyme, which catalyzed the reversible hydrolysis of ceramide with an optimal pH around neutral, was also demonstrated. The formed sphingosine was efficiently absorbed and metabolized in the gut to palmitic acid and incorporated into chyle triacylglycerols.

Progr Colloid Polym Sci (1998) 108:119–128
© Steinkopff Verlag 1998

The intestinal alkaline SMase in rat intestinal mucosa and contents has recently been further characterized [90]. The level of alkaline SMase was low in the duodenum and reached its highest levels in the middle of the small intestine. The enzyme was also present in colon and rectum. This distribution pattern differs distinctly from that of acid SMase, and of alkaline phosphatase and other brush border enzymes as disacharidases, which have highest activity in the upper part of the small intestine. Furthermore, a novel alkaline SMase activity was identified in human bile [56]. This was a surprising discovery; although bile is important for the absorption of lipids it is not considered a source of digestive enzymes. Pancreatic juice, which is the important supplier of lipolytic enzymes, seems to be of no significance for the digestion of SM; no SMase activity could be detected in human pancreatic juice, except for a low level of acid SMase.

The bile alkaline SMase seems to be specific to humans and was not demonstrated in other species that were studied (rat, guinea pig, hamster, pig, cow, sheep, baboon) [91]. The function of the bile SMase is unknown. The release of bile from gallbladder to intestine in response to a meal, and the rather high activity indicates, however, that the enzyme participates in the digestion of dietary SM. A hypothesis is that it also may have some function in the intestinal absorption of cholesterol. Together the mucosal and bile alkaline SMases could play important roles in digestion of dietary SM.

Human milk contains about 150 mg SM per litre [92], accounting for about 40% of total milk phospholipids. The dietary intake of SM in the breast-fed infant is thus around 60–150 mg/day. Since the digestion of milk triglycerides in human neonates differs from that in adults, due to the presence of bile salt stimulated lipase (BSSL) in human milk and due to the limited capacity of the pancreas in the neonate, the question was raised whether also the digestion of SM is different in suckling infants. It has recently been shown that human milk contains two enzyme activities which may participate in the degradation of SM [93]. First, an acid, not bile salt dependent SMase is present. Although this enzyme should be suited to operate in gastric contents, the finding may reflect the presence of some lysosomal enzymes in milk and its physiological function has not been proven. Furthermore alkaline SMase is present also in the newborn [91]. Second, milk contains ceramidase activity with neutral to alkaline pH-optimum, which cleaves ceramide to free sphingoid base and fatty acid. This ceramidase activity could be assigned to BSSL, which is important for digestion of milk triglycerides in newborn infants [94]. Although the detailed course of SM digestion in the newborn has so far not been clarified, there are thus enzymes that can generate poten-

tially bioactive sphingolipid metabolites from milk SM in the gut. We have speculated that these metabolites might be of importance for optimal development of the gut mucosa in the newborn, breast-fed infant.

The digestion and absorption of glycosphingolipids is not well characterized. Earlier studies showed [95] that digestion of glucosylceramide is slow and incomplete but follows a pattern similar to that for SM. No pancreatic enzyme hydrolyzing glycosphingolipids was found, whereas intestinal mucosa contain both an acid enzyme, probably lysosomal, and an enzyme with neutral pH-optimum, that hydrolyzes the glycosidic bond of glycosylceramides [96].

The course of SM digestion

Early studies [54, 55] indicated that the course of SM digestion was slower and more incomplete than for the glycerolipids. In a recent series of in vivo experiments on rats, the digestion of SM at different levels of the intestinal tract was examined, after oral administration of radioactive SM [97]. The conclusions from these experiments are that SM digestion is extended all over the whole small intestine and occurs mainly in the middle and lower parts. This coincides with the distribution of the alkaline SMase, which further indicates that this enzyme is important for digestion. The capacity of SM digestion was limited. Also after administration of small amounts, all of the small intestine and colon are exposed to SM and its bioactive metabolites, which could have a number of effects on the exposed intestinal cells. Recent experiments by Schmelz et al., in which radiolabeled SM was placed in intestinal loops of mice [98], also indicated that, although limited, the ability to hydrolyze SM is extended over the whole small intestine.

Exposure of colon to biologically active sphingolipid metabolites

It has recently been suggested that ceramide derived from dietary SM may influence tumour development in the colon. Addition of milk-SM to a normal diet in mice reduced the promotion phase in the development of colon tumours [99, 100], induced by the chemical carcinogen dimethylhydrazine. Although the inhibitory factor has not been established, ceramide may be responsible, due to its known ability to regulate cell growth. It has been suggested, as a possible mechanism, that tumour cells have defects in the activation of SMase, and that this defect could be by-passed by dietary sphingolipid metabolites [101].

The activity of SMase in human colon tumours has recently been examined in our group. The results showed [102] that the activity was decreased in the carcinoma, compared to normal colon tissue. The difference was most significant for the alkaline SMase activity, which was decreased by 75%. We have suggested that this enzyme has a regulatory role in cell proliferation in the colon mucosa and that dietary SM may be a nutritional factor that influences the events. If the suggestion is correct, that ceramide may suppress development of colon cancer, it is of interest to increase the amount of dietary SM and ceramide that reaches the colon. When rats were fed a mixture of cholesterol and radioactive SM, the uptake of SM in the rat intestine was further reduced (Nyberg et al., unpublished). The percentage recovery of radioactivity in faeces was more than twice as high when SM was given with cholesterol than without. One third of the radioactivity in faeces was present as ceramide.

Several studies have shown that there is a strong interaction between cholesterol and SM in cell membranes [103]. The critical factor is thought to be the high degree of saturation in the very long acyl chains of SM, which allows cholesterol to interact along its entire length. The strong interaction might also influence the absorption of these molecules in the rat intestinal tract. In most dietary sources of SM, the two compounds occur associated with each other in membrane structures, as for example in the milk fat globule and in meat.

SM also influences the absorption of cholesterol. On feeding rats dispersions with cholesterol and milk SM or soybean PC, a significantly higher cholesterol absorption was obtained with soybean PC than with milk SM (Nyberg et al., unpublished).

Future perspectives

SM hydrolysis initiates an intracellular signalling pathway, which has been shown to play important roles in a variety of signal transduction systems. The extent to which this signalling system is used in inflammation, immune responses and apoptosis is not known but it might be a commonly employed pathway, that could be exploited therapeutically [7].

As also stressed in recent reviews by Merrill and coworkers, sphingolipids are dietary constituents with potential biological effects in the gastrointestinal tract. Of particular interest is to explore the possibility that effects of exogenous ceramide in colon may influence the regulation of cellular growth in a way that affects cancer development. This question involves studies of the factors that regulate the exposure of colon to sphingolipid metabolites. To clarify the physical as well as enzymological factors that regulate the degradation is of major importance. For instance, the interactions between sterols and SM may be exploited both for inhibition of cholesterol absorption and to influence the exposure of colon to sphingolipid metabolites. The possibility that metabolites formed from milk SM in the neonates influence the maturation of the gut mucosa needs to be clarified. Additional ways that sphingolipids has been suggested to be of nutritional interest are as a source of dietary choline [86], and in altering serum lipid levels [104].

References

1. Thudicum JLW (1884) In: Bailliere, Tindall and Cox (eds) A Treatise on the Chemical Constitution of the Brain. Fascimile Edition 1962, Archon Books, Connecticut
2. Barenholz Y, Thompson TE (1980) Biochim Biophys Acta 604:129–158
3. Merrill AH Jr, Sweeley CC (1996) In: Vance DE, Vance JE (eds) Biochemistry of Lipids, Lipoproteins and Membranes. Elsevier, Amsterdam, pp 309–339
4. Hannun YA, Linardic CM (1993) Biochim Biophys Acta 1154:223–236
5. Pick L, Bielschowsky M (1927) Klin Wochenschr 6:1631–1637
6. Shapiro D, Flowers HM (1962) J Am Chem Soc 84:1047–1050
7. Kolesnick RN (1991) Prog Lipid Res 30:1–38
8. Bielawska A, Crane HM, Liotta D, Obeid LM, Hannun YA (1993) J Biol Chem 268:26226–26232
9. Malmsten M, Bergenståhl B, Nyberg L, Odham G (1994) JAOCS 71:9, 1021–1026
10. Yeagle PL, Hutton WC, Huang CH, Martin RB (1976) Biochemistry 15:2121–2124
11. Stults CLM, Sweeley CC, Macher BA, (1989) Methods Enzymol 50:167–214
12. Kobayashi T, Mitsuo K, Goto I (1988b) Eur J Biochem 171:747–752
13. Merrill AH Jr, Wang E, Mullins RE, Jamison WCL, Nimkar S, Liotta DC (1988) Anal Biochem 171:373–381
14. Sweeley CC (1991) In: Vance DE, Vance JE (eds) Biochemistry of Lipids, Lipoproteins and Membranes. Elsevier, Amsterdam, pp 327–360
15. Shipley GG, Avecilla LS, Small DM (1974) J Lipid Res 15:124–131
16. Sánchez-Yagüe J, Cabezas JA, Llanillo M (1988) Biochem International 16:5, 809–814
17. Huang C, Mason JT (1986) Biochim Biophys Acta 864:423–470
18. Shinitsky M, Barenholz Y (1978) Biochim Biophys Acta 515:367–394
19. Barenholz Y (1984) In: Shinitsky M (ed) Physiology of Membrane Fluidity, Vol 1. CRC Press, Boca Raton, pp 131–173
20. Cooper RA, Durocher JR, Leslie MH (1977) J Clin Inv 60:115–121
21. Landh T (1994) J Phys Chem 98:8453
22. Landh T, Larsson K (1992) Patent application no. PCT /SE92/00692
23. Ljusberg-Wahren H, Nyberg L, Larsson K (1996) Chimica Oggi/Chemistry Today, June

Progr Colloid Polym Sci (1998) 108:119–128
© Steinkopff Verlag 1998

24. Davenas E, Caviatti M, Nordoy A, Renaud S (1984) Biochem Biophys Acta 793:278–286
25. White DA (1973) In: Ansell GB, Hawthorne JN, Dawson RMC (eds) Form and Function of Phospholipids. Elsevier, Amsterdam, pp 441–482
26. Merrill AH Jr (1991) J Bioenergetics Biomembranes 23:1, 83–104
27. Chapman MJ (1986) Methods Enzymol 128:70–143
28. Esko JD, Raetz CRH (1983) In: Boyer PD (ed) The Enzymes, Vol XVI. Academic Press, New York, pp 207–253
29. Barenholz Y, Gatt S (1982) In: Hawthorne JN, Ansell GB (eds) Phospholipids, Ch 4. Elsevier Biomedical Press, Amsterdam, p 129
30. Borochov H, Zahler P, Wilbrandt W, Shinitzky M (1977) Biochim Biophys Acta 470:382–388
31. Eisenberg S, Stein Y, Stein O (1969) J Clin Invest 48:2320–2329
32. Seth SK, Newman HAI (1975) Circ Refs 36:294–299
33. Rouser G, Kitchevsky G, Yamamoto A (1972) Adv Lipid Res 10:261–336
34. Brady RO, Koval GJ (1958) J Biol Chem 233:26–31
35. Merrill AH Jr, Jones D (1990) Biochim Biophys Acta 1044:1–12
36. Mandon EC, Van Echten G, Birk R, Schmidt RR, Sandhoff K (1991) Eur J Biochem 198:667–674
37. Williams RD, Wang E, Merrill AH Jr (1984) Arch Biochem Biophys 228:282–291
38. Karlsson KA (1970) Chem Phys Lipids 5:6–43
39. Stoffel W, Le Kim D, Sticht G (1968) Hoppe-Seyler's Z Physiol Chem 349:1637–1644
40. Merrill AH Jr, Wang E (1986) J Biol Chem 261:3764–3769
41. Akanuma H, Kishimoto Y (1979) J Biol Chem 254:1050–1056
42. Singh I (1983) J Neurochem 40:1565–1570
43. Moser HW, Moser AB, Chen WW, Schram W (1989) In: Scriver CR, Beaudent AL, Sly WS, Valle D (eds) The Metabolic Basis of Inherited Diseases, 6th Edition, Vol II, Ch 65. McGraw-Hill, New York, pp 1645–1655
44. Riley RT, Norred WP, Bacon CW (1993) Annu Rev Nutr 13:167–189
45. Sweeley CC (1985) In: Vance DE, Vance JE (eds) Biochem Lipids and Membranes Benjamin Cummings, Menlo Park, pp 361–403
46. Barenholz BG, Roitman A, Gatt S (1966) J Biol Chem 241:3731–3737
47. Fowler S (1969) Biochem Biophys Acta 191:141–146
48. Koval M, Pagano RE (1991) Biochim Biophys Acta 1082:113–125
49. Ballou LR, Lauderkind SJF, Rosloniec EF, Raghow R (1996) Biochim Biophys Acta 1301:273–287
50. Pentchev PG, Kruth HS, Comly ME, Butler JD, Vanier MR, Wenger DA, Patel S (1986) J Biol Chem 261:16 775
51. Okazaki T, Bielawska A, Domae N, Bell RM, Hannun YA (1994) J Biol Chem 269:4070–4077
52. Das DVM, Cook HW, Spence MW (1984) Biochim Biophys Acta 777:339–342
53. Desmukh GD, Radin NS (1985) J Neurochem 44:1152–1155
54. Nilsson Å (1969) Biochim Biophys Acta 176:339–347
55. Nilsson Å (1968) Biochim Biophys Acta 164:575–584
56. Nyberg L, Duan R-D, Axelsson J, Nilsson Å (1996) Biochim Biophys Acta 1300:42–48
57. Duan R-D, Nilsson Å (1997) Hepatology, in press
58. Hassler DF, Bell RM (1993) In: Advances in Lipid Research, Vol 26
59. Gatt S, Gottesdiner TJ (1975) J Neurochem 26:421–422
60. Sugita M, Connolly P, Dulaney JT, Moser HW (1973) Lipids 8:401–406
61. Yada Y, Higuchi K, Imokawa G (1995) J Biol Chem 270:12 677–12 684
62. Stoffel W, Melzner I (1980) Hoppe-Seyler's Z Physiol Chem 361:755–771
63. Slife CW, Wang E, Hunter R, Wang S, Burgess C, Liotta DC, Merrill AH Jr (1989) J Biol Chem 264:10 371–10 377
64. Ghosh S, Strum JC, Bell RM (1997) FASEB J 11:45–50
65. McGovern UB, Jones KT, Sharpe GR (1995) Br J Dermatol 132:892–896
66. Berridge MJ (1983) Biochem J 2:849–858
67. Hannun YA, Obeid LM (1995) Trends Biochem Sci 20:73–78
68. Hannun YA, Obeid LM, Dbaibo GS (1996) In: Bell RM (ed) Handbook of Lipid Research, Vol 8. Lipid Second Messengers, Ch 5, Plenum Press, New York, pp 177–204
69. Hannun YA, Loomis CR, Merrill AH Jr, Bell RM (1986) J Biol Chem 261:12 604–12 609
70. Kolesnick RN (1987) J Biol Chem 262:16 759–6 762
71. Spiegel S, Foster D, Kolesnick R (1996) Current Opinion in Cell Biology 8:159–167
72. Hannun YA, Bell RM (1989) Science 243:500–507
73. Okazaki T, Bell RM, Hannun YA (1989) J Biol Chem 265:70–75
74. Kim M-Y, Linardic C, Obeid L, Hannun YA (1991) J Biol Chem 266:484–489
75. Jayadev S, Linardic CM, Hannun YH (1994) J Biol Chem 269:5757–5763
76. Bell RM, Hannun YA, Merrill AH Jr (eds) (1993) In: Advances in Lipid Research: Sphingolipids and their Metabolites, Vols 25 and 26. Academic Press, Orlando, FL
77. Saba JD, Obeid LM, Hannun YA (1996) Phil Trans R Soc Lond B 351:233–244
78. Linardic CM, Hannun YA (1994) J Biol Chem 269:23 530–23 537
79. Merrill AH Jr, Wang E (1992) Methods Enzymol 209:427–437
80. Liu J, Mathias S, Yang Z, Kolesnick RN (1994) J Biol Chem 269:3047–3052
81. Dobrowsky RT, Hannun YA (1992) J Biol Chem 267:5048–5051
82. Hannun YA (1994) J Biol Chem 269:3125–3128
83. Bielawska A, Crane HM, Liotta D, Obeid LM, Hannun YA (1993) J Biol Chem 268:26 226–26 232
84. Raines MA, Kolesnick RN, Golde DW (1993) J Biol Chem 268:14 572–14 575
85. Wu J, Spiegel S, Sturgill TW (1995) J Biol Chem 270:11 484–11 488
86. Desai NN, Zhang H, Olivera A, Mattie ME, Spiegel S (1992) J Biol Chem 267:23 122–23 128
87. Zeisel SH, Char D, Sheard NF (1986) J Nutr 116:50–58
88. Blank ML, Cress EA, Smith ZL, Snyder F (1992) J Nutr 122:1656–1661
89. Nyberg L, Burling H (1993) Patent SE-501697
90. Duan R-D, Nyberg L, Nilsson Å (1995) Biochim Biophys Acta 1259:49–55
91. Duan R-D, Hertervig E, Nyberg L, Hauge T, Sternby B, Lillienau J, Farooqi A, Nilsson Å (1996) Digest Dis Sci 41:1801–1806
92. Zeisel SH (1990) J Nutr Biochem 1:332–349
93. Nyberg L, Farooqi A, Duan R-D, Nilsson Å, Hernell O (1997) J Pediatr Gastroenterol Nutrition, submitted
94. Hernell O, Bläckberg L (1994) J Pediatr 125:556–561
95. Nilsson Å (1969) Biochim Biophys Acta 187:113–121
96. Leese HJ, Semenza G (1973) J Biol Chem 248:8170–8173
97. Nyberg L, Duan R-D, Lundgren P, Nilsson Å (1997) J Nutr Biochem 8:12–118
98. Schmelz E-M, Crall KJ, Larocque R, Dillehay DL, Merrill AH Jr (1994) J Nutr 124:702–712
99. Dillehay DL, Webb SK, Schmelz E-M, Merrill AH Jr (1994) J Nutr 124:615–620
100. Schmelz EM, Bushnev AS, Dillehay DL, Liotta DC, Merrill AH (1997) Nutr Cancer 28:81–85

101. Merrill AH Jr, Schmelz E-M, Wang E, Schroeder JJ, Dillehay DL, Riley RT (1995) J Nutr 125:1677S–1682S

102. Hertervig E, Nilsson Å, Nyberg L, Duan R-D (1997) Cancer 79:448–453

103. Slotte JP, Bierman EL (1988) Biochem J 250:653–658

104. Imaizumi K, Tominaga A, Sato M, Sugano M (1992) Nutr Res 12:543–548

105. Wurtman JJ (1979) In: Barbeau A, Growdon JH, Wurtman JJ (eds) Nutrition and the brain, Raven Press, NY. Vol 5, pp 73–81 Raven Press, NY

106. Long C (ed) (1961) In: Biochemists' Handbook, E and FN Spon Ltd, London

Progr Colloid Polym Sci (1998) 108:129–138
© Steinkopff Verlag 1998

A.S. Fogden
M. Stenkula
C.E. Fairhurst
M.C. Holmes
M.S. Leaver

Hexagonally perforated lamellae with uniform mean curvature

A.S. Fogden (✉) · M. Stenkula
Physical Chemistry 1
Center for Chemistry
and Chemical Engineering
University of Lund
P.O. Box 124
S-221 00 Lund
Sweden

C.E. Fairhurst · M.C. Holmes · M.S. Leaver
Department of Physics
Astronomy and Mathematics
University of Central Lancashire
Preston PR 1 2HE, Lancashire
United Kingdom

Abstract An increasing variety of self-assembling amphiphilic systems are found to exhibit a striking type of intermediate phase which comprises a one-dimensional stacking of two-dimensional networks, or perforated lamellae, often termed a mesh phase. This study focuses on idealized mesh models with the highest attainable symmetry and interfacial smoothness, namely hexagonal arrays of perforations with uniform mean-curvature throughout. The exact surfaces are constructed numerically and their geometries are analyzed in detail as a basis for assessing the deviations of observed mesh structures. The simplest network, of connectivity 3, is also compared to its 6-connected dual.

Key words Perforated lamellae – hexagonal array – uniform mean curvature – mesh phase

Introduction

Our present understanding of the aggregation of amphiphilic molecules is based upon a unified picture of liquid-crystalline structures laid out by Luzzati and coworkers in the late 1960s in a series of papers on the polymorphism of lipids [1–3]. Small angle X-ray scattering from the anhydrous high-temperature phases of divalent cation soaps revealed a novel family of structures, in all of which the polar groups pack into rod-like elements organized into networks embedded in the continuous matrix of their molten paraffin chains. In the body-centred cubic phase the rods lie along two-fold axes, connected three by three to form two interwoven, three-dimensional labyrinths [1]. The other two kinds are optically aniso-tropic (birefringent) and built from planar networks, in the one case a hexagonal mesh, again three-connected, with the layers stacked in the rhombohedral symmetry R3̄m, and the second structure comprising four-connected square meshes [2]. Further, the high temperature phases of lecithins, with a small amount of water, displayed ana-logous cubic and mesh phases [3]. In these models the rod segments, although relatively short, are assumed to be perfectly cylindrical and meet with sharp intersections. Although such cusped junctions are somewhat unphysical, the proposed structures matched the observed X-ray in-tensities and the predicted head-group areas were consis-tent across the different polymorphs.

Around a decade later the Luzzati picture was adapted by Kåre Larsson and others to describe the cubic phases formed in lipid/water systems, and in particular for monoolein [4, 5]. At sufficiently high dilution it becomes more appropriate to describe the pair of water-swollen networks of monolayer-coated rods as a single smooth bilayer sheathing the central partition of the cell. Using the Helfrich theory [6] of curved bilayer free energy and the concept of the shape parameter [7], the midsurface is assigned zero mean-curvature and so becomes a triply periodic minimal surface. The cubic model of Luzzati thus corresponds topologically to the gyroid surface, with the other two varieties of lyotropic cubic phase commonly encountered then being ascribed to the diamond and primitive surfaces. Although these diluted lipid systems do not appear to display equilibrium mesh phases, Larsson also considered structures related to these, by taking slices through infinite minimal surfaces to model cooperative phenomena involving proteins in biological membranes [5].

Investigations of phase behavior of surfactants in water revealed an analogous set of intermediate structures of the opposite type, i.e. built with apolar tunnels in a water continuum. These studies, combined with curvature intuition, suggested that mesh phases are favored over cubic (V_1) if the hydrocarbon chain is long or switched to fluorocarbon and thus reduced in flexibility. As one example, the lithium-perfluorooctanoate/water system was found to display a tetragonal mesh phase [8]; the interface was modelled simplistically using cylindrical rods meeting at cube junctions to verify that volume fractions and self-diffusion coefficients were consistent with this topology. Surprisingly, even the most commonly used surfactant, sodium dodecyl sulfate, was found to pass, down a concentration gradient, through four intermediate phases (again including a square mesh) over a narrow composition region [9]. NMR diffusion on such systems also indicated that bilayer perforations are not merely a feature of equilibrium meshes, but can also be prevalent as defects in lamellar phases, becoming correlated on approaching the transition.

These trends in the stabilization of surfactant meshes are not restricted to charged systems. They are also manifested in the phase behavior of nonionic surfactants of the general type C_nEO_m, thus providing the opportunity for systematic studies in which the interactions can be directly tailored by adjusting the length of the alkyl chain and/or poly(oxyethylene) head-group without the complication of long-ranged electrostatics. Recent investigations by some of the current authors on the $C_{30}EO_9$/water system [10] revealed an extensive intermediate phase region, which was probed with optical, spectroscopic and small-angle scattering techniques, including neutron diffraction on heating of a shear-aligned hexagonal phase [11]. This combination of techniques indicated a rhombohedral mesh structure, although analysis using rod/box models could not conclusively pin down its topology; indeed a six-connected (triangular) mesh appeared to yield a better fit than the three-connected structure of Luzzati. On switching to $C_{16}EO_6$ the reduction in alkyl chain length effected a near balance in preference for mesh and cubic; on cooling from the lamellar phase the binary system exhibited in turn a random mesh phase (disrupted lamellar L_α^H), a metastable ordered mesh and a gyroid cubic phase [12]. The ordered mesh displayed characteristics similar to those for $C_{30}EO_9$, thus posing the same structural ambiguities.

In diblock copolymer melts the two chemically bonded, immiscible chains A and B microseparate on cooling the disordered phase to form A- and B-rich domains with a variety of possible interfacial geometries. The morphologies and phase sequences observed by the groups of Bates [13] and Hashimoto [14, 15] are quite similar to those encountered with surfactants or lipids; cubic and three-connected mesh phases are commonplace. The task of visualizing topologies is expedited by the combination of scattering techniques with transmission electron microscopy, although the latter can lead to misinterpretations since images of cast films vary with section thickness and angle. Further, the preparations may require long equilibration times and are prone to kinetic trapping of metastable states. Relative to the situation for surfactants in water, the melts of conformationally symmetric, linear diblock copolymers offer increased possibilities for theoretical description of the self-assembly by adapting standard treatments of Gaussian polymers. The framework of a self-consistent mean-field theory, incorporating entropic chain stretching, incompressibility and immiscibility, has been established and a diversity of approximate solution methods have been developed to span the extremes of weak and strong segregation [16]. Importantly, progress has been made in increasing the geometrical sophistication of descriptions of complex topologies by expanding the segment spatial distributions as Fourier series for the given symmetry class and working in their reciprocal space.

As expected by the delicate free energy balance of the intermediate structures, the theoretical phase diagrams for block copolymers depend upon their particular assumptions. Recent analysis suggests that the three-connected mesh structure is nearly stable in the pure melt, and thus could be a long-lasting transient, with apparently little preference for a two-layered (ab...) over a triple (abc...) stacking repeat [16]. Although diblock/homopolymer blends have received less theoretical attention, the existing studies point to stabilization of the mesh by addition of homopolymer [17]. The synthesis of complex molecular architectures, involving star-, graft-, tapered- and multicomponent multi-blocks [15], poses renewed challenges to modelling. Although the general principles of the mean-field theory admit of such adaptations, complete treatments covering the convoluted microstructures observed would be an awkward task.

Comparing the above-mentioned developments in our understanding of the aggregation of lipids and surfactants with those for block copolymers, especially with regard to modelling mesh structures, it could be concluded that the former are in need of a more sophisticated geometrical description beyond the rod/box constructions. For this purpose we can borrow from the techniques of Fourier decomposition employed in block copolymer theories [16]. Conversely, the latter could benefit, at least in the

Progr Colloid Polym Sci (1998) 108: 129–138
© Steinkopff Verlag 1998

short term, by a departure from self-consistent field theories to simpler prescriptions aimed at unifying the underlying geometrical trends for copolymer molecules of varying architecture and blends. In both cases the equilibrium interfaces are commonly understood to depend to a large degree on their spontaneous mean curvature. Thus one possibility is to consider surfaces with uniformly prescribed mean curvature, i.e. possessing the spontaneously preferred value at every point. Such surface families have already proven useful for bicontinuous cubic phases, generalizing the central minimal surface to models of the hydrophobic/hydrophilic interfaces on either side [18].

In the current study we take the obvious step of modelling mesh topologies using uniform mean curvature. That such surfaces exist was proven already in 1970 by Lawson [19], however their explicit construction has remained largely unaddressed [20]. A uniformly curved model of meshes can be justified on a number of counts. The Helfrich bending free energy for surfactant monolayers, together with the bare surface tension of block copolymers, both suggest a tendency for constant mean curvature. In neither case though can this interfacial ideality be claimed to be the sole driving force for aggregation. The geometrical approach decouples the mesh layers from their neighbors in the stack, although this may be a reasonable assumption if the interlayer interactions are weak relative to intrinsic contributions. More serious is the neglect of molecular packing requirements (the incompressibility constraint) for the hydrocarbon chains enclosed in a type 1 mesh, as the layers have not been observed to take up any oil. Thus we must regard the uniformly curved meshes constructed here as a structural basis for assessing the importance of layer correlations and global packing demands. In the following section we address the commonly occurring three-connected mesh and also, to illustrate the effect of increased topological complexity, consider the six-connected mesh due to its possible relevance to the nonionic surfactant systems. In a subsequent section these smooth meshes will be briefly compared to the corresponding rod/box models.

Uniformly curved meshes of connectivity 3 and 6

The general structure of a single such mesh layer of connectivity 3 and 6 is given in Figs. 1 and 2, respectively, to clarify the layout and definitions used for presentation of the exact results in Figs. 3–8. The underlying symmetries are identical for both cases. The parts A are plan views of a rectangular piece comprising 4 of the p6m translational cells (shaded, with side length a), which in turn are subdivided by the vertical mirror planes into 12 triangles or asymmetric units. The unit highlighted in parts A is shown

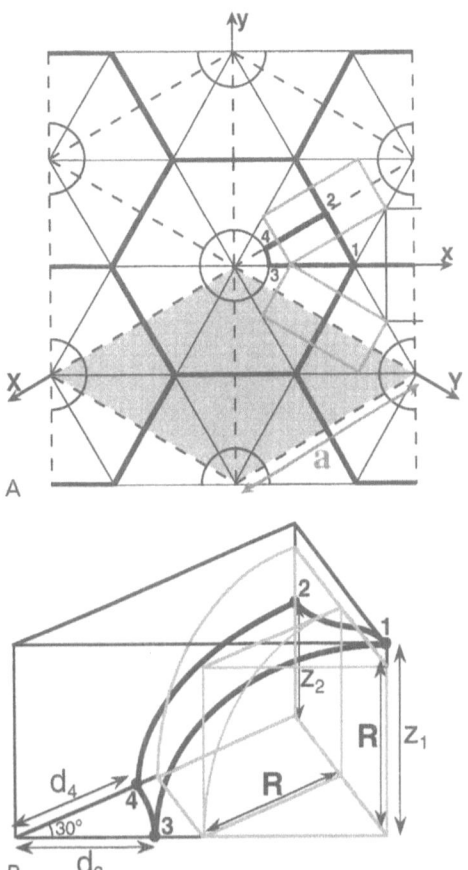

Fig. 1 Diagram of one sheet of a three-connected regular mesh, showing (A) the overall plan view, and (B) an asymmetric unit in perspective. The two types of mesh interface considered are illustrated in both, with the rod/box in light gray and the smooth surface in black with vertices numbered

in perspective in parts B; reflection over the bottom face (the mesh midplane) gives the lower half.

Onto this common symmetry scheme the two distinct topologies are imposed, namely the skeletons of a regular hexagonal (three-connected) and triangular (six-connected) network of mesh tunnels shown in dark gray in Figs. 1A and 2A, respectively. The circles indicate the pattern of layer perforations thus created. Two types of surface model for clothing these labyrinths are included in both cases. The simplest rod/box model is illustrated in light gray, built from cylinder segments of radius R meeting at equilateral triangular (Fig. 1) or hexagonal (Fig. 2) prisms with this same height [10–12]. A generalized, smoothened version of this is indicated by the mirror plane boundary curves in black, in order to represent a uniform mean-curvature mesh. Note that no symmetry has, as yet, been attached to the upper plane bounding Figs. 1B and 2B; we merely choose it to coincide with the maximum

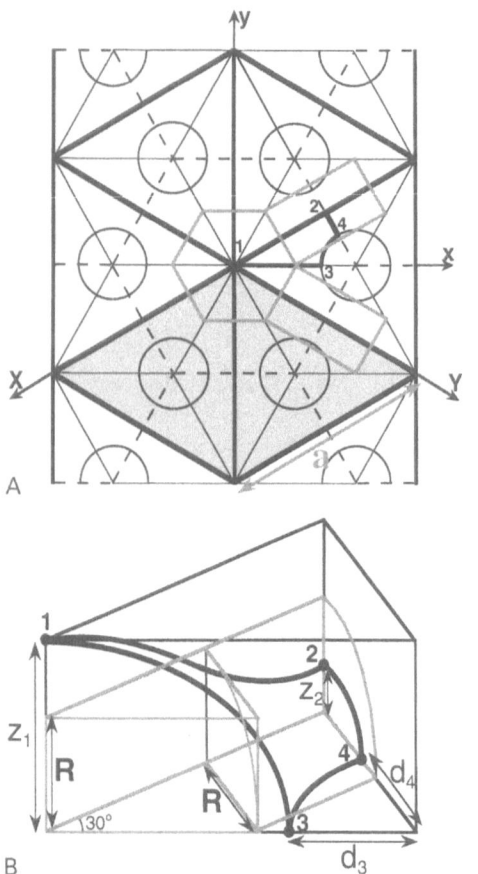

Fig. 2 Diagram of one sheet of a six-connected regular mesh, applying the same descriptions as in Fig. 1

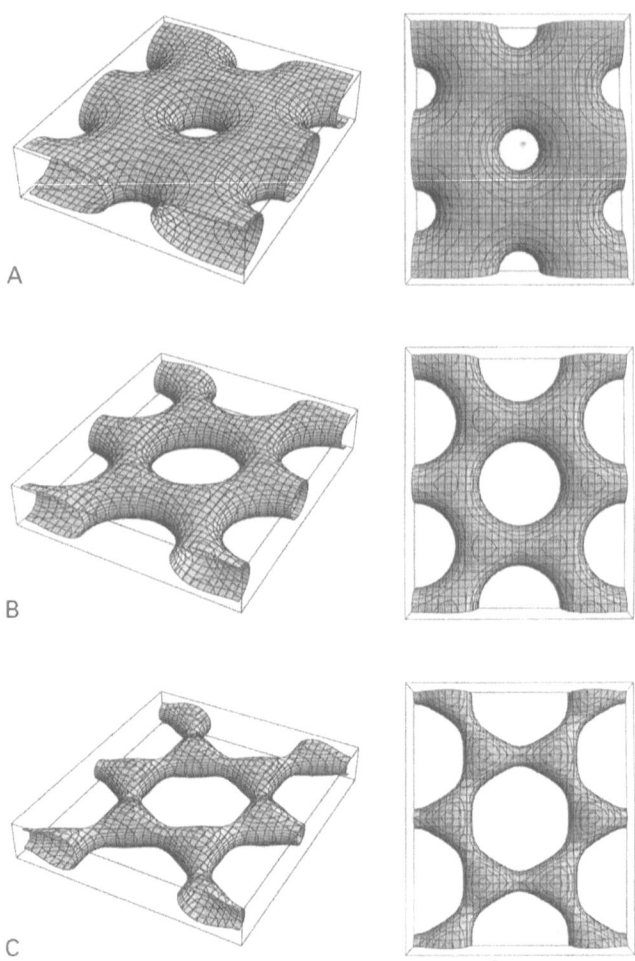

Fig. 3 Three examples of three-connected meshes bearing uniform mean curvature, with values (A) $H^* = 0.695$, (B) $H^* = 2.11$ and (C) $H^* = 3.71$. Each is viewed from the same two perspectives

height z_1 of the isolated mesh layer. The extremal dimensions of the holes are labelled d_3 and d_4, while those of the mesh neck are z_2 and $a/2 - d_4$ (three-connected) or $a/(2\sqrt{3}) - d_4$ (six-connected). For this asymmetric unit we define the volume enclosed between the surface and the midplane as V, the surface area S, and its projection onto the midplane, i.e. the shadow area, $S_\perp$.

The determination of these geometrical characteristics for the rod/box models, as functions of their R, is completely trivial; indeed this is the main reason for their use. In contrast, calculation of the characteristics for the uniformly curved models corresponding to each fixed value of the mean curvature H, is far from straightforward. In spite of their long established existence [19], little is known of the quantitative details of these surfaces, with regard to their embedding in three-dimensional Euclidean space, the range of H values attainable with this topology, and the topological transformations which terminate the mesh family. The mathematical task is more complicated than that for the minimal (zero mean curvature) surfaces [20]

since the extension of their complex-plane parametrization to surfaces of non-zero H involves non-analytic functions [21]. Anderson and coworkers [18] employed an on-surface finite-element method to numerically generate the constant mean-curvature companions of cubic minimal surfaces. For meshes the computational difficulties are compounded by the absence of such a (non-degenerate) analytic starting point for perturbation in H.

For the above reasons we choose to switch from these parametric descriptions and tackle the solution by Fourier decomposition. For a surface defined implicitly by the equation $f(x, y, z) = 0$, the mean curvature at a point (x, y, z), i.e. the average of the two local principal curvatures c_1 and c_2, is given by

$$H = \frac{1}{2}(c_1 + c_2) = \frac{1}{2}\nabla \cdot \left(\frac{\nabla f}{|\nabla f|}\right). \qquad (1)$$

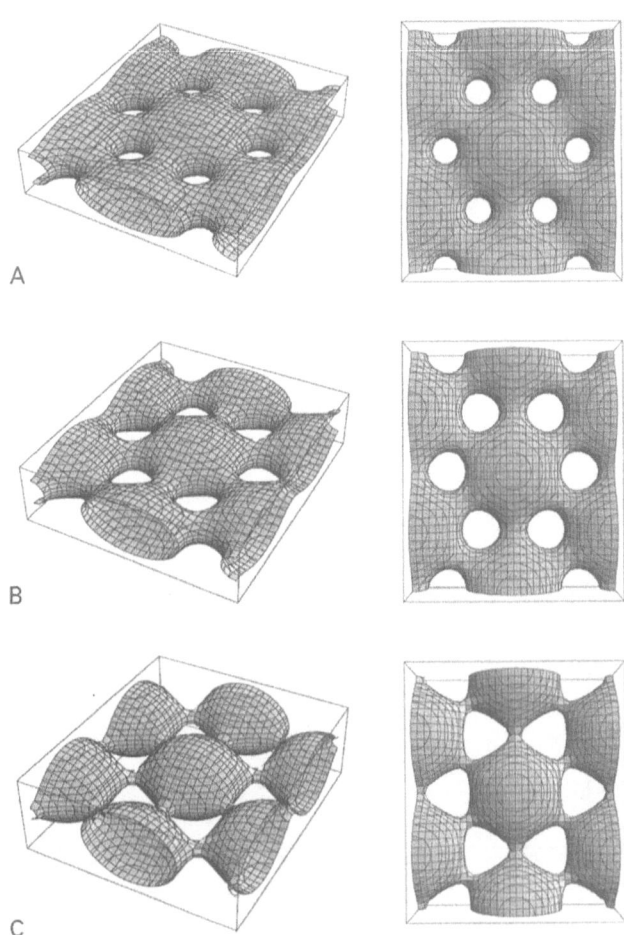

Fig. 4 Three examples of uniformly curved six-connected meshes, corresponding to (A) $H^* = 0.907$, (B) $H^* = 1.49$ and (C) $H^* = 2.10$

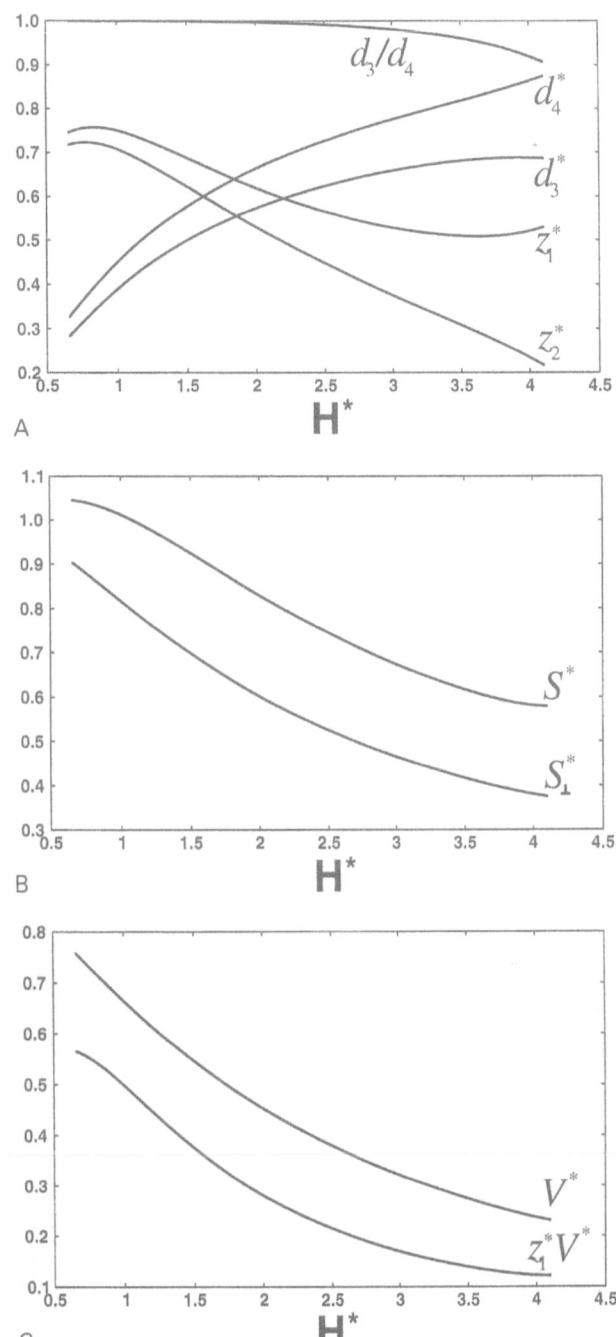

Fig. 5 Plots of the geometric characteristics of the uniformly curved three-connected meshes, showing (A) distances, (B) areas and (C) volumes, vs. H^*. See the text for details

Any periodic function f can be expressed uniquely as a Fourier series:

$$f(x, y, z) = \sum_{(hkl) = -\infty}^{\infty} |A(hkl)| \cos[2\pi(hX + kY + lZ) - \alpha(hkl)] \quad (2)$$

where the wave-vector indices and their structural moduli and phases are subject to the symmetry invariance of the particular space group [22]. For a fully ordered mesh phase of rhombohedral or hexagonal symmetry, the space-group coordinates, scaled with the horizontal and vertical periods a and c (see Figs. 1 and 2), are defined in terms of the Cartesians as

$$X = -\frac{1}{\sqrt{3}}\frac{x}{a} - \frac{y}{a}, \qquad Y = \frac{1}{\sqrt{3}}\frac{x}{a} - \frac{y}{a}, \qquad Z = \frac{z}{c}. \quad (3)$$

Our present study is restricted to an isolated mesh layer, so the series reduces to that for the two-dimensional, maximum symmetry, space-group p6m. Throughout the following discussion we use the dimensionless mean curvature

$$H^* = Ha \quad (4)$$

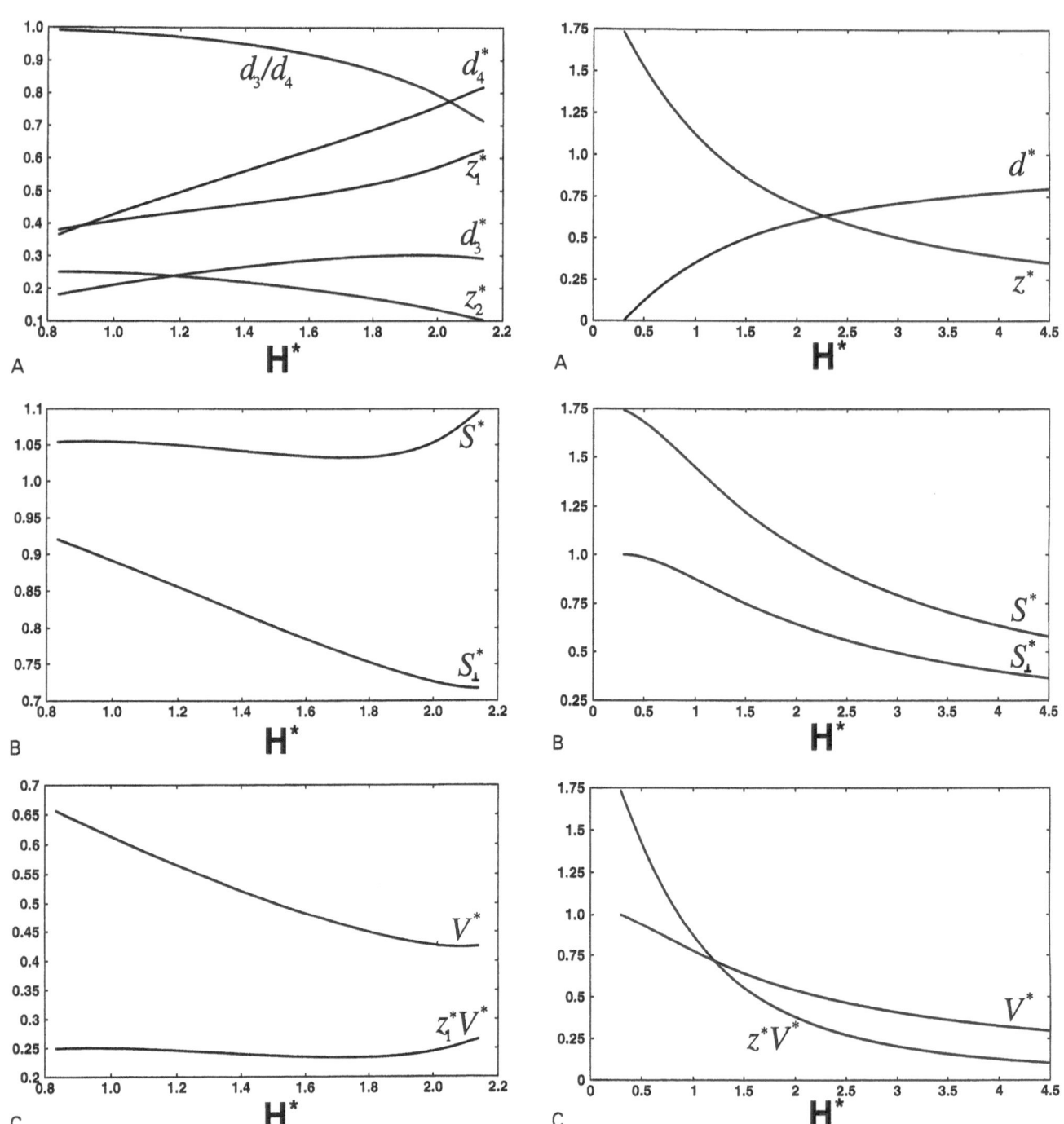

Fig. 6 Plots of (A) distances, (B) areas and (C) volumes, vs. H^* for the uniformly curved six-connected meshes

Fig. 7 Plots of the geometrical characteristics of the three-connected rod/box models, to be compared with Fig. 5. Note that the abscissa H^* here represents the average curvature over the rod/box surface, given in Eq. (11) (see the text for further explanation)

and also scale all the geometrical characteristics mentioned above with this arbitrary a.

The uniformly curved meshes are numerically generated using a finite number of the Fourier invariants from Eq. (2); the coefficients are iterated to minimize the surface variation of the mean curvature using a least-squares approach. The method is similar in part to that used by

Mackay for minimal surfaces [23]. We have obtained the set of three-connected meshes corresponding to each individual H^* value taken in steps of 0.03 across the interval $0.66 < H^* < 4.1$, and the six-connected meshes at the same density of H^* values between 0.83 and 2.14. In all

Progr Colloid Polym Sci (1998) 108:129–138
© Steinkopff Verlag 1998

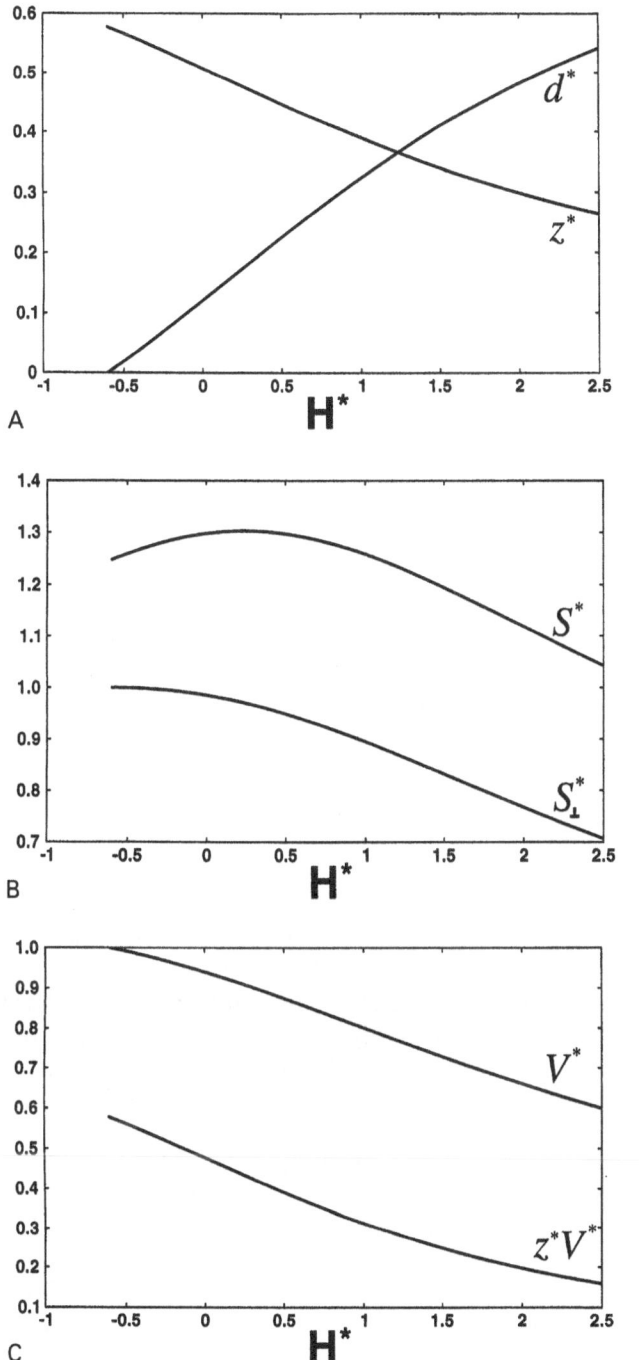

Fig. 8 Plots for the six-connected rod/box models, to be compared with Fig. 6. Again the abscissa is the surface-averaged value in Eq. (13)

Gaussian curvatures and/or trans-surface distances. Such inhomogeneities pose problems for our method since the number of invariants required for uniform convergence rises rapidly, but at the same time these features render the surfaces highly unrealistic as interfacial candidates for lyotropic liquid crystals.

Figure 3 displays three representatives of the family of uniformly curved three-connected meshes thus generated. The (x, y) region is the rectangle in Fig. 1A and each surface is shown both in oblique view and from above. In Fig. 3A $H^* = 0.695$ and the mesh takes on the appearance of non-overlapping catenoidal-like rings, approaching complete rotational symmetry, and connected with relatively flat sections. On increasing the mean curvature to $H^* = 2.11$ in Fig. 3B the catenoidal and planar elements merge by inflation to give a more homogeneous mesh. The enlarged perforations remain quite circular and the mesh thickness decreases to expose greater variations in local height. These trends continue to Fig. 3C, at which $H^* = 3.71$. The holes are now noticeably hexagonal, leading to a significant narrowing of the mesh necks and suggesting approach to a topological transition.

Three examples from the six-connected mesh family are shown in Fig. 4, retaining the same viewpoints as throughout Fig. 3. The mesh in Fig. 4A, for $H^* = 0.907$, is analogous to that in Fig. 3A although the "catenoids" overlap more strongly and the large hub regions are noticeably cupped. This difference, while due in part to the slightly higher curvature, is a direct result of the increased genus, through the reduction of the perforation symmetry to three-fold and raising of the hub symmetry to six-fold. On driving the mean curvature up to $H^* = 1.49$ (Fig. 4B), the holes increase in width and now display their triangular aspect. Although this case is close to the middle of the H^* range investigated, its Gaussian curvature distribution is significantly less homogeneous than its counterpart in Fig. 3B; the hubs which dominate the mesh now grow in height with increasing H^*. These inhomogeneities are even more apparent in Fig. 4C, where $H^* = 2.10$, i.e. almost exactly that value in Fig. 3B. The dimension d_4 of the perforations (see Fig. 2B) grows at the expense of d_3, the necks become narrower and a topology change is imminent.

To present the quantitative plots of the variations in distances, areas and volumes across this span of uniform H^* surfaces, we first introduce our definitions of the corresponding starred quantities which are independent of the length a. We begin with the three-connected meshes, referring to Fig. 1B. The widths d_3 and d_4 are naturally scaled with their triangle edge lengths

$$d_3 = \frac{1}{\sqrt{3}} a d_3^* , \qquad d_4 = \frac{1}{2} a d_4^* . \tag{5}$$

cases 8 to 24 Fourier invariants are sufficient to reduce the surface-averaged deviation of H to small fraction of 1%. We have not attempted to cover the full range of accessible H^* for either topology. Uniform meshes certainly exist for values above and below our endpoints, however these display increasingly large variations in the distribution of

In the absence of a vertical reference we define the heights z_1 and z_2 relative to the length of the third triangle edge, thus

$$z_1 = \frac{1}{2\sqrt{3}} a z_1^* , \qquad z_2 = \frac{1}{2\sqrt{3}} a z_2^* . \qquad (6)$$

The surface area S and its shadow area $S_\perp$ are scaled against that of the midplane triangle

$$S = \frac{1}{8\sqrt{3}} a^2 S^* , \qquad S_\perp = \frac{1}{8\sqrt{3}} a^2 S_\perp^* . \qquad (7)$$

Finally, we define the enclosed volume V relative to the volume of the triangular prism of height z_1 in Fig. 1B. Thus

$$V = \frac{1}{48} a^3 z_1^* V^* . \qquad (8)$$

This set of dimensionless characteristics of the three-connected meshes are plotted over the range $0.66 < H^* < 4.1$ in Fig. 5. Figure 5A displays the four distances in Eqs. (5) and (6), together with the ratio d_3/d_4 of the perforation width extrema. This last quantity decreases very slowly from unity, so the holes are close to circular across the majority of this range. At both extremes, i.e. curvatures lower than Fig. 3A or higher than Fig. 3C, some of the trends reverse. At the low end z_1^* and z_2^* pass through maxima and begin decreasing (i.e. thinning) and converging (flattening). Although we have not sought to generate surfaces below this value, it is clear that as H^* approaches zero the four starred distances converge pairwise to zero. The limiting (minimal) surface is then infinite, completely isolated, catenoids. At the high curvature end of the range, z_1^* and d_3^* pass through a minimum and maximum, respectively, and the transitions are more subtle. Again we have not investigated the behavior beyond these values, since the necks become prohibitive; the qualitative trends are, however, clear. The topological termination is the necking-off to an array of touching spheres (these are the only closed, handle-free surfaces of uniform mean curvature), of radius $a/(2\sqrt{3})$. Thus z_2^* becomes zero, z_1^* and d_4^* both reach unity, and d_3^* tends to $1/2$. Note that this mean-curvature value, $H^* = 2\sqrt{3}$, is not the endpoint of the range though, indeed the sphere array is less curved than the mesh in Fig. 3C. Thus the plots in Fig. 5A continue up to some maximum attainable H_{max}^* (which we have not determined), at which point the solution family bifurcates, giving a second branch returning to the sphere terminus. Two distinct mesh embeddings exist for uniform mean-curvatures in this subrange $2\sqrt{3} < H^* < H_{max}^*$. This bifurcative behavior is expected, since the surface evolution must trace back through itself during the necking process, and has been analyzed for the transitions of the

bicontinuous cubics [18]. Figure 5B shows the pair of dimensionless areas in Eq. (7), while Fig. 5C gives the curves for both the dimensionless volume V^* and, more importantly, the product $z_1^* V^*$ entering into Eq. (8). All of these curves will loop around at H_{max}^* and return to their values for the touching spheres.

Since these three-connected meshes only possess the single degree of freedom H^*, their fitting to the results of small angle scattering from a liquid-crystalline mesh phase (or suspected one) is a simple matter. The value of $z_1^* V^*$ is specified by the composition, i.e. the volume fraction of hydrocarbon Φ_{hc} enclosed in a type 1 mesh, together with the hexagonal unit cell ratio c/a. Assuming, for the sake of definiteness, that this cell comprises three mesh layers (a rhombohedral abc stacking),

$$z_1^* V^* = \frac{1}{\sqrt{3}} \frac{c}{a} \Phi_{hc} . \qquad (9)$$

From the lower, monotonic curve in Fig. 5C the corresponding value of H^*, and hence the mean curvature, can be read-off unambiguously. Using the value of S^* in Fig. 5B, the interfacial area per surfactant head-group a_{hg} is then expressed in terms of the volume per surfactant chain v_{hc} as

$$S^* = \frac{1}{6} c \, \Phi_{hc} \frac{a_{hg}}{v_{hc}} . \qquad (10)$$

Given knowledge of the head-group area and mean curvature through the adjoining phase regions (cylindrical, cubic or lamellar), the sensibility of this uniform H mesh model of the intermediate phase can be directly assessed. Further, the results in Fig. 5A can be used in conjunction with Eqs. (5) and (6) to check whether the mesh heights are too large or their necks too small for packing the given chain length. Provided the four extremal dimensions are physically reasonable, the mesh at this H^* value is probably quite close to the true interfacial structure.

To present the companion results for the six-connected meshes, we use a set of starred definitions entirely analogous to those in Eqs. (5)–(8), only modified by the interchange of triangle edge lengths. The formulae for d_3, S and $S_\perp$ are unchanged by this, d_4 in Eq. (5) is divided by $\sqrt{3}$, while z_1 and z_2 in Eq. (6) and V in Eq. (8) are all multiplied by $\sqrt{3}$. The corresponding sets of curves are given in Fig. 6 over the range $0.83 < H^* < 2.14$. Many of the trends are quite similar to those in Fig. 5, although certain differences are significant. In particular, z_1^* now increases over the whole range investigated, and the ratio d_3/d_4 deviates more strongly from unity, as expected by the reduction in the internal symmetry of the holes from $6m$ to $3m$. On the other hand, S^* and the product $z_1^* V^*$ vary very little. The lower endpoint of the family is the same as for the

Progr Colloid Polym Sci (1998) 108:129–138
© Steinkopff Verlag 1998

three-connected meshes, namely as H^* drops below 0.83 and approaches zero all four starred distances in Fig. 6A tend to zero, resulting in perfect catenoids. Likewise the topological terminus at the opposite extreme is an array of touching spheres (necking off from Fig. 4C), now with radius $a/2$ and so $H^* = 2$. Again z_2^* vanishes, z_1^* and d_4^* both reach unity, and now d_3^* tends to $1 - \sqrt{3}/2$. A similar type of bifurcation occurs at some maximum $H_{max}^* > 2.14$ so, for example, there exists a six-connected partner to the mesh in Fig. 4C with identical curvature.

The application of this information for structural fitting to six-connected meshes follows, in principle, the above prescription, using the altered definitions of the starred quantities. The necessary modification to Eq. (9) is to divide the right-hand side by $\sqrt{3}$; Eq. (10) is unchanged. Note though, that the lack of variation of S^* and $z_1^*V^*$ in Fig. 6 renders the fitting either impossible or inconclusive. The uniform mean-curvature meshes only fit a very narrow window of compositions (irrespective of surface area or dimensions); within this window a definite value of H^* cannot reasonably be inferred without additional criteria. These features indicate that, if six-connected meshes are manifest in lyotropic systems [10–12], uniformity of curvature is not their driving force.

Comparison with primitive rod/box models

We complete the current study by returning to the comparison of these families of uniformly curved meshes with those of the simplest rod/box constructions as illustrated in Figs. 1 and 2. These two types of model, while sharing the same topology, differ dramatically in their mean-curvature distributions and their appearance. By comparing the set of geometrical characteristics in Figs. 5 and 6 with their analogues for the discrete model, the degree of distinction when restricted to the less sensitive measures used in elementary structural fitting, can be ascertained. The rod/box formulae for the dimensionless quantities defined in Eqs. (5)–(10) (together with their modifications for connectivity 6 mentioned above) are trivially obtained in terms of the single free variable R, and will not be given here. To facilitate the assessment we switch to plotting these rod/box characteristics vs. the surface average of their non-uniform mean-curvature $\langle H \rangle = \int H \, dS / \int dS$. The average is calculated by rounding all exposed edges, i.e. smoothly collaring the adjoining facets, then shrinking these down to zero width.

For the three-connected rod/box this gives

$$\langle H^* \rangle = \frac{1 - (2 + \sqrt{3}\pi - 10/\sqrt{3})r}{2r[1 - (2 + \sqrt{3} - 2/\pi - 4\sqrt{3}/\pi)r]} \quad (11)$$

where we again use the definition in Eq. (4), and

$$r = R/a . \quad (12)$$

The results are presented in Fig. 7 over the subrange $0.2992 < \langle H^* \rangle < 4.5$, where the lower limit corresponds to the endpoint $r = 1/2$. Certain distinctions from the uniform mesh curves in Fig. 5 are immediately apparent. The scaled distances d_3^* and d_4^* are now identical, and labelled d^* in Fig. 7A; similarly $z_1^* = z_2^* \equiv z^*$ there. Also, the behavior at the low curvature extremes of each are certainly different, and the rod/box does not display a bifurcation at the opposite end. However, in the central parts of the range, of greatest relevance to lyotropic systems, the curves in Figs. 5 and 7 are quite similar, and definite conclusions as to the geometry of the mesh may require accurate data and deeper theoretical input.

For the six-connected rod/box the smoothing gives

$$\langle H^* \rangle = \frac{1 - (2\sqrt{3} + \pi - 8/3)r}{2r[1 - (1 + 2\sqrt{3} - 4/\pi - 2\sqrt{3}/\pi)r]} . \quad (13)$$

Note that this averaged curvature is not strictly positive – it passes through zero at $r = 0.2539$ and drops to -0.5979 at the maximum radius $r = 1/(2\sqrt{3})$. The scaled quantities are plotted in Fig. 8 up to $\langle H^* \rangle = 2.5$. The distinction with the results for uniform meshes in Fig. 6 is now more pronounced, as expected from the increase in topological complexity for connectivity 6. Most notable is the difference in the curves for $z_1^*V^*$ and S^*; the former curve in Fig. 8C now decreases monotonically to admit a large window of compositions for structural matching [10–12].

Extensions

In this study we have numerically constructed the uniformly curved mesh families for the regular networks of connectivity 3 and 6. Their analogue for connectivity 4 also exists, and could be generated by applying the same basic algorithm within the two-dimensional space-group p4m. Although square-mesh phases have seldom, if ever, been observed or predicted in block copolymer systems, the body-centred tetragonal symmetry can be difficult to distinguish from the rhombohedral on the basis of small-angle scattering [10], so this third family would provide a useful partner for a fuller comparison.

In forthcoming studies we shall use the uniformly curved meshes to aid in the resolution of ambiguities arising from the three- and six-connected rod/box models for the intermediate phases of the nonionic surfactants $C_{30}EO_9$ and $C_{16}EO_6$. These synthetic systems offer ideal test cases, due to their chemical purity and monodispersity of chain lengths; moreover the head-group areas measured in the hexagonal and lamellar regions are very similar, suggesting only a narrow range of allowable values across the intermediate. Further, at the lower volume fractions of surfactant the inter-mesh correlations (due to head-group overlap repulsion) disappear, while the in-plane organization persists, providing a regime in which our geometrical assumption of layer decoupling is valid. Even in the presence of correlations, it is frequently inferred that the layer structure is largely unperturbed. In particular, puckering of three-connected meshes (in which the connections tilt alternately up and down around the six-rings as in the parallel layers of the diamond lattice) is apparently never encountered, although admissible within the global symmetry R3m.

Any possible failure in fitting these nonionic systems to our three-connected meshes (given that the six-connected option is immediately precluded by Fig. 6 for the given compositions) would then point directly to the importance of the internal incompressibility constraint. To accordingly generalize our model of uniform mean curvature, while retaining maximal simplicity, the ideal vehicle would be the surfactant shape parameter. The algorithm could be readily adapted to minimize a combination of deviations in the local mean-curvature and packing shape (which can be regarded as independent quantities by virtue of the breakdown of the parallel surface construction for meshes). This would then provide a two-variable surface family for comparison, and moreover for prediction of mesh formation given the basic molecular architecture.

References

1. Luzzati V, Spegt PA (1967) Nature 215:701
2. Luzzati V, Tardieu A, Gulik-Krzywicki T (1968) Nature 217:1028
3. Luzzati V, Gulik-Krzywicki T, Tardieu A (1968) Nature 218:1031
4. Hyde ST, Ericsson B, Andersson S, Larsson K (1984) Z Kristallogr 168:213
5. Larsson K (1989) J Phys Chem 93:7304
6. Helfrich W (1973) Z Naturforsch c 28:693
7. Israelachvili J, Mitchell DJ, Ninham BW (1976) J Chem Soc Faraday Trans 72:1525
8. Kekicheff P, Tiddy GJT (1989) J Phys Chem 93:2520
9. Kekicheff P (1991) Mol Cryst Liq Cryst 198:131
10. Burgoyne J, Holmes MC, Tiddy GJT (1995) J Phys Chem 99:6054
11. Fairhurst CE, Holmes MC, Leaver MS (1996) Langmuir 12:6336
12. Fairhurst CE, Holmes MC, Leaver MS (1997) Langmuir 13:4964
13. Föster S, Khandpur AK, Zhao J, Bates FS, Hamley IW, Ryan AJ, Bras W (1994) Macromolecules 27:6922
14. Hashimoto T, Koizumi S, Hasegawa H, Izumitani T, Hyde ST (1992) Macromolecules 25:1433
15. Hasegawa H, Hashimoto T, Hyde ST (1996) Polymer 37:3825
16. Matsen MW, Bates FS (1996) Macromolecules 29:1091
17. Matsen MW (1995) Phys Rev Lett 74:4225
18. Anderson DM, Davis HT, Scriven LE, Nitsche JCC (1990) Adv Chem Phys 77:337
19. Lawson HB (1970) Ann Math 92:335
20. Fogden AS (1996) Phil Trans R Soc Lond A 354:2159
21. Gackstatter F (1990) J Phys Colloq 51:163
22. Henry NFM, Londsdale K (eds) (1969) International Tables for X-Ray Crystallography Kynoch, Birmingham, UK
23. Mackay AL (1994) Chem Phys Lett 221:317

Progr Colloid Polym Sci (1998) 108:139–152
© Steinkopff Verlag 1998

S.T. Hyde
A. Fogden

Hexagonal mesophases: honeycomb, froth, mesh or sponge?

S.T. Hyde (✉)
Applied Mathematics Department
Research School of Physical Sciences
Australian National University
Canberra 0200
Australia

A. Fogden
Physical Chemistry 1
Chemical Center
Lund University
S-22100 Lund
Sweden

Abstract The mesostructure of hexagonal phases in lyotropic liquid crystals and some inorganic derivates is reconsidered in topological terms, and it is argued that related rhombohedral sponge and mesh structures may well form in the vicinity of a conventional hexagonal phase. The standard signature of a hexagonal phase from small-angle scattering patterns is insufficient evidence for the hexagonal honeycomb arrangement of one-dimensional channels believed to characterize the mesostructure, and an alternative swelling analysis is suggested to offer a more detailed probe of film topology in hexagonal systems.

Key words Lyotropic liquid crystals – mesophase topology –bicontinuous structures

Introduction

The identification of mesostructure in "soft" materials, which exhibit both temporal and spatial structural fluctuations over many length and time scales, is a subtle issue. Mesocrystallinity – translational order at supramolecular length scales – of such materials under certain thermodynamic conditions, is evident from, for example, small angle scattering, which reveals pseudo-Bragg reflections – correlation peaks in scattering intensity, due to an underlying lattice in the material. However, that lattice is far less rigid than the lattice of atomic crystals. One rarely obtains more than ten reflections in small angle spectra of lyotropic liquid crystals of amphiphiles in water, at least an order of magnitude fewer than routinely found in diffraction patterns of conventional crystals. The application of conventional crystallographic techniques – developed for atomic crystals – must then be treated with some caution. Even if

the observed intensity peaks suffice to pin down the underlying space group symmetry, matching to model structure factors of uniform electron density partitions can at best only distinguish the most likely candidate within the working set of interfaces assumed. Complementary probes of structure, particularly those that are characteristic of the *topology* of the mesostructure, are needed.

In this paper we address these issues by focusing on ubiquitous hexagonal phases formed in amphiphile–water mixtures and related materials. The mesostructure of hexagonal phases is commonly accepted to be relatively simple: two-dimensional cylinders, packed with hydrophobic chains, arranged in a hexagonal lattice embedded in a polar continuum, for type I phases, and vice versa for type II phases (see Fig. 1). This structure can be viewed as a hexagonal "honeycomb" of one species filled with rods of the other [1] (such as the honeycomb surface running between the chain ends of aligned, close-packed, worm-like reverse micelles).

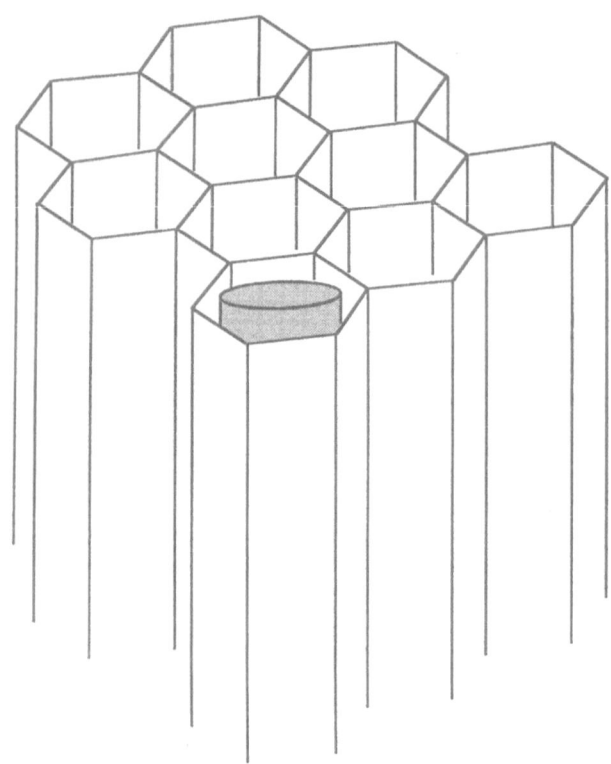

Fig. 1 Cartoon of the mesostructure of a conventional hexagonal mesophase. The shaded rods represent the hydrophobic/polar domains in type I/II H phases, respectively

Alternative structural topologies include bicontinuous rhombohedral "sponge" structures, and rhombohedral "mesh" structures. Sponges, meshes and honeycombs are examples of three-, two- and one- dimensional channels on one side of each monolayer surface, or on both sides of a central bilayer. Zero-dimensional, discontinuous volumes are a further possibility, found for example in discontinuous cubic (and possibly hexagonal) mesophases, or disordered arrangements, of reverse micelles (I_{II} and L_2, respectively). Here the surface separating the free-chain ends of opposed monolayers resembles a (reversed) closed cellular froth.

The analysis highlights the importance of mesostructural topology, often the poor relation of crystallographic symmetry in characterizing quasi two-dimensional soft materials. There are indications – to be discussed below – that the abstract tools of topology and differential geometry required to distinguish froths, honeycombs, meshes and sponges, shed some light on the structures of real systems, including lyotropic liquid crystals and related mesoporous inorganic materials formed by "templating" with amphiphiles.

A brief account of sponge topology and geometry

Both mesh and sponge mesostructures are known to occur in amphiphilic systems: Luzzati et al. first reported tetragonal and rhombohedral mesh structures in 1960s (in lyotropic and thermotropic systems) [2] and cubic rod networks, topologically identical to crystalline sponges in a number of lyotropic systems.

The identification of these structures with smoothly curved mesh and sponge surfaces came later, largely due to the work of Kåre Larsson and associates [1, 3, 4]. This identification has proved crucial to further understanding of the genesis and relative stability of these mesostructures, and we owe a significant debt to Larsson for that insight.

A geometrical investigation of mesh structures can be found elsewhere in this volume (see the article by Fogden et al.); here we shall discuss sponges in more detail. The generic form for a sponge is best illustrated by a remarkable engraving by Robert Hooke (see Fig. 2), drawn from observations of sea-sponges in the earliest optical microscopes. The sponge contains two interwoven labyrinths, and the structure consists of two continuous domains (one solid, the other pore), it is "bicontinuous". (This is not an essential restriction for sponges – the remarkable self-intersecting crystallographic surfaces of Koch and Fischer contain up to eight separate tunnel systems [5]!)

The simplest bicontinuous morphologies are generated by a single-sheeted surface, so that any two points separating the labyrinths, both lying in the surface, can be connected by a path that lies everywhere in the surface. The dividing surface is at least on average *hyperbolic*, i.e. its

Fig. 2 Engraving by Robert Hooke of a sea-sponge

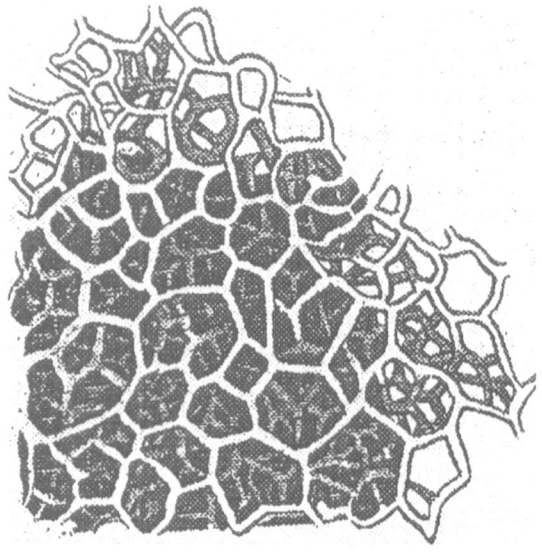

Progr Colloid Polym Sci (1998) 108:139–152
© Steinkopff Verlag 1998

surface-averaged Gaussian curvature:

$$\langle K \rangle \equiv \iint\limits_{\text{surface}} K \mathrm{d}a / \iint\limits_{\text{surface}} \mathrm{d}a \tag{1}$$

is negative. The Gaussian curvature is the product:

$$K = k_1 k_2 \tag{2}$$

of the principal curvatures, k_1 and k_2, of the most curved planar arcs, formed by orthogonal cross-sections through the surface. The geometrically simplest examples of hyperbolic surfaces are *minimal surfaces*, which are everywhere equally concave and convex, so that $k_1 = -k_2$ and the arithmetic mean, the mean curvature (H), vanishes.

The only minimal surfaces known whose surface-averaged Gaussian curvature is negative are the triply periodic examples, which have an underlying three-dimensional lattice. (Two-, one- and non-periodic minimal surfaces exist, but these have asymptotically flat ends that result in $\langle K \rangle$ vanishing.) Disordered minimal surfaces formed by "melting" triply-periodic examples, remain unknown, although the possibility that such structures exist in three-dimensional Euclidean space cannot be definitely ruled out. (We suspect that translationally and orientationally disordered examples are unlikely to be found, although quasi-periodic – and likely self-intersecting – minimal surfaces may emerge.)

An ever-swelling inventory of triply periodic minimal surfaces exists. The surfaces are distinguished by (i) whether or not they are free of self-intersections ("embedded"); (ii) their space group symmetry, often listed together with the subgroup symmetry resulting from the removal of all elements that exchange sides of the surface, known as the *black–white* subgroup (for embedded cases the symmetry of one labyrinth of the surface); (iii) the surface topology. The last measure of surface form conventionally refers to the topology of a translational unit cell of the surface, which is "compactified" by gluing opposite faces of the cell (separated by a lattice translation vector), thereby forming a boundary-free surface containing a number of distinct handles. The *genus* of the surface per unit cell refers to the number of handles in the compactified object [6].

Due to some flexibility in the choice of unit cell, a number of measures of surface genus can be found in scientific literature. From a differential geometric perspective, the genus of the surface is equal to that of the compactified surface contained within a primitive unit cell found in the black–white subgroup, referred to as the "lattice fundamental region" [5]. This genus must be at least three for all triply periodic hyperbolic surfaces (and at least two for doubly periodic mesh surfaces, which can be attained with uniform, but now non-zero, H). For structural studies of condensed materials decorating minimal surfaces, one is usually more interested in the genus per conventional unit cell of the full space group of the surface. Within a single space group (and, less commonly, a single group/subgroup pair), a number of triply periodic minimal surfaces can exist, distinguished by their topologies. Thus, even within this special class, identification of the symmetry is insufficient to characterize the structure. One must, in addition, determine the topology of the surface within that unit cell (in very rare cases this genus can also be manifested in multiple forms). In principle, the surface genus can be experimentally measured from swelling studies of the material, outlined later in this paper.

The simplest, and most celebrated, triply periodic minimal surfaces are then the cubic cases of genus three (per lattice fundamental region): the P, D and gyroid surfaces (of space group – black/white subgroup symmetries Im$\underline{3}$m-Pm$\underline{3}$m, Pn$\underline{3}$m-Fd$\underline{3}$m and Ia$\underline{3}$d-I4$_1$32 respectively). The labyrinths on either side of the P and D surfaces are identical (and six- and four-connected, respectively), while the (three-connected) labyrinths on either side of the gyroid are enantiomeric – one is right handed, the other left handed, as shown in Fig. 3.

Lower symmetry relatives of these cubic surfaces have been characterized. The one-variable families comprise rhombohedral and tetragonal gyroids, rG (R$\underline{3}$c-R$\underline{3}$2) and tG (I4$_1$/acd-I4$_1$22), respectively, and rhombohedral and tetragonal relatives of the P and D surfaces, rPD (R$\underline{3}$m-R$\underline{3}$m), tP (I4/mmm-P4/mmm) and tD (P4$_2$/nnm-I4$_1$/amd). Further degradations to orthorhombic and monoclinic relatives are also known [7–10].

Genus-three tetragonal (CLP; P4$_2$/mcm-P4$_2$/mmc) and hexagonal (H; P6$_3$/mmc-P$\underline{6}$m2) surfaces, whose labyrinth configurations are topologically distinct from those of the P, D or gyroid, have also been elucidated.

The simplest genus-four example is the cubic I-WP surface (Im$\underline{3}$m-Im$\underline{3}$m). This illustrates an important feature of "unbalanced" triply periodic minimal surfaces; the labyrinths on either side of the surface are quite different (one labyrinth is eight-connected, the other four-connected), and, unlike all the cases mentioned above, the I-WP surface does not divide space into equal sub-volumes, rather the partitioning is approximately in the ratio 0.536: 0.464 [11].

The Gaussian curvature of minimal surfaces necessarily varies over the surface, although it is everywhere strictly non-positive. These curvature variations, and differences in the global form of surfaces, mean that they differ from each other in their space-filling properties. A simple global measure of the space-filling features of these triply periodic sponges is the dimensionless surface to volume ratio, which we call the *homogeneity* index:

$$h \equiv A^{3/2}/(2(-\pi(1-g))^{1/2}V) \tag{3}$$

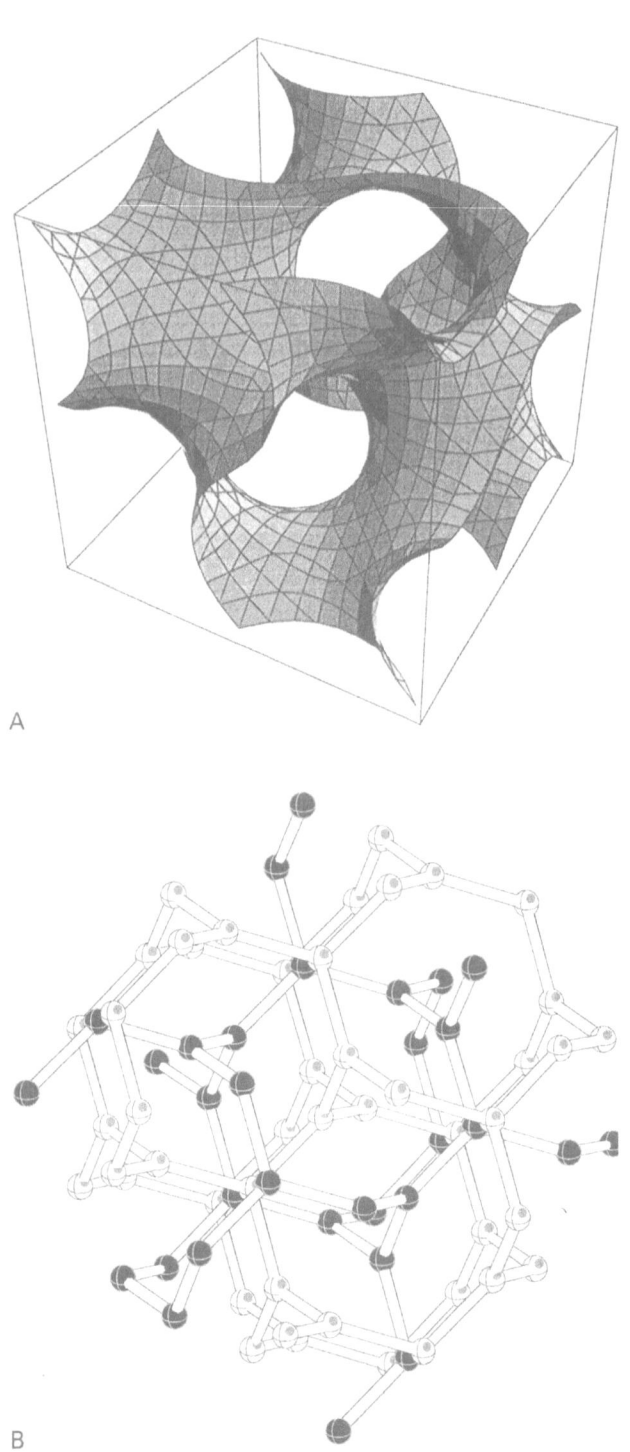

A

B

Fig. 3 (A) Unit cell of the cubic gyroid triply periodic minimal surface, (B) labyrinth graphs on either side of the gyroid (one with white nodes, the other black, showing several unit cells viewed almost down the [1 1 1] direction)

where A denotes the area of the surface within a repeat unit of volume V and genus g. This number affords a crude measure of the sponge geometry, since an ideal "homogeneous" sponge – with a single radius of (convex and concave) curvature at all points (which is unrealizable in three-dimensional Euclidean space[1]) – has a homogeneity index of exactly 3/4. (Note however, that one can only infer that the sponge is inhomogeneous from this index; if h is equal to 3/4 the sponge may be inhomogeneous to varying degrees, but with the same fortuitous balancing of surface areas and associated volumes.) Its values for the P, D and gyroid are 0.716346, 0.749844 and 0.766668, respectively.

The concept of homogeneity is a useful one, particularly relevant to the energetics of mesophase formation in soft materials (described in the next section). A direct measure of the inhomogeneity of Gaussian curvatures is given by their second moment over the sponge:

$$\mu_2 \equiv \langle K^2 \rangle / \langle K \rangle^2 \tag{4}$$

A key conjecture, which is appearing more reasonable as data accumulates, couples the Gaussian curvature inhomogeneity to the surface topology of the lattice fundamental region and the surface symmetry. We suggest that, for a given embedded triply periodic minimal surface, the inhomogeneity (μ_2) increases with increasing genus (adding handles), and "decreasing" symmetry (decreasing space group order, i.e. number of equivalent sites of a point in general position within a primitive unit cell of the group). Among these ordered surfaces, the genus-three cubic examples – the P, D and gyroid – are the most homogeneous ($\mu_2 = 1.218755$ for all three cases by virtue of their Bonnet relation), and can be described as the least frustrated immersions of a minimal surface sponge in our three-dimensional space.

Although a multitude of cubic minimal surfaces, of genus exceeding three, have been discovered in the past decade, the exact parametrizations have been established for only a few cases, including the I-WP, F-RD and C(P) surfaces, with $g = 4$, 6 and 9, respectively. The corresponding values of μ_2 are 1.48375, 1.61557 and 2.80449 [10], and relative to the P surface, exhibit the rise expected on increasing genus at fixed symmetry of the primitive unit cell (Pm$\bar{3}$m). The four values are plotted vs. the (negated) Euler characteristic $\chi = 2(1 - g)$ in Fig. 4, which also displays these higher topologies in one-eighth of their full unit cells. Three of the surfaces (sharing the space group Im$\bar{3}$m) lie almost exactly along the same line. That the F-RD falls significantly below (actually corresponding to the value for $\chi = -7$ on the line) may be a reflection of its

---

[1] Hilbert established that a singularity-free immersion of the pseudosphere, a surface of constant negative Gaussian curvature into three-dimensional euclidean space is impossible. It can, however, be achieved in spherical three-dimensional space and analytically in seven-dimensional flat (euclidean) space!

Progr Colloid Polym Sci (1998) 108:139–152
© Steinkopff Verlag 1998

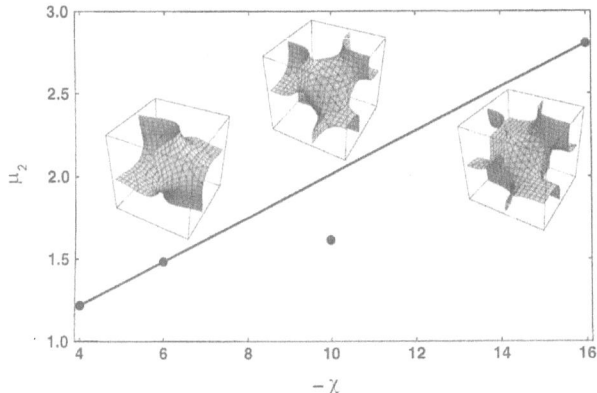

Fig. 4 Gaussian curvature inhomogeneity (see Eq. (4)) for the known examples of cubic minimal surfaces

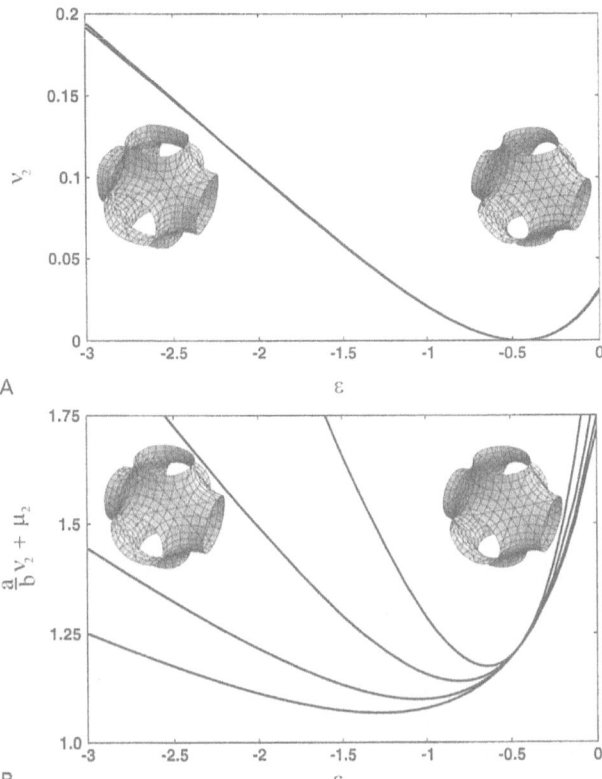

Fig. 5 Plots of (A) mean curvature, and (B) combined mean and Gaussian curvature, indices (see Eqs. (4) and (5)) over the one-variable family of surfaces in Eq. (6), as explained in the text

true face-centered symmetry (Fm3m). However, in the absence of exact mathematical prescriptions for other cubic examples (most notably the C(D) surface [12]), it is difficult to explore these deeper trends.

The trends in Gaussian curvature inhomogeneity on reducing symmetry, with genus fixed at its minimum value of three, are analyzed in the following section. Although μ_2 increases, again as expected, on distortion of the cubic P, D and gyroid surfaces, the heirachy and interrelations of these one-variable families become quite intricate. We do not yet know the relative importance of increasing genus vs. decreasing symmetry. It is interesting though, to note that these three simplest examples can be continuously transformed into one another, using only the rhombohedral and tetragonal degradations (preserving topology and vanishing H throughout), without ever exceeding the μ_2 level of the I-WP surface.

Before addressing the issue of sponge stability, a short note on the assumption of zero mean curvature is in order. The μ_2 value common to the three simplest minimal surfaces is by no means an absolute (or even a local) minimum for triply periodic hyperbolic partitions, although they probably represent the space groups and topologies within which the lowest values are attainable. To analyze these possibilities, some generalized measure of "curvature frustration", extending Eq. (4) to cases of non-zero H, must be introduced. In keeping with the use of dimensionless and intensive indices, one choice for balanced partitions (i.e. $\langle H \rangle = 0$) is the linear combination

$$av_2 + b\mu_2$$

where

$$v_2 = \langle H^2 \rangle / \langle -K \rangle \tag{5}$$

with appropriate (non-negative) weighting factors a and b.

As an illustration for the space group Im3m, consider zero-level surfaces of the general Fourier series for this symmetry, truncated at the two terms of lowest frequency, (1 0 0) and (1 1 1),

$$\cos 2\pi x + \cos 2\pi y + \cos 2\pi z$$
$$+ \varepsilon \cos 2\pi x \cos 2\pi y \cos 2\pi z = 0 . \tag{6}$$

Over the parameter interval $-3 < \varepsilon < 1$ the (balanced) surfaces share the P topology [13]. Figure 5a displays a unit cell of the surfaces for $\varepsilon = -3$ and 0, and the variation of v_2 intermediate to these. The minimum value is 4×10^{-6} at $\varepsilon = -0.452$, for which Eq. (6) is an exceedingly accurate approximation to the exact P surface. The homogeneity index at this minimum agrees with the P value to all six significant figures given above. Moreover, the rise in both of these indices out to the endpoints of the plot remain closely correlated; the second, almost coincident, curve in Fig. 5A is the approximation

$$v_2 \approx 8.85(h - 0.716346) . \tag{7}$$

These simple results reemphasize a clear warning, namely that surfaces can appear close to minimal judging by relatively insensitive measures such as the homogeneity index (i.e. area per amphiphile headgroup), or from electron microscope textures, and yet carry substantial deviations in mean curvature.

Figure 5B plots the variation in μ_2 (the lowest curve) over the same parameter range. This index decreases from 1.219 at $\varepsilon = -0.452$ (again very close to the P value) to reach a minimum of 1.068 at $\varepsilon = -1.278$ (at which $h = 0.721$ and $v_2 = 0.041$). The figure also shows, in ascending order, the combination in Eq. (5) for weighting ratios $a/b = 1, 3.71$ (for which values from the v_2- and μ_2-minimizing surfaces are equal) and 10. Although this ratio would be hoped to be quite large (to avoid bringing into question the validity of a curvature expansion formalism for the free energy), finite deviations from the P minimal surface are expected. The values of the associated membrane bending constants, k_C and κ_3, are rarely known in practice, but for V_{II} phases of ionic amphiphiles in excess 1:1 electrolyte the electrostatic contributions can be derived from the Poisson–Boltzmann theory [14, 15]. For a genus-three structure with lattice parameter α, the ratio a/b varies between $0.16(\alpha/\lambda_D)^2$ at low surface charge (in which case the intrinsic, non-electrostatic, bilayer contributions would dominate) and $0.07(\alpha/\lambda_D)^2$ for high charge, where λ_D is the usual Debye screening length.

Figure 5B also displays the surfaces corresponding to these μ_2- and v_2-minimizing values of ε (at left and right, respectively). The two surfaces are, by eye, almost indistinguishable from each other, or from any surface in the interval $-1.278 < \varepsilon < -0.452$ in which the minimum of their linear combination must reside. This problem of resolution is not surprising since the symmetry embeds the same framework of straight lines in each surface; indeed, this is the reason for the success of the two-term approximation to the P surface. Preliminary investigations reveal that addition of the next two Fourier terms ((2 1 0) and (2 2 1)) to Eq. (6) only serves to slightly reduce the minimum attainable μ_2, to around 1.063, with a concominant increase in v_2 to around 0.046. Accordingly, this new optimal surface appears identical, for all practical purposes, to that in Fig. 5B. Thus truncations such as Eq. (6) can offer simple, yet quite accurate, representations for global optimization of H and/or K distributions within the space group Im$\bar{3}$m. The two-term analogue for Pn$\bar{3}$m symmetry can be readily constructed and yields equally accurate approximations to the D surface.

Determination of the triply periodic hyperbolic partition(s) bearing the most homogeneous distribution of Gaussian curvatures, and the strength of deviations in v_2 from zero as μ_2 approaches unity, pose interesting problems which shall not be further addressed here. Indeed, such problems can give rise to a multitude of extensions, related to the choice of integration measure. For

example, if the definitions were to be generalized to the monolayer surface average $\langle |H - H_0|^m \rangle$ or $\langle |K - K_0|^m \rangle$, the minimizing solution could switch from a smooth to a piecewise-smooth surface (cylindrical struts intersecting sharply along junction lines) as the power m is decreased below unity.

Given that some, albeit probably not severe, departures of ordered sponge partitions from zero mean curvature are inevitable, it could be argued that minimal surface models should be replaced by the general tool of Fourier decomposition for the given space group. The latter approach is especially suited to problems which involve space-filling foliations, such as detailed calculations of chain segment distributions in block copolymer melts or director fields in thermotropic liquid crystals. However, for lyotropic liquid crystals, and related systems, the natural language is differential geometry (even in the presence of direct interactions across solvent domains) and the natural basis for sponges are the triply periodic minimal surfaces. Although the mathematical elegance of an exact parametrization in terms of complex analytic functions is a property unique to minimal surfaces, Gackstatter derived an analogous (but non-analytic) representation for, in principle, any surface [16]. While this new formalism involves non-linear integrability conditions, these can be linearized and readily solved by perturbing from the analytic minimal surface. Moreover, for relatively slight deviations from minimality, only the first-order corrections are required (e.g. the slope of the linear dependence of volume fraction on uniform H close to zero). Similar schemes have also been considered by Nitsche [17] in the course of optimising surface functionals of the form

$$\iint (\alpha + \beta H^2 - \gamma K + \delta K^2)\, da \qquad (8)$$

(with or without external constraints). The minimal surfaces were found to define the central backbone about which the stationary states can be accessed by such perturbations, provided δ is not too large.

Thus, the central importance of triply periodic minimal surfaces of genus three (per lattice fundamental region), and cubic, tetragonal and rhombohedral symmetries, to hyperbolic forms is clear.

Relative stability of sponges

Assuming the generality of the trends illustrated above, it follows that, for a variety of free energy functionals based on elastic bending energies about a preferred curvature, or preferred function of curvatures, the most stable hyperbolic configurations can be represented by the P, D and gyroid minimal surfaces, together with their rhombohedral and tetragonal distortions. We ignore entropic contributions, largely because their relative input to free energies remains unknown, apart from some preliminary results [18]. Indeed, the hypothesis that these highly symmetric surfaces give access to an optimal solution affords a simple explanation of the mesocrystallinity in many soft systems. Their long-range order need not be induced by global interactions, rather as a result of local curvature requirements alone, which are best satisfied by crystalline sponges, since molten bicontinuous geometries appear to form non-minimal surfaces. (Note that molten sponges are believed to occur in dilute "sponge" (L3) mesophases. In that case, the membrane curvature is very low, due to the high solvent fraction, and there is negligible difference between curvature variations in periodic minimal surfaces, and related disordered hyperbolic surfaces, so the latter

presumably form due to their higher entropic content. The disordered surfaces are nevertheless similar to minimal surfaces [19].)

The idea that hyperbolic mesostructured materials form in response to local curvature requirements is rather more than Platonic idealization of the complex physics governing self-assembly of these materials. That is suggested by an apparent underlying common relation between lattice parameters of isotropic mesocrystalline molecular materials (and some covalent atomic materials), tabulated in Table 1. The ratio of cell edges measured in neighboring phases of some silicates, lyotropic and thermotropic liquid crystals (including blue phases, exhibiting only orientational order) in this variety of soft molecular systems is precisely that expected assuming the phases consist of

Table 1 Distinct phases of some liquid crystals related by the Bonnet transformation of underlying two-dimensional hyperbolic surfaces

Material	Phase	Symmetry	α (Å)	# Atoms/ molecules per unit cell	Related sponge	Ratio	Bonnet ratio	Difference
Lyotropic liquid crystals								
Monoolein–water (lipid)	Bicontinuous cubic	Pn$\bar{3}$m Ia$\bar{3}$d	84.5[a] 140.8[a]	~800 ~3600	D gyroid	1.66	1.5757	5%
DDAB-cyclohexane-water (Surfactant)[b]	Bicontinuous cubic	Im$\bar{3}$m Pn$\bar{3}$m	116[c] 92[c]	~600 ~800	P D	1.26	1.2793	2%
Thermotropic liquid crystals								
R-biphenyl carboxylic acid[d]	Smectic D	Ia$\bar{3}$d Pn$\bar{3}$m[e]	170.8[f] 111.9[f]	1650 3150	gyroid D[e]	1.53	1.5757	3%
CB15-E9[g]	Blue Phase I Blue Phase II	I4$_1$32 P4$_2$32	4970[h] 3120[h]	~3 × 10^8[i] ~7 × 10^7[i]	gyroid[j] D[j]	1.59	1.5757	1%

[a] Hyde ST, Andersson S, Ericsson B, Larsson K (1984) Z Kristallogr 168:213.
[b] DDAB = didodecyldimethyl ammonium bromide, a double-chained cationic surfactant.
[c] Barois P, Hyde ST, Ninham BW, Dowling T (1990) Langmuir 6:1136.
[d] The molecules forming the Ia$\bar{3}$d and Pn$\bar{3}$m phases differ slightly: $X = NO_2$ for the Ia$\bar{3}$d phase, and CN for the Pn$\bar{3}$m phase. The generic acid is:

$$H_{37}C_{18}-O-\overset{\overset{X}{|}}{C_6H_3}-C_6H_4-\overset{\overset{O}{\|}}{C}-OH$$

[e] Levelut and Clerc index this phase to the Im$\bar{3}$m space group; an identification that leads to a number of anomalies reported in their paper. Here we *assume* Pn$\bar{3}$m symmetry, consistent with the D-surface, which yields excellent accord with the ratio of lattice parameters expected by the Bonnet transformation.
[f] Levelut A-M., Clerc M, Structural investigations on the "Smectic D" and related mesophases, preprint, 1997.
[g] CB15 is a strongly polarizable chiral molecule:

$$CH_3-CH_2-\overset{\overset{CH_3}{|}}{C}{}^*H-CH_2-C_6H_4-C_6H_4-CN;$$

E9 is a (nematic-forming) additive.
[h] Estimated from "Bragg wavelengths" reported in: Heppke G., Jérôme B, Kitzerow H-S, Pieranski P (1989) J Phys (France) 50:2991; assuming the refractive index is unity.
[i] Calculated assuming a specific volume of $1 \, cm^3 \, g^{-1}$.
[j] There has been some doubt about the identity of the BP structures, although this analysis makes it clear that the BP's can be described by the gyroid and D-surface, proposed by Pansu and Dubois-Violette (1990) In Dubois-Violette E., Pansu B (eds) The Geometry of Interfaces Colloque de Physique, Vol C-7, Editions de Physique: pp C7-281, C7-296.

interfaces whose Gaussian curvature (strictly $\langle K \rangle$), and average radius of curvature, is conserved.

A simple measure of the average radius of curvature of a single-sheeted cubic sponge, characterized by a lattice parameter of α, a homogeneity index, h, and Euler characteristic, χ, is, from Eq. (3):

$$R \equiv \langle -K \rangle^{-1/2} = \left(\frac{h}{-2\pi\chi} \right)^{1/3} \alpha \ . \tag{9}$$

If there is a preferred radius of (hyperbolic) curvature, conserved between parent and product phases, the resulting scaling of lattice parameters is dependent on $h^{1/3}$, which is itself dependent on the details of the surface geometry, and the intrinsic topology, $(-\chi)^{-1/3}$, according to this equation. In the case of the isometric P, D and gyroid surfaces, this "hidden variable" of curvature has been recognized earlier, due to the special, curvature conserving, Bonnet relation that links all three surfaces [4].

The examples, listed in Table 1, are all modelled by these three ubiquitous surfaces, and their relative lattice dimensions can be at once explained by an underlying requirement of preservation of Gaussian curvature in parent and product phases at each point on the interface. Without this locality assumption, there is little reason to expect universal lattice parameter ratios. Were the interfaces not close to minimal surfaces, this scaling of lattice parameters would also fail to follow (excepting an unlikely coincidental equivalent scaling of surface-to-volume ratios).

These examples demonstrate the validity of the notion that self-energies of assembled amphiphiles in lyotropic liquid crystals can be modelled – at least to a first approximation – by an elastic bending energy, introduced by

Helfrich [20], minimized at some "preferred" curvature. The exact form of this bending energy functional remains uncertain. One possible form is that based on the recognition that constituent amphiphiles in the assembly have a preferred molecular shape, determined by a balance of competing interactions in the aggregate, and simply quantified by the molecular shape parameter:

$$s \equiv v/al$$

introduced by Israelachvili et al. [21]. The shape parameter is coupled to the geometry of the resulting aggregate, and the mean and Gaussian curvatures of an interface running between hydrophobic and hydrophilic regions are related to this parameter by:

$$s = 1 + Hl + Kl^2/3 \ . \tag{10}$$

The equality holds provided the amphiphiles are packed so that their director lies normal to the interface, and there is no collective tilt of the molecules (as sometimes is found in semi-crystalline "gel" phases, and "ripple" phases), and the monolayer is of constant thickness (l). This relation allows a generic "phase diagram" to be constructed, relating the shape parameter to the amphiphile (more strictly, the hydrophobic portion of the amphiphile) volume fraction for the simplest homogeneous geometries of zero, positive and negative curvatures. This diagram is shown in Fig. 6 for a range of idealized structures.

In particular, the shape parameter of an amphiphilic type II bilayer, folded onto a minimal surface, is coupled to the volume fraction by the approximate expression:

$$\varphi = 4\sqrt{3}hs \sqrt{\frac{s-1}{(3s-1)^3}} \ , \tag{11}$$

Fig. 6 Plots of the variation of shape parameters with amphiphile volume fraction for spherical, cylindrical, mesh (shaded region) and sponge mesophases (type I/1 and II/2 structure)

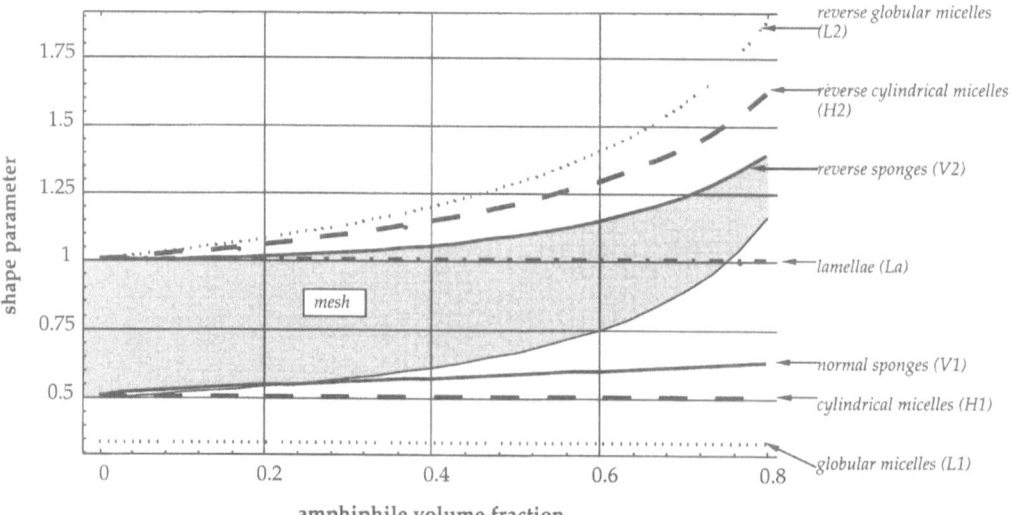

which is exact if the bilayer is of uniform thickness ($2l$), so that the interfaces running through the head groups on either side of the minimal surface are parallel. The assumptions of interfacial homogeneity and the parallel surface construction are less suited to some of the other phase types, namely V_I, mesh, H_{II} and L_2.

Neglecting, for now, energy costs associated with inhomogeneities, which may preclude the formation of a particular geometry, it follows that the sequence of phases expected for amphiphiles whose shape parameter exceeds unity is, on water dilution, given by the appropriate horizontal cut (at constant s) through Fig. 6. The generic sequence on increasing dilution is:

sponge monolayer → mesh → sponge bilayer (i.e. V_{II} if cubic) → rod-shaped reverse micelles (H_{II} if ordered into the least frustrated hexagonal array) → reverse globular micelles (I_{II} if cubic, L_2 if molten).

More detailed predictions rely on the form of the bending energy functional within this model assuming a preferred molecular shape $s = s_p$. Supposing an elastic form, with modulus κ', the bending energy per unit area of the membrane is

$$E_{bend} = \frac{\kappa'}{l^2 A} \iint (s - s_p)^2 \, da, \qquad A \equiv \iint da. \tag{12}$$

For a bilayer sheathing a minimal surface, Eq. (12) can be rewritten, assuming the parallel surface model in Eq. (10), to read [22]:

$$E_{bend} = \frac{\kappa}{Al^2} \iint (Kl^2 - K_p l^2)^2 \, da \tag{13}$$

for small curvatures ($\langle -Kl^2 \rangle \ll 1$), with K and K_p now denoting the Gaussian curvature, and its preferred value, for the bilayer midsurface (and $4\kappa'/9$ relabelled as κ). This can be recast to give a minimum bending energy per amphiphile (of chain volume v_s):

$$e_{min.bend} \approx \kappa \frac{v_s}{l^3} (K_p l^2)^2 \left(1 - \frac{1}{\mu_2}\right). \tag{14a}$$

This energy minimum is realized at amphiphile fraction:

$$\varphi_{min.bend} \approx 2h \sqrt{\frac{-K_p l^2}{\mu_2}}. \tag{14b}$$

This equation limits the variety of sponges expected to occur in sponge-shaped bilayers – depending on the magnitude of the elastic modules (κ), only those geometries which exhibit small curvature frustrations may occur. Central to these are the cubic genus three surfaces (which are well-known in lyotropic systems as the bicontinuous cubic phases [2, 23]), and perhaps tetragonal and rhombohedral relatives. The fine details of relative stabilities are dependent on the variation of membrane thickness with volume fraction. Possible variations are canvassed more fully elsewhere [10].

For simplicity, assume that the thickness remains fixed, and the curvature remains small. Plots of the variations of this elastic bending energy minimum with composition are shown for the exact minimal surface geometries in Fig. 7. In order of increasing water content, the model implies the following phase sequence, assuming sufficiently small κ (cf. thermal energy, kT) to enable the least frustrated structures to form:

T and R **mesh** structures whose 2-D correlations for a turbostratic stacking of layers give hexagonal and square lattices, or possible three-dimensional crystalline stackings of layers to give the R$\underline{3}$m and I422 symmetries characteristic of the T and R mesophases [1, 24]

    gyroid (cubic) - >
    tG OR rG (the latter is marginally less favored under the assumptions adopted here) - >
    D (cubic) - >
    rPD - >
    P (cubic) - >
    tP.

Once the homogeneity index drops below about 0.6, the formation of a classical H_{II} mesophase, consisting of a hexagonal packing of infinite cylindrical, discrete, reverse micelles – a hexagonal honeycomb of amphiphilic chains – is possible for volume fractions of amphiphile exceeding about 30% (type II phases).

These calculations are noteworthy in two respects. Firstly, all "ring-like" minimal surfaces (e.g. rPD, tP, H) exhibit a pair of solutions for a range of values of the axial ratios of lattice parameters (c/a) [25]. Further, for all families here, a pair of surfaces, with distinct axial ratios, exists over a range of $h/\sqrt{\mu_2}$ values (and thus both candidates can be realized in sponge-like bilayers at a given volume fraction of amphiphile). Usually, one solution is more frustrated (larger μ_2) than the other, and the optimal solution for the bilayer is a single well-defined structure. However branches of the rG surface are only marginally different in this respect, while the branches of the tD surfaces show almost identical curvature variations, so both cases lead to the possibility of distinct structures – of the same symmetry and topology but different axial ratios – being equally favorable. Secondly, in the vicinity of the cubic surfaces, all lower symmetry degradations show relatively shallow variation of h and μ_2 with axial ratio. Since the (assumed fixed) volume fraction of amphiphile scales with $h/\sqrt{\mu_2}$, fluctuations in the axial ratio are expected, which can occur without inducing compositional variations, or bending energy costs. In particular, we may

volume fraction of amphiphile (scaled by curvature)

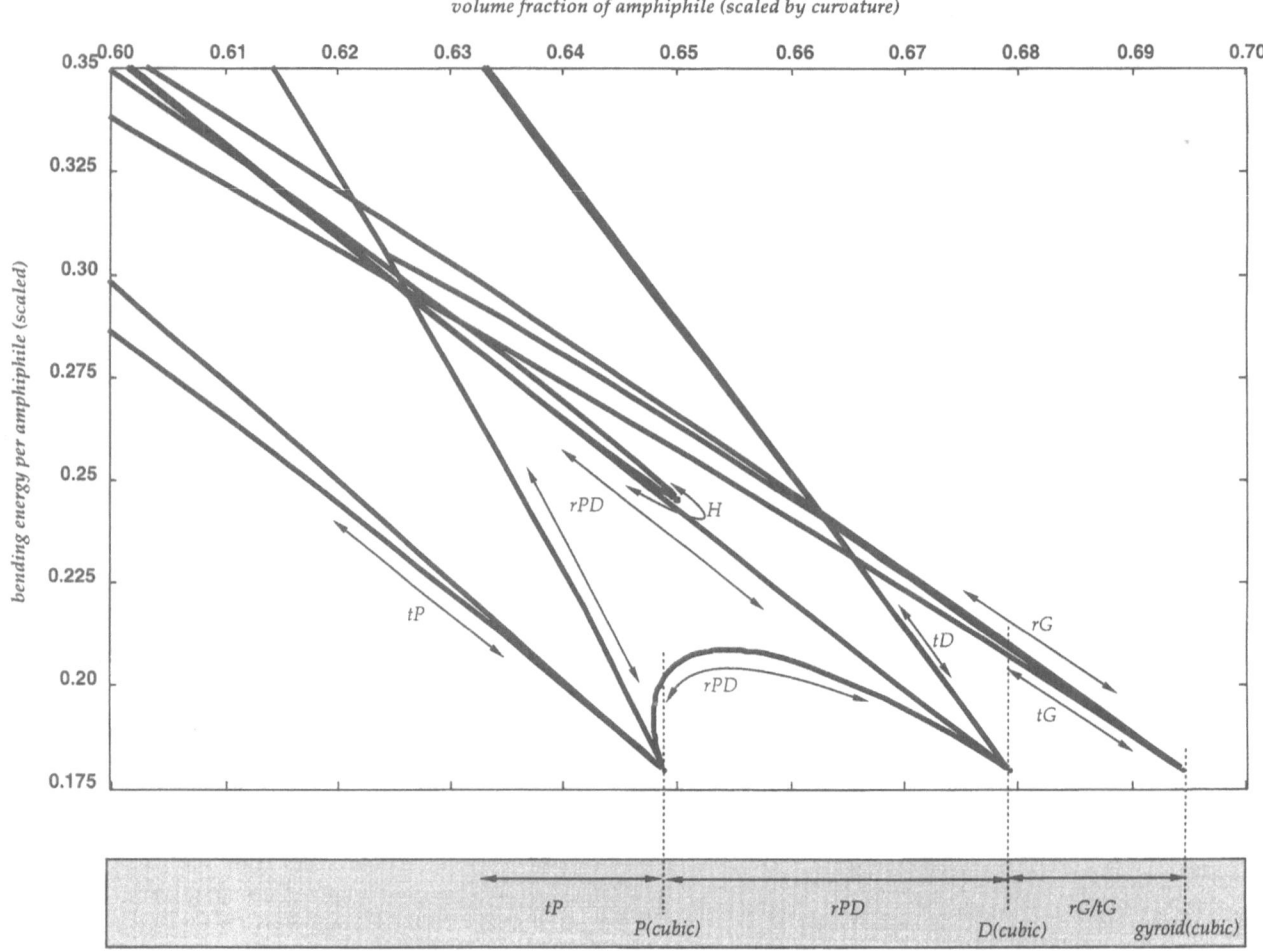

Fig. 7 Plots of the scaled elastic bending energy minimum, $(1 - 1/\mu_2)$ vs. scaled amphiphile volume fraction $(h/\sqrt{\mu_2})$ (see Eqs. (14a) and (14b)) for the most homogeneous examples of triply periodic minimal surfaces: the cubic P, D, and gyroid, rhombohedral distortions of the gyroid (rG – two almost superposed branches, which diverge at higher bending energies) and the P and D surfaces (rPG – two branches from the cubic D, two from the cubic P), tetragonal distortions of the gyroid (tG – two branches, the upper almost coinciding with rG), P (tP – two diverging branches) and D surfaces (tD – two superposed branches), together with the hexagonal H surface (one γ-shaped loop). All other known minimal surfaces have higher second moments. The shaded box shows the sequence of bicontinuous amphiphilic sponge mesophases expected on dilution, assuming their stability is governed by the Gaussian curvature variations of the bilayer alone, the bilayer thickness remains constant, and the curvature is small, deduced from the main plot. These sponges are flanked by mesh and honeycomb (classical hexagonal) structures at higher and lower amphiphile fractions respectively

expect such fluctuations to occur in tG and rG members in the vicinity of the cubic gyroid phase, and likewise for the rPD in the vicinity of the cubic P and D phases.

This model is primitive, in that it assumes phase behavior is set solely by a preferred molecular shape. Nevertheless, it does suggest that anisotropic bicontinuous phases of rhombohedral and tetragonal symmetries (perhaps washed out by fluctuations) may be expected at concentrations approaching those where the isotropic bicontinuous cubic phases are found.

An alternative model for mesophase stability has been constructed, which relies on the amphiphilic aggregate

being subject to an elastic bending energy per unit area involving the mean and Gaussian curvatures (H_n and K_n, respectively) measured at a "neutral" surface (not the central surface, but displaced some distance l_n towards the head-groups) of the form [26]:

$$E_{\text{bend}} = \frac{1}{A}\left(\kappa \iint (H_n - c_p)^2 \, da + \kappa^* \iint (H_n^2 - K_n) \, da\right). \quad (15)$$

This model differs from that outlined above, since it assumes a preferred (linear) curvature (c_p) for an interface traced along the inextensible neutral surface, located somewhere between the head groups and the chain ends in the amphiphilic sheets. At first sight, this model may be expected to yield quite different phase behavior from the previous one. For example, hyperbolic structures suffer an energy cost. Closer analysis shows this expectation to be false. Indeed, some manipulation allows the minimum for Eq. (15) to be written (assuming μ_2 is close to one, valid for a dilute (type II) system, or a homogeneous concentrated system, and parallel neutral surfaces):

$$e_{\text{min·bend}} \approx \kappa \frac{v_s}{l^3} \left(\frac{l}{l_n}\right)^2 \left((K_p l_n^2)^2 - \frac{(\kappa^*/\kappa)^2}{4(1 + \kappa^*/\kappa)} \frac{1}{\mu_2}\right). \quad (16)$$

Here l and K_p are defined as in Eq. (14a), and μ_2 can likewise be taken as its minimal surface value (since the deviation from its neutral surface value is negligible for sufficiently low amphiphile concentrations, and moreover, should not be retained without a consistent extension of Eq. (15) to include the corresponding higher-order curvature terms). Comparing Eqs. (14a) and (16), both models lead to similar scaling within this dilute regime. So the requirement of a preferred molecular shape is functionally equivalent to that of preferred linear curvature, aside from the offset due to the hyperbolic penalty of the latter. This model too ensures that the phase sequence lies within the structural sequence outlined above, viz. bicontinuous bilayer sponges (V_{II} phases), followed by a (hexagonal) honeycomb (H_{II} phase) and inverse micellar froth, (e.g. cubic discontinuous, I_{II}) at higher water contents.

Experimental observations of hexagonal and related mesophases

The theoretical scenario outlined in the previous section, is a direct consequence of the adoption of a preferred intrinsic average geometry for the amphiphiles. If it is valid, the identification of reversed hexagonal mesophases with a hexagonal honeycomb may be faulty, given the range of low (bending) energy rhombohedral sponge and mesh structures whose scattering spectra are not dissimilar to that of a two-dimensional classical hexagonal phase.

Alternatively, if these identified systems are indeed classical H_{II} phases, local bending energy models do not adequately capture the physics underlying mesophase formation in these lyotropic systems. That possibility cannot be ruled out, although a wealth of evidence can be found that is consistent with the curvature approach.

There are, however, already clues that some hexagonal mesophases are topologically more interesting than honeycombs.

Kåre Larsson has published a report of two adjacent H_{II} phases in the monoolein lipid–water system [27].

While theoretical models can undoubtedly be advanced demonstrating double minima in the energies of hexagonal arrays of cylinders at distinct lattice spacings (as has been done to explain coexisting lamellar mesophases), distinct hexagonal phases are equally likely to be the result of distinct topologies within each phase.

A number of amphiphile–water systems display the sequence $H_{II} \to V_{II}$ on dilution, some mixed systems have the remarkable sequence $H_{II} \to L_\alpha \to V_{II}$ [26]. The latter sequence is "reentrant" if one accepts the conventional topological prescriptions for the H_{II} and L_α mesophases, changing from inverse curvature, to zero, to inverse. An alternative explanation of these phases, such as identification of the hexagonal and lamellar phases as mesh or sponge structures of underlying (fluctuating) rhombohedral or tetragonal symmetries, would remove the theoretical challenges these observations pose. In that case their phase behavior would remain consistent with the expectations of the curvature models. (It has been claimed that the honeycomb (H_{II}) $\to$ sponge (V_{II}) sequence is consistent with the functional in Eq. (15) [26]. We are unconvinced by this argument. One can generate these transitions on increasing dilution, but only for very low homogeneity indices for the V_{II} phases (less than ca. 0.6), in which case the bending energy is likely to be high due to Gaussian curvature inhomogeneities.)

A further indirect clue to the uncertainty surrounding a honeycomb description of "hexagonal" phases is afforded by epitaxial data for $V_{II} \to H_{II}$ transitions, analyzed collectively in Ref. [28]. These data can be compared with expected lattice parameter ratios assuming the systems have similar shape parameters in parent and product phases. (This analysis is a simple generalization of the technique described by Eq. (9) for sponge–sponge epitaxial relations.) The epitaxial data is inconsistent with this hypothesis for type II systems, and further analysis shows that a significant density of negative disclinations – inducing negative Gaussian curvature, and handles between the one-dimensional channels of the classical hexagonal honeycomb – are consistent with the epitaxial lattice parameter ratios [29].

How then can one expect to identify more correctly the topology of these mesophases, given that their symmetry may be a poor measure of fluctuating assemblies? One technique that is available, and has long been used, is swelling measurements. Those data, capturing the functional variation of diffraction spacings (d) with concentration (φ), can be used to directly probe the interfacial topology of the scattering film, although care must be exercised. It is untrue, for example, that the conventional swelling dependence of smectic phases:

$$d \sim \varphi^{-1} \quad (17)$$

is a signature of defect-free lamellar interfaces. The same relation applies to swelling of reversed phases of *any* topology – including sponge, mesh, honeycomb and froth – provided the membrane swells while the cross-sectional area per amphiphile (measured at the free chain ends) remains fixed [19].

It is convenient to generalize the shape parameter to characterize any AB composite structure, whose A and B domains are assumed to be bounded by a reasonably homogeneous (and sharp) interface. Two parameters can be defined, pointing towards either side of the dividing surface: one for each domain. These parameters relate the volumes in A and B domains, v_A and v_B, that can be ascribed to a patch of the interface, of area a, with radii of curvature l_A and l_B towards the A- and B-sides of the interface:

$$s_A \equiv \frac{v_A}{al_A} \text{ and } s_B \equiv \frac{v_B}{al_B} .$$

Suppose the composite is close to homogeneous (and thus well-characterized by a single length scale). If the composite changes its fraction of A, curvatures of the interface must scale with volume fraction as [30]:

$$\varphi_A^{s_A} \sim \frac{l_A}{(l_A + l_B)} \sim \frac{l_A}{d} \text{ and } \varphi_B^{s_B} \sim \frac{l_B}{d}, \tag{18}$$

where the exponents, s_A and s_B, are precisely equal to the shape parameters of A and B domains, whose dependence on topology and volume fraction (assuming homogeneity) is plotted in Fig. 6. It is evident that this swelling exponent is just larger than unity, regardless of topology, for reversed (type II) phases (except in very concentrated lyotropes), whereas it varies dramatically with topology for "normal" (type I) phases, for which the shape parameter describes the "inner" side of the interface. In order to determine the interfacial topology, it is preferable to use the equation for the analogues of type I systems, which means one needs to know the thickness ($\sim l$) of the "inner" component that contains the centers of curvature of the interface [30].

This relation cannot be applied with certainty unless the two independent length scales, l_A and d can be determined. That is often impossible in amphiphilic systems, where scattering studies yield d only. Indeed, the *assumed* mesostructure is usually used to estimate the other length l, equal to the membrane thickness. Were the membrane thickness to be determined independent of structural assumptions, the swelling technique would offer an unequivocal topological signature, and allow direct determination of "defect" densities in these systems. (In some cases, it can be determined, leading to measures of homogeneity index and membrane topology per unit volume for sponges, both cubic and disordered [30].)

This technique is not confined to lyotropic amphiphilic assemblies. It is applicable to any two-phase composite, provided a sharp interface between the phases can be traced within the material. For example, porous materials can be considered as composites (of solid and pore), and the analysis offers some hope of characterizing the pore topology, of central importance to these materials.

Mesoporous inorganic materials, which are synthesized in the presence of "templating" surfactant–water liquors, are closely related to lyotropic systems. These materials exhibit a variety of mesostructures, including hexagonal, lamellar, mesh and cubic sponge morphologies [31–34], and their form at the mesoscopic scale is likely to be moulded by the templating amphiphiles. Temporal fluctuations found in soft matter are suppressed in these covalently polymerized materials, and a number of new mesophases have been detected from diffraction alone, including novel mesh phases, the cubic gyroid phase, and a possible rhombohedral relative (rG) in the midst of a $V_I \rightarrow H_I$ transition [35]. A large proportion of these products are however less ordered (with frequently less than five small-angle "diffraction" peaks), and are routinely described as lamellar and honeycomb phases.

That description often relies on electron microscopy (a technique so far largely inaccessible to soft materials). Despite the ease with which these solid mesostructured materials can be studied using EM techniques, mesostructural topology remains (in our opinion) uncertain, (Like diffracting X-rays, EM operators invariably select those regions of highest geometric correlations, and ignore other, less correlated regions, which may in fact offer more structural information!) The swelling analysis affords an explicitly topological probe, distinct from structural snapshots offered by conventional EM images or isolated small-angle scattering spectra.

A good example of this structural subtlety is afforded by "hexagonal" FSM materials, synthesized with a suite of chemically similar surfactants of varying chain lengths from C8 to C18 [36]. In all cases, TEM images are suggestive of a honeycomb morphology, and small-angle powder patterns reveal 2–4 peaks, consistent with the two-dimensional hexagonal lattice. The thickness of the inorganic film (l) in these materials can be estimated by electron microscopy, and the film is usually sufficiently correlated to yield small-angle diffraction peaks (d). Conventional porosity measurements together with the silica membrane solid density (close to that of fused quartz) allow estimation of φ. These data, plotted in Fig. 8, fit well the power-law swelling form of Eq. (18), provided the exponent is ca. 0.8. That means these FSM materials are more likely to be sponges, or possibly meshes (see Fig. 6), than honeycombs (whose exponent is 1/2), or froths (1/3).

Progr Colloid Polym Sci (1998) 108:139–152
© Steinkopff Verlag 1998

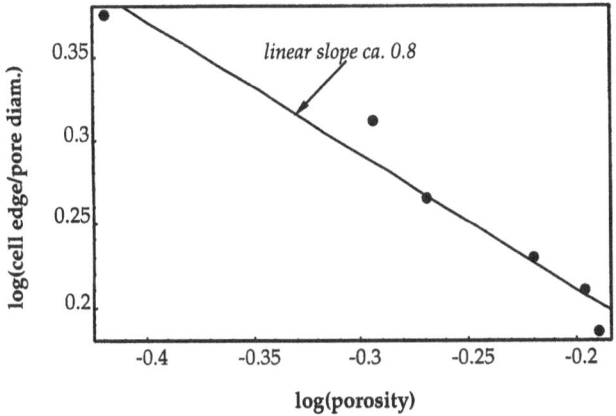

Fig. 8 Log–log plot of scaled repeat spacings vs. pore volume fractions for a range of mesoporous silica "hexagonal" materials (FSM-16), synthesized by Inagaki et al. from kanemite layer-silicate precursors, using cationic single-chain surfactants whose chains contain 8, 10, 12, 14, 16 and 18 carbon atoms [36]. The fit implies an average shape parameter characteristic of sponges (cf. V_I curve, Fig. 6), rather than a honeycomb (H_I, $s = 0.5$)

tures require only small changes in amphiphile molecular dimensions to form [19]. Hexagonal mesophases, of space group $P6_3/mmc$, have been reported recently in a non-ionic amphiphile–water system [37], and in a mesoporous silica film grown in an ionic amphiphile–water system [38]. This symmetry is characteristic of a hexagonal closed packed array of globular micelles, however it is also that of relatively low energy triply periodic H surface (see Fig. 7). Both reports have assumed the former structure, but the possibility of a bicontinuous structure partitioned by the H surface cannot be a priori excluded.

To date only a few examples of hexagonal phases that do not form classical honeycombs are characterized. However, given the variety of distinct topologies available to these materials in the vicinity of honeycombs, including froth, mesh and sponge morphologies, it is likely that more will be found. One can generalize this concern, which extends to all mesophases. Similarly, the identification of one-dimensional smectic order need not imply the phase to be a classical lamellar structure. In particular, the possibility of fluctuating rhombohedral (rPD, rG) and tetragonal (tP, tD, tG) sponges, and mesh structures, exhibiting hexagonal or lamellar-like diffraction patterns may lead to incorrect simplification of amphiphile mesostructure.

The issue of mesostructure in these materials, subject to fluctuations, is most importantly a topological one. Over-reliance on geometric correlations, and a pseudo-crystallographic approach, may be dangerously misleading, and disguise a plethora of fascinating structures available to Nature.

Conclusions

We restate the possibility that sponges (discrete) micellar phases, and froths may exhibit similar diffraction patterns. Significant topological differences can arise with little change apparent from conventional structural probes such as scattering. Bicontinuous and discontinuous cubic struc-

References

1. Larsson K (1989) J Phys Chem 93:7304
2. Luzzati V, Gulik-Krzywicki T, Tardieu A (1968) Nature 218:1031
3. Larsson K, Fontell K, Krog N (1980) Chem Phys Lipids 27:321
4. Hyde ST, Andersson S, Ericsson B, Larsson K (1984) Z Kristallogr 168:213
5. Fischer W, Koch E (1996) Phil Trans Roy Soc Lond A 354:2105
6. Hilbert D, Cohn-Vossen S (1952) Geometry and the Imagination. Chelsea, New York
7. Fogden A, Hyde ST (1992) Acta Cryst A48:442
8. Fogden A (1992) Acta Cryst A48:575
9. Fogden A, Hyde ST (1992) Acta Cryst A48:575
10. Fogden A, Hyde ST, in preparation
11. Anderson DM, PhD thesis, University of Minnesota

12. Koch E, Fischer W (1990) Acta Cryst A46:33 (For the C(D) surface, the flat point along $3m$ is not of order 4 as listed in their Table 1, but rather of first order, surrounded by a further three first-order flat points. This generalization, while complicating the parametrization, is necessary for the minimal surface to globally lock into its space group.)
13. Fogden A, Lidin S (1994) J Chem Soc Faraday Trans 90:3423
14. Fogden A, Daicic J, Kidane A (1997) J Phys II France 7:229
15. de Vries R (1996) J Chem Phys 103:6740
16. Gackstatter F (1990) Colloque de Physique C-7:163
17. Nitsche JCC (1993) In Friedman JN, Davis T (eds), Statistical Thermodynamics and Differential Geometry of Microstructured Materials, Springer, New York

18. Marcelja S (1995) Fizika B4:197
19. Hyde ST (1995) Colloids Surfaces A: Physicochem Eng Aspects 103:227
20. Helfrich W (1973) Z Naturforsch 28c:693
21. Israelachvili JN, Mitchell DJ, Ninham BW (1976) J Chem Soc Faraday Trans 2, 72:1525
22. Fogden A, Hyde ST, Lundberg G (1991) J Chem Soc Faraday Trans 87(7):949
23. Fontell K (1990) Colloid Polym Sci 268:264
24. Luzzati V, Tardieu A, Gulik-Krzywicki T (1968) Nature 217:1028
25. Lidin S (1988) J Phys France 49:421
26. Templer RH, Seddon JM, Duesing PM, Winter R, Erbes J, J Phys Chem, in press
27. Larsson K (1988) J Coll Interf Sci 122:298
28. Luzzati V (1995) J Phys (France) II 5:1649

29. Hyde ST (1996) Curr Opinion Solid State Mat Sci 1:653
30. Hyde ST (1997) Langmuir 13(4):842
31. Huo Q, Margolese D, Ciesla U, Feng P, Gier T, Sieger P, Leon R, Petroff P, Schüth F, Stucky G (1994) Nature 368: 317
32. Beck JS, Vartuli J, Roth W, Leonowicz M, Kresge C, Schmitt K, Chu C, Olson D, Sheppard E, McCullen S, Higgins J, Schenkler J (1992) J Am Chem Soc 114: 10835
33. Inagaki S, Koiwai A, Suzuki N, Fukushima Y, Kuroda K (1996) Bull Chem Soc Japan
34. Stucky G, Monnier A, Schüth F, Huo Q, Margolese D, Kumar D, Krishnamurty M, Petroff P, Firouzi A, Jamicke M, Chmelka B (1994) Mol Cryst Liq Cryst 240:187
35. Landry CC, Monnier A, Norby P, Hanson JC, Hyde ST, Chmelka BF, Stucky GD, in preparation
36. Inagaki S, Fukushima Y, Kuroda K (1993) Chem Comm 8:680
37. Clerc M (1996) J Phys II (France) (6):961
38. Tolbert SH, Schäffer TE, Feng J, Hansma PK, Stucky GD (1997) Chem Mater (9):1962

Progr Colloid Polym Sci (1998) 108:153–160
© Steinkopff Verlag 1998

F. Caboi
M. Monduzzi

On microstructural transitions
of lamellar phase forming surfactants

Review article in honour of Kåre Larsson

F. Caboi · M. Monduzzi (✉)
Dipartimento Scienze Chimiche
Università di Cagliari
Via Ospedale 72
I-09124 Cagliari
Italy
E-mail: monduzzi@vaxca1.unica.it

Abstract In binary or ternary surfactant systems, if a complete segregation at an interface between the hydrophilic and lipophilic domains is assumed, the microstructure is related to the interfacial geometry. This is determined by two factors: the local interfacial curvatures set by the balance of molecular forces at the interface, and the interfacial topology and degree of connectivity that is imposed by the need to satisfy global packing constraints. The local constraint is characterized by the packing parameter of the surfactant, v/al. For lipids and membrane mimetic systems this is close to unity.

Several microemulsions and liquid-crystalline phases formed from surfactants with surfactant parameter close to unity are shown to exhibit peculiar structural transitions. This is demonstrated by NMR and conductivity experiments. This feature, $v/al = 1$, shared by lipids, seems to allow quite diverse flexibility in microstructure. This is despite the very different molecular structures of the surfactants studied (DDAB, AOT, MO, PFPE). Microemulsions and liquid-crystalline phases exhibit drastic structural changes on addition of a very small amount of a further new component, or with minimal variation in the composition of the system.

Other properties exhibited by such systems are shared and quite general: on increasing the volume fraction of the hydrophobic domain in binary surfactant/water system, or upon water and oil dilution, for microemulsions, an evolution of microstructure from a continuous water network towards closed water domains is observed.

Key words Surfactant – packing parameter – microstructural transition

Introduction

Lipids, the focus of much of Kåre Larsson's work, are usually surfactants that form bilayers essential for the existence of cell membranes. In the language of self-assembly their surfactant parameter is close to unity. Such lipids form a variety of bicontinuous cubic phases which have been well studied. Biomembranes, which besides proteins and other components, contain mixtures of many lipids, exhibit enormously more flexibility in their self-assembly capacity. The question then arises whether such flexibility is available to the lipids themselves, or is a property of more complicated biological systems alone. We here attempt to address this question by studying some model systems with surfactant parameter close to unity.

In the last few decades a better understanding of surfactant self-assembly has emerged. Models firmly based on thermodynamic and kinetic principles appear to account for the microstructure of liquid-crystalline (LC) or microemulsion phases and to predict their structural evolutions [1, 2]. Such theories are well known and need only minimal rehearsal here.

The self-assembled aggregate is described by the geometry of the interface dividing the hydrophobic and hydrophilic domains [1]. This interface has two attributes: the interfacial curvatures, related to the local geometry, and the interfacial topology, that describes the global geometry in terms of the degree of interfacial connectivity. The local packing constraint on the curvatures at the interface is measured by the packing parameter of the surfactant, v/al (v is the hydrophobic chain volume, a is the head group area and l is the chain length [2]).

The surfactant parameter is a useful molecular characteristic to predict which phases form with a given surfactant. It connects molecular properties to the preferred curvatures of the aggregate interface. The parameter v/al is related to the mean (H), and the Gaussian (G) curvatures, through the relation [1]

$$v/al = 1 + Hl + Kl^2/3 . \qquad (1)$$

Generally, if v/al is less than 1/3 the surfactant forms spherical micelles; between 1/3 and 1/2 cylindrical micelles are preferred. Between 1/2 and 1 a much more complex situation is obtained, which has been explored in Ref. [1]. Generally, bilayers now form. (But whether they form – even in dilute systems – single-walled vesicles or multibilayers or cubic phases depends on chain stiffness. Packing constraints on the inner and outer sides of a curved bilayer can be very different). When v/al is around 1, lamellar aggregates are favored, and if $v/al > 1$, aggregates with inverse curvature form. Here water is enclosed by a surfactant monolayer and the hydrophobic part is the continuous medium [3].

In a system with a third component like oil, an effective packing parameter is a natural generalization. The v/al ratio can change as a consequence of temperature variations, oil or water addition or through the introduction of a new component. For instance, the effective surfactant chain volume increases if a highly penetrating oil is added. The area of the polar head changes by adding salt or with a change in temperature (different hydration or steric interaction). So the composition can itself change v/al. Then microstructure and also interfacial topology can alter. Alternatively, structural modifications can occur as a result of curvature variations, which do not alter v/al [1], and simply result from global packing constraints imposed by mass conservation of components.

We here focus on microstructures, obtained by various experimental techniques, for different types of surfactant systems. They have as common feature a surfactant parameter close to 1. It will be shown how some such liquid-crystalline phases and microemulsions are particularly affected by minimal perturbations of the system. These can be: a small addition of a new component, an increase of the chain length of the surfactant in a homologous series, or a variation of the composition. The main focus is on microstructural transitions from a water continuous network to a closed water domain. These closed domains can assume the shape of a water-in-oil (w/o) droplet structure, or an inverse hexagonal (H_2) liquid-crystalline organization.

The transitions occurring in ternary systems are along water dilution lines with increasing ϕ_w/ϕ_s ratio, or along oil dilution lines with increasing ϕ_o. In our systems, the latter is shown to imply an additional transition of the surfactant interface: from a bilayer to a monolayer organization.

For surfactant/water binary systems the curvature variation needed to induce analogous evolutions, are obtained either by increasing the length of the hydrophobic chain, or by solubilizing a small amount of a hydrophobic species.

Figure 1 shows the surfactant molecules which are here considered: the single-chain surfactants perfluoropolyether ammonium carboxylates (PFPE–NH$_4$), ($v/al \approx 1.0$) and monoolein (MO), ($v/al \approx 1.0$), and the double-chain surfactants di-dodecyl–dimethyl–ammonium bromide (DDAB), ($v/al = 0.82$), and di-octyl–sulfossuccinate sodium and calcium salts (NaAOT, CaAOT), ($v/al \approx 1.12$). Table 1 lists the surfactant systems investigated. In parallel are listed the dilution criteria used for study, and modifications induced by additives. The observed structural transitions (cf. Fig. 2) are also listed in Table 1.

Ternary microemulsions

The microemulsion regions formed by DDAB, PFPE, NaAOT and CaAOT surfactants are shown in Fig. 3. The water and oil dilution lines studied are indicated. Note that a bicontinuous cubic phase is present for all. Transitions from a bicontinuous microstructure towards partially, or totally closed water domains have been confirmed by NMR self-diffusion and conductivity measurements. This transition, and its evolution is well described in terms of the DOC model [4, 5]. This simple geometric (Disordered Open Connected) model was first proposed by Hyde et al., to predict the microstructure and phase boundaries of the DDAB system in different alkanes. It depends on only a single parameter v/al_{eff} which measures

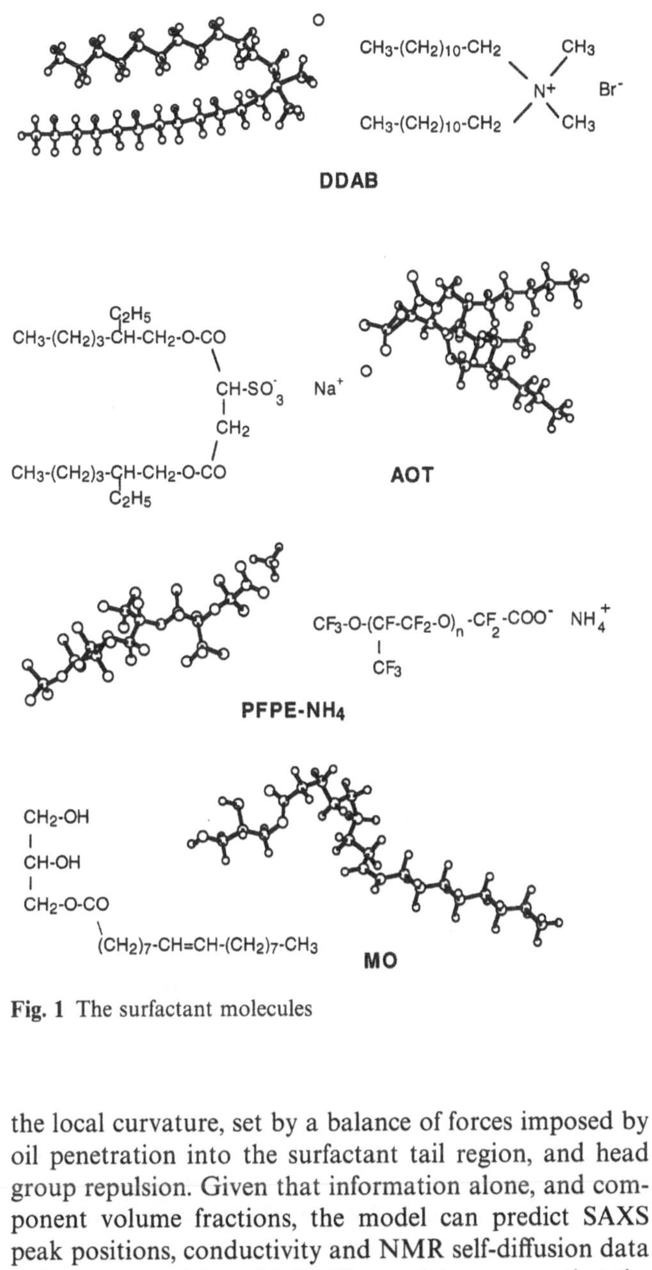

Fig. 1 The surfactant molecules

the local curvature, set by a balance of forces imposed by oil penetration into the surfactant tail region, and head group repulsion. Given that information alone, and component volume fractions, the model can predict SAXS peak positions, conductivity and NMR self-diffusion data in L_2 microemulsions [4, 5]. The model presumes that the surfactant resides substantially at the oil–water interface. The analytical approximation invoked to describe a surface of constant average curvature of the DOC model is a simple one. It uses a network of spheres connected to Z neighbors through cylinders that satisfy the same curvature constraints imposed by the v/al ratio. The interface topology is not fixed but it is dependent on the composition and the microstructure can vary from interconnected structures to monodisperse spheres, for instance, upon water dilution [1, 4, 5].

The DOC model captures the phase behavior and microstructure of a range of DDAB/alkane microemulsions quantitatively. It relies on local and global packing

constraints, the only two parameters of consequence, and therefore should have general validity. This has indeed been confirmed by comparing the DDAB/W/HEX microemulsion with different PFPE–NH$_4$ (S2)/W/PFPE oils microemulsions [6]. Upon water dilution the water self-diffusion data for the DDAB/HEX system are correctly reproduced with an effective surfactant parameter $v/al = 1.53$. The connectivity parameter Z, determined by volume fractions of components, varies from 6.5 to 0.5, as shown in Fig. 4A. This corresponds to the structural transitions E–B–F in Fig. 2. The surfactant PFPE–NH$_4$ (S2) was investigated in the presence of three PFPE oils that differ in molecular weight. These are: PFPE, MW = 700 ($O1$); PFPE, MW = 800 ($O2$); PFPE, MW = 1000 ($O3$). The behavior is exactly as for the DDAB system. Figure 4B shows water self-diffusion data reproduced with $v/al = 1.22$, with Z varying from 10 to 0.35 for the PFPE ($O1$) system. The data for the highest MW PFPE ($O3$) oil are reproduced with $v/al = 1.14$. Only a slight decrease of connectivity is observed (Z ranging from 10.30 to 8.5). Due to the low degree of oil penetration here, the system maintains a partially connected water network along the entire dilution line, implying a E–B (Fig. 2) structural transition.

In ternary phase diagrams of DDAB with short-chained (strongly penetrating) oils, the microemulsion region extends up to the oil corner. With tetradecane the microemulsion phase separates at an intermediate oil content ($\phi_0^{max} = 0.59$) [7]. Along the oil dilution line at s/w = 0.43, the NMR water self-diffusion coefficients decrease with increasing oil volume fraction as shown in Fig. 5. Here we notice and remark on the weak dependence of diffusion coefficients on oil chain length. At low oil content all systems form bicontinuous cubic phases with a multiply connected bilayer of the surfactant interface (see phase diagrams of Fig. 3).

On the other hand, at very high oil content, and typically, for the decane and dodecane systems, the ratios D_w/D_o have values consistent with a spherical w/o droplet microstructure. The V_2–L_2 transition implies a bilayer–monolayer evolution of the surfactant interface organization (C–B–F transition in Fig. 2) together with a gradual disconnection of the water network. For the tetradecane system the low penetration of the oil into the surfactant chains does not favor a negative curvature of the interface. A phase separation occurs (C–B transition in Fig. 2). Nevertheless, a bilayer–monolayer transition, which brings about a partially disconnected water network, is still suggested.

We have found very similar transitions in a variety of systems where a reversed curvature (w/o) microstructure exists. This is so, e.g., for the AOT microemulsion at high oil content. These systems are usually considered to be

Table 1 Summary of the various surfactant systems and structural transitions involved upon different composition changes

Surfactant[a]	Perturbation	Transition[b]
DDAB/W/OIL		
HEX	Water dilution	Bicontinuous → w/o (E–B–F)
C10	Oil dilution	Bilayer → monolayer (A–C–B–F)
C12	Oil dilution	Bilayer → monolayer (A–C–B–F)
C14	Oil dilution	Bilayer → monolayer (A–C–B)
PFPE–NH$_4$/W/PFPE Oil	Water dilution	Bicontinuous → w/o
(S2) (O1, O2, O3)		O1, O2 (E–B–F)
		O3 (E–B)
AOT/W/ISO	Water dilution	No transition (F)
AOT/W/ISO + HSA	Water dilution	Bicontinuous → w/o
[HSA]/[AOT] = 5.8 × 10^{-5}		(E–B–F)
CaAOT/W/C10	Water dilution	Bicontinuous → w/o
PFPE–NH$_4$/W		(E–B–F)
S1 (MW = 434)	ϕ_s Increase	(A–B)
S2 (MW = 722)	ϕ_s Increase	(A–C)
S3 (MW = 923)	ϕ_s Increase	No transition (D)
S1 → S2 → S3	Chain length increase	(B–C–D)
MO/W/VK$_1$	Increase of VK$_1$ content	C–D

[a] Abbreviations. (HEX) hexane, (C10) decane, (C12) dodecane, (C14) tetradecane, (ISO) isooctane. (PFPE) perfluoropolyether, (PFPE OIL). O1 MW = 700, O2 MW = 800, O3 MW = 1000. (DDAB) didodecyldimethyl ammonium bromide. (AOT) dioctylsulfossuccinate. (HSA) human serum albumin. (MO) monoolein. (VK$_1$) Vitamin K$_1$.
[b] Cf. Fig. 2.

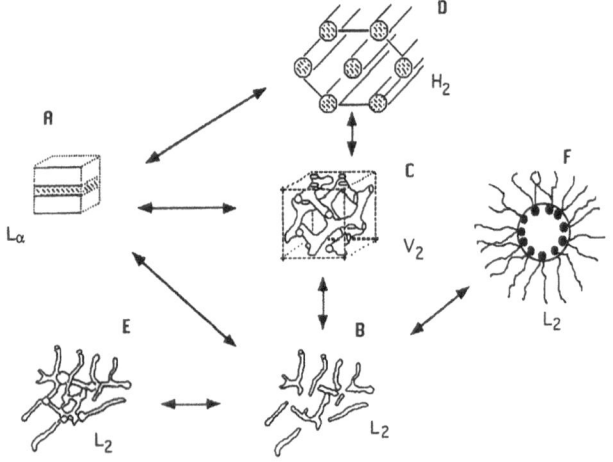

Fig. 2 Schematic picture of the structural transitions involved in the microemulsion and liquid-crystalline phases

simply reverse micelles. However, on introduction of a low amount of the globular protein human serum albumin (HSA) (HSA/AOT = 5.8 × 10^{-5}) into the microemulsion formed by AOT/W/ISO [8] there are changes in structure. The protein induces a significant increase of the water self-diffusion coefficients at low water content (see Fig. 6). Diffusion data are now consistent with the existence of

a continuous water network, exactly as for the DDAB system. With increasing water content the trend parallels that observed in the absence of the protein, therefore suggesting a gradual disconnection toward closed w/o droplets (E–B–F transition in Fig. 2). It seems likely that, at low water content, the globular protein rearranges its disposition at the interface to relocate within the water domain, thus favoring a significant clustering of the water droplets which, in turn, creates a water continuous network. When a sufficient amount of water allows restoration of the original globular shape, the disconnected microstructure is again favored [8].

An interesting example of this type of transition is found in the CaAOT/W/C10 system [9], where conductivity and self-diffusion data, measured in the whole microemulsion region, along several water dilution line at 25 °C, show a percolation behavior.

In terms of standard percolation theory [10–12], conductivity data clearly indicate the occurrence of a static percolation whose threshold is dependent on the volume fraction of the dispersed phase, as shown in Fig. 7. We remark that the maxima of conductivities as a function of ϕ_w increase with increasing s/o ratio, and lie on a straight line moving from line a (s/o = 2/8) up to line d (s/o = 1). The E–B–F transitions of Fig. 2 appear to be strongly suggested. Along the line e (s/o = 7/3), at very high

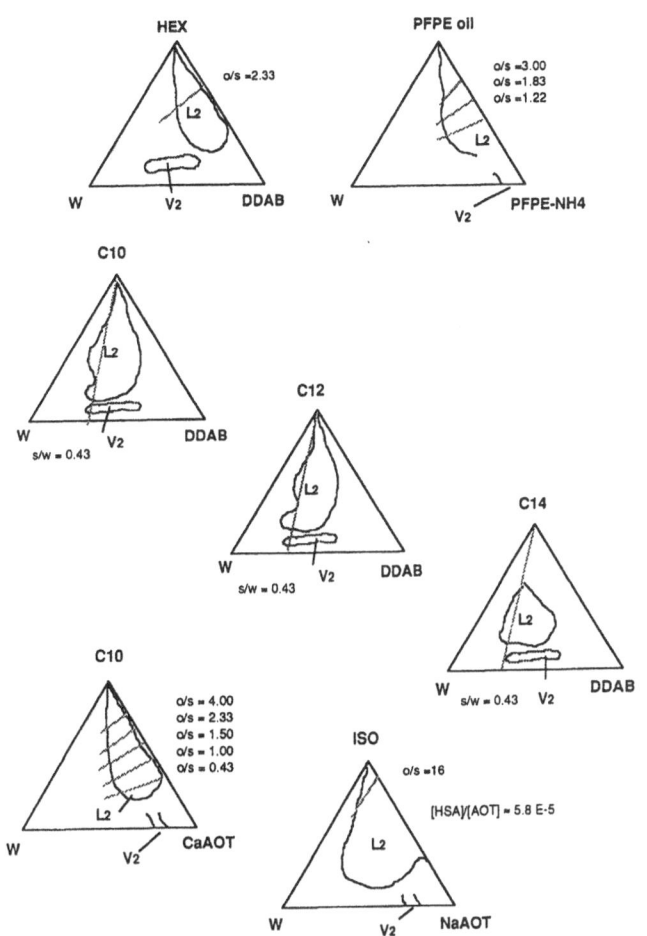

Fig. 3 Microemulsion regions (L_2) together with the investigated oil and water dilution lines in the DDAB, PFPE–NH$_4$, NaAOT and CaAOT ternary phase diagrams. V_2 cubic regions are also shown

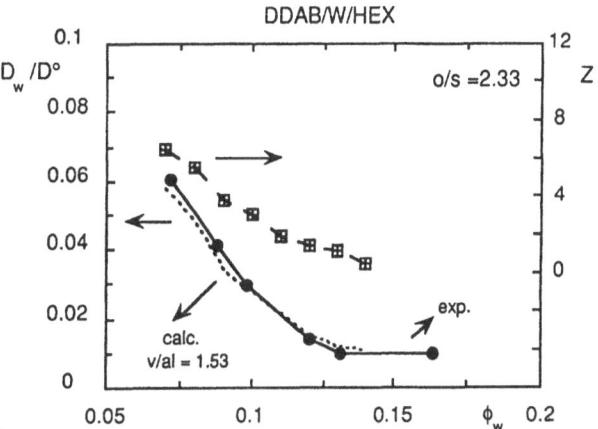

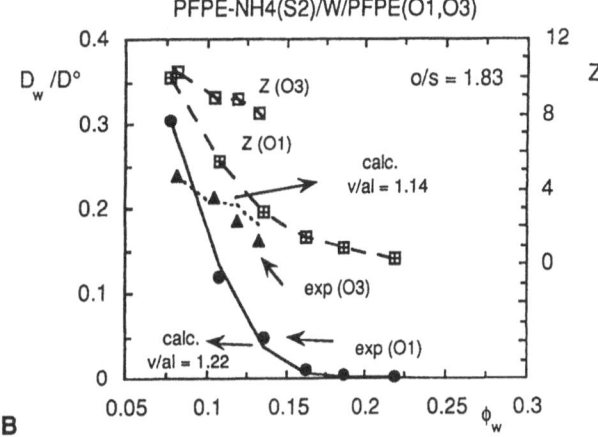

Fig. 4 Relative water self-diffusion coefficients, D_w/D_o (experimental and calculated values) and connectivity factors (Z) from DOC model vs. the water volume fraction ϕ_w [6]. (A) DDAB/W/HEX and (B) PFPE–NH$_4$ ($S2$)/W/PFPE ($O1$, $O2$) microemulsions

surfactant concentration, the conductivity reaches a noticeable maximum. It does not decrease to very low values before phase separation occurs. This can be related to very strong interaggregate attractive interactions. In terms of the DOC model [5] the degree of connectivity must be rather high. Surfactant interfacial organization locally approaches a bilayer microstructure (due to low water content: $\phi_w \ll \phi_s$) thus preventing the total disconnection to w/o droplets (E–B transition in Fig. 2).

Surfactant/water binary systems

Turning attention to surfactant/water binary systems, we remark first on the phase behavior observed at high surfactant concentration (>80 wt%) with a homologous series of three PFPE–NH$_4$ surfactants. These have short ($S1$, MW $=434$), intermediate ($S2$, MW $=722$) and long ($S3$, MW $=923$) PFPE chains. The phase diagrams are shown in Fig. 8 [13, 14].

At an intermediate concentration, $S1$ and $S2$ form a lamellar phase, while $S3$ shows a hexagonal phase. With increasing surfactant concentration $S1$ and $S2$ evolve to a partially water connected L_2 phase (A–B transition) and to a bicontinuous V_2 phase (A–C transition), respectively. A mean spontaneous curvature around zero seems to account well for their packing molecular constraints. Only the $S3$ surfactant, which retains a H_2 LC phase, almost independently of the water content, fullfills the packing requirements suitable to produce a stable inverse curvature of the interface. Notice that with increasing surfactant chain length, the B–C–D structural evolution is observed.

It should be remarked that in the case of $S1$ the occurrence of non-spherical-shaped aggregates, partially connected by small water channels is a plausible hypothesis. This follows from the results obtained through the DOC model for the PFPE microemulsions [6]: self-diffusion

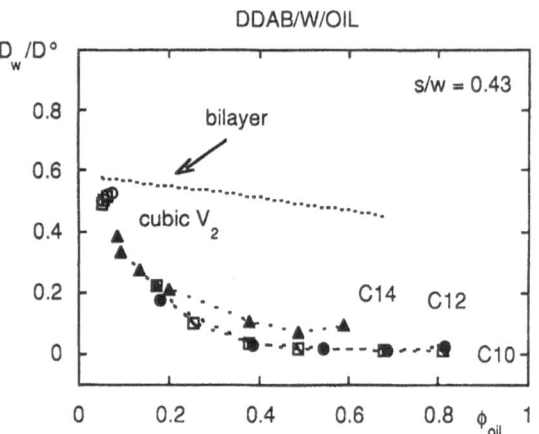

Fig. 5 Dependence of the relative water self-diffusion coefficients, D_w/D_o, on the oil volume fraction along an oil dilution line (s/w = 0.43) of the DDAB/W/Oil microemulsion [7]. L_2 region. (●) DDAB/W/C10, (□) DDAB/W/C12, (▲) DDAB/W/C14. V_2 cubic phase: (○) DDAB/W/C10, (■) DDAB/W/C12. Calculated D_w/D_o are also reported for a continuous bilayer surfactant organization which coincides with the V_2 cubic LC phase

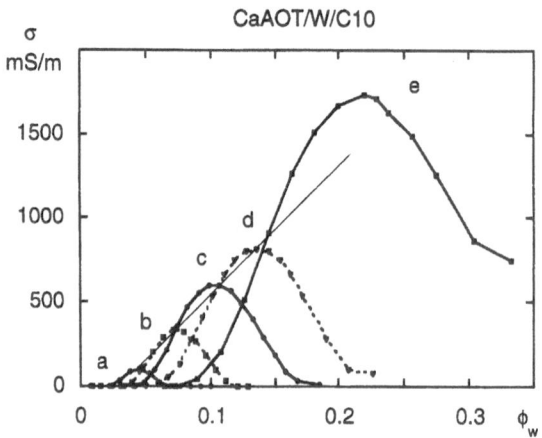

Fig. 7 CaAOT/W/C10 microemulsion [9]. Conductivity data, σ (ms/m), as a function of the water volume fraction (ϕ_w) for different water dilution lines, at constant s/o (wt%) ratio: (a) s/o = 2/8, (b) s/o = 3/7, (c) s/o = 4/6, (d) s/o = 1, (e) s/o = 7/3. The conductivity curves are for visualization purpose. A straight line connects the maxima of conductivity for a–d curves

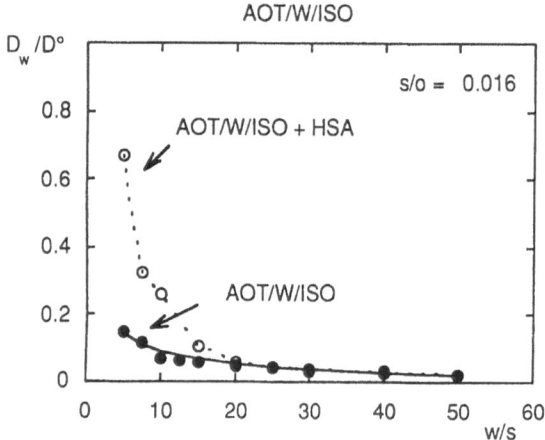

Fig. 6 Relative water self-diffusion coefficients, D_w/D_o, vs. water/surfactant molar ratio (w/s) in the AOT/W/ISO (●) and AOT/W/ISO + HSA (○) microemulsions [8], at the constant surfactant/oil molar ratio s/o = 0.0165 at 25 °C. [HSA]/[AOT] = 5.8 × 10⁻⁵. The solid line of AOT/W/ISO system is calculated from composition

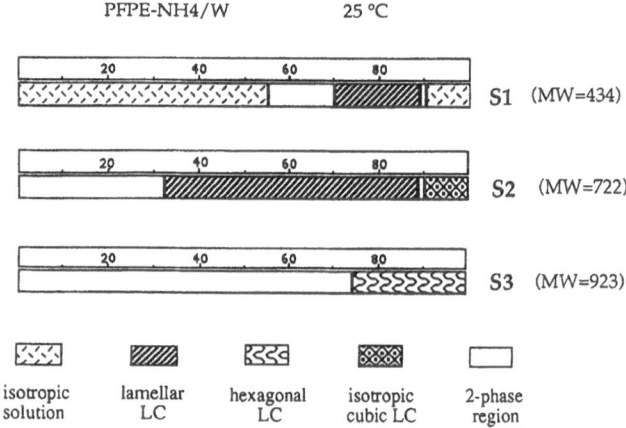

Fig. 8 Phase diagrams at 25 °C of PFPE–NH₄ surfactants (S1, S2 and S3) in D₂O, as a function of surfactant wt% redrawn from [13,14]

coefficients are reported similar to those found in the (S1)/W binary system at S1 > 90 wt%.

As an additional example of a continuous–disconnected water network transition, it is interesting to compare with the case of monoolein. This surfactant has a capacity to form cubic phases over a wide range of the binary phase diagram in water. Bicontinuous cubic structures are the subject of increasing interest mainly due to their importance in biological systems, and because of their use as host systems. Recent investigations have demonstrated the usefulness of MO cubic phases for controlled drug release in pharmaceutical applications [1, 15].

The phase diagram of MO/W system has been reinvestigated, at 25 °C, by NMR self-diffusion and SAXS, in the presence of small amounts of the lipophylic vitamin K_1 [16]. The L_2 and LC (L_α, C_G, C_D) phases of the MO/W system are not modified by 1 wt% of VK₁. However, at a VK₁ content of around 4 wt%, significant structural modifications are observed. This is shown in Fig. 9. The C_G phase, at a water content around 18 wt%, becomes a reverse hexagonal phase H₂. By contrast, in the pure MO/W system this phase appears at high temperature (80 °C) only. The surfactant parameter for the C_G and the H₂ phases are about 1.3 and 1.7, respectively [17]. Since the VK₁/MO ratio is 1/25, the VK₁ molecules, if uniformly dispersed in the surface region, can hardly change the packing parameter of monoolein.

Progr Colloid Polym Sci (1998) 108:153–160
© Steinkopff Verlag 1998

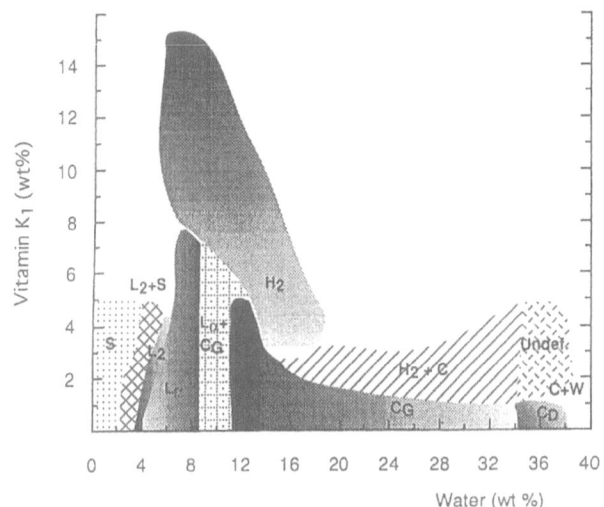

Fig. 9 Partial phase diagram of MO/W/VK₁ at 25 °C redrawn from [16]

This is surprising. It remains that a very small quantity of VK₁ has a large effect on the phase behavior of the MO/water system. A plausible explanation is that these effects can be accounted for only by assuming that the vitamin molecules segregate locally within the bilayer. These small aggregates, located in peculiar edges of the bilayer of the C_G phase, must then induce local changes of the curvature, which promote the transition toward a hexagonal structure (C–D transition in Fig. 2).

Concluding remarks

From the examples reported above, the peculiar microstructural transition from continuous to disconnected water network, observed upon small perturbations, seems to be almost a general behavior for all surfactants of the same surfactant parameter. The phenomena are related to a surfactant parameter ranging approximately from 0.8 to 1.2. This becomes greater than unity (1.3–1.8) upon composition changes. This is in agreement with the observation that the topology of the interface can change rapidly in surfactant/water binary systems [18], with small variation of the hydrophobic volume fraction, if the surfactant parameter is slightly higher than unity. In our systems we

have seen the effect of chain length and increase in surfactant concentration in the PFPE–NH₄ systems; or the effect of a hydrophobic species addition, in the MO/W system. In the case of the PFPE–NH₄ systems, it is reasonable that an increase of v/al occurs moving from S1 to S3, but only local curvatures are likely to change with increasing concentration of each surfactant, as in the case of MO/W system.

These transitions occur in microemulsions, upon water or oil dilution. This implies a continuous monolayer → w/o droplet or a bilayer → monolayer transition, respectively. In our systems, the latter transition involves also a gradual disconnection of the water network. In terms of DOC model a microstructure of water closed droplets prevails at high oil content, when the connectivity degree Z drops below 1.1. At low oil content, when the internal volume fraction is very high, a partial disconnection may occur as in the case of DDAB/W/C14 and PFPE (S2)/W/PFPE (O3).

Ultimately, both types of transitions promote the formation of spheres, which is a configuration of maximum stability and implies a decrease of connectivity to zero.

It is not always clear if the effective v/al changes during the observed transitions. Certainly, the process involves a variation of the Gaussian curvature from the negative values of a continuous bilayer or monolayer (A, C and E structures in Fig. 2) to the positive values of the spherical droplet through a value for the mean Gaussian curvature of zero. That circumstance characterizes an intermediate microstructure of partially connected cylinders (structure B in Fig. 2) having a sponge-like morphology.

The phenomena revealed by our studies are characteristic of delicately poised systems. These are those which form lamellar or reverse phases in water. Their surfactant parameter is close to or greater than unity. The ease with which surfactant interfacial curvature can be modified is suggestive. One expects that in more complex natural systems (formed by lipids, proteins, water and electrolytes) important structural evolutions induced by small perturbations and occurring at a microscopic level, can play a crucial role to determine the macroscopic performance of living systems.

Acknowledgements Thanks are due to B.W. Ninham and S. Hyde for stimulating discussions and advice. MURST (Italy), CNR (Italy), Consorzio Sistemi Grande Interfase (CSGI-Firenze) and Assessorato Igiene Sanita' (Sardinia Region) are acknowledged for support.

References

1. Hyde S, Andersson S, Larsson K, Blum Z, Landh T, Lidin S, Ninham BW (1997) The Language of Shape, Chs 1–5. Elsevier, Amsterdam
2. Mitchell DJ, Ninham BW (1981) J Chem Soc Faraday Trans II 77:601
3. Israelachvili JN (1992) Intermolecular and Surface Forces, 2nd edn. Academic Press, San Diego
4. Hyde ST (1989) J Phys Chem 93:1458
5. Hyde ST, Ninham BW, Zemb T (1989) J Phys Chem 93:1464
6. Monduzzi M, Knackstedt MA, Ninham BW (1995) J Phys Chem 99:17 772

7. Monduzzi M, Caboi F, Larche' F, Olsson U (1997) Langmuir 13:2184
8. Monduzzi M, Caboi F, Moriconi C (1997) Colloid and Surface 129:327
9. Caboi F, Capuzzi G, Baglioni P, Monduzzi M (1997) J Phys Chem 101:10205
10. Knackstedt MA, Ninham BW, Monduzzi M (1995) Phys Rev Lett 75:653
11. Feldman Y, Kozlovich N, Nir I, Garti N (1995) Phys Rev E 51:478, and references therein
12. Feldman Y, Kozlovich N, Nir I, Garti N, Archpov V, Idiyatullin Z, Zuev Y, Fedotov V (1996) J Phys Chem 100:3745
13. Monduzzi M, Chittofrati A, Boselli V (1994) J Phys Chem 98:7591
14. Caboi F, Chittofrati A, Monduzzi M, Moriconi C (1996) Langmuir 12:6022
15. Larsson K (1994) Lipids – Molecular organization, physical functions and technical applications, Vol 5. The Oily Press, Dundee
16. Caboi F, Nylander T, Razumas V, Talaikyté Z, Monduzzi M, Larsson K (1997) Langmuir 13:5476
17. Larsson K (1989) J Phys Chem 93:7304
18. Hyde S (1992) Pure Appl Chem 64:1617

Progr Colloid Polym Sci (1998) 108:161–165
© Steinkopff Verlag 1998

P. Billsten
M. Wahlgren
H. Elwing

Adsorption of human carbonic anhydrase II onto silicon oxides surfaces: The effects of truncation in the N-terminal region

P. Billsten (✉)
IFM Laboratory of Applied Physics
Linköping University
S-58183 Linköping
Sweden

M. Wahlgren
Department of Food Technology
Uuniversity of Lund
P.O. Box 124
S-22100 Lund
Sweden

H. Elwing
Department of General
and Marine Microbiology
University of Göteborg
Lundberg bld. Medicinaregatan 9c
S-41390 Göteborg
Sweden

Abstract The adsorption of human carbonic anhydrase II pseudo-wild type ($HCAII_{pwt}$) and an N-terminally truncated version thereof onto silica surfaces were studied. The amount adsorbed and the adsorption kinetics were measured using in situ ellipsometry. A substantial difference was seen between the two proteins. The adsorbed amount of the truncated version ($2.53\ mg/m^2$) indicates an end-on orientation, while the $HCAII_{pwt}$ seems to adsorb side-on ($1.84\ mg/m^2$). It is suggested that the orientation effects arise from the truncation. The truncation is known to unfold the two most N-terminal helical segments, which could inhibit adsorption with the N-terminal region facing the surface, due to steric repulsion.

Key words Protein adsorption – ellipsometry – carbonic anhydrase – mutant proteins – solid surfaces

Introduction

Human carbonic anhydrase II (HCAII) is an enzyme that catalyses the interconversion of carbon dioxide into bicarbonate and protons. We have previously used this protein and mutants thereof to study the structural changes that occur upon adsorption to silica nanoparticles [1]. It was found that truncated species of the protein obtained a structure similar to that of a folding intermediate of the enzyme, while the changes of the $HCAII_{pwt}$ were smaller [1]. The aim of this work was to link these studies to measurements of adsorbed amount and adsorption kinetics. The benefit of using HCAII in studies of conformational changes upon adsorption is that its unfolding and folding properties in solution have been extensively studied [2–7]. Furthermore, the protein is well characterized and its three dimensional structure is known to a resolution of 1.54 Å from X-ray diffraction [8]. It is a 29 300 Da large monomeric α/β-protein containing 259 amino acids. The overall size of the enzyme is 39 by 42 by 55 Å (Fig. 1).

To understand how proteins are affected by adsorption to solid surfaces, it is desirable to compare proteins that differ only in one aspect. One way of achieving similar proteins is to compare site-specific mutants that only vary in one or a few amino acid positions. There are several mutants of human carbonic anhydrase II available. The two mutants used in this study were human carbonic anhydrase pseudo wild type ($HCAII_{pwt}$) and a truncated version of this protein. The $HCAII_{pwt}$ differs from the wild type protein by a replacement of the only naturally occurring cystein in position 206 with a serine. In the truncated protein (trunc 5) the additional change is a removal of the four most N-terminal amino acid residues. The $HCAII_{pwt}$ is reported to have almost identical properties as the wild type enzyme. Trunc 5 has a decreased structural stability by 4–5 kcal/mol compared to

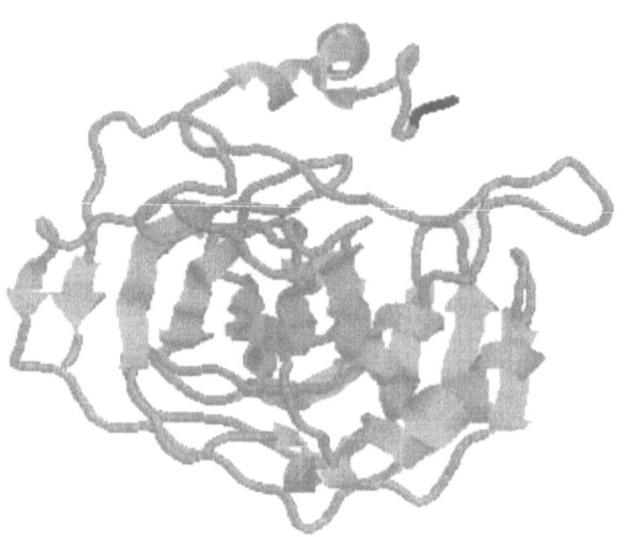

Fig. 1 The structure of carbonic anhydrase II from X-ray crystallography, with a resolution of 1.88 Å. The data were obtained from Brokhaven protein data bank [8]

HCAII$_{pwt}$, as measured with guanidine hydrochloride (GuHCl) denaturation. Furthermore the enzyme activity is approximately 30% compared to HCAII$_{pwt}$ [2].

Figure 1 shows the native enzyme and the amino acids removed due to the truncation (shown in black). The N-terminal region of the protein forms an arm which has very few critical contacts with the rest of the molecule. This arm can be removed to the 24th amino acid without disrupting the native structure of the remaining enzyme. However, circular dichroism and intrinsic tryptophan fluorescence measurements suggest that removal of the four most N-terminal amino acids is sufficient to unfold the two N-terminal helical segments made up by amino acids 13–18 and 21–24. Furthermore, the Trp residue located in position 16 becomes more solvent exposed in the trunc 5 mutant [2]. These observations indicate unfolding of the N-terminal region.

Materials and methods

The water used was distilled, passed through an ion exchanger and active charcoal (Millipore Corporation, Bedford, USA), and then distilled twice in a glass still. All chemical were of analytical grade. The silica surfaces, from Okmetic OY, Finland (resistivity 1–20 Ω cm (100) orientation, type P, boron doped), were kindly provided by Stefan Welin-Klintström (Linköping, Sweden) and prepared as described in [9] to give a 300 Å thick oxide layer. The surfaces were cleaned at 80 °C for 5 min with NH$_4$OH: H$_2$O$_2$: H$_2$O (1:1:5) (v:v:v) and HCl: H$_2$O$_2$: H$_2$O (1:1:1)

(v:v:v), respectively, and then washed with water and ethanol [9]. The surfaces were stored in ethanol were treated, prior to use, for five minutes in low pressure air plasma (0.2–0.3 Torr) using an r.f. glow discharge unit (Harrick PDC 3 XG, Harrick Scientific Corp., Ossining, NY).

The mutants used were HCAII pseudo wild type (HCA$_{pwt}$) and trunc 5. HCAII$_{pwt}$ has the naturally occurring cystein in position 206 replaced by a serine using site-directed mutagenesis. The spectroscopic properties and activity of this mutant are apparently identical to that of the wild type. Trunc 5 has its four most N-terminal amino acids removed, as well as the cystein in position 206 replaced by a serine using site-directed mutagenesis. The different mutant genes of HCAII were inserted in a pACA plasmid [10], which was then transfected into Escherchia coli BL21/DE3 [11].

Each strain was grown in a medium containing 20 g/l peptone, 20 g/l NaCl, 10 g/l yeast extract, 0.5 mM ZnSO$_4$ and 560 μg/ml ampicillin at 37 °C. When the optical density of the culture reaches $A_{660} = 0.45$ the synthesis of the desired protein was induced by the addition of isopropyl-β-D-thiogalactopyranoside (IPTG) to a final concentration of 0.5 mM.

The cultures were harvested after 16 h, and the cells were lysed by French pressing. After removal of larger cell fragments by centrifugation at 10 000 rpm for 45 min, the cell extracts were loaded onto an affinity chromatography column and purified in a single step [12].

Full details of mutagenesis, production and purification procedures are described in Aronsson et al. [2]. The protein concentrations were determined by measuring the adsorbance at 280 nm. The extinction coefficients used were determined according to the method of Gill and von Hippel [13].

The adsorbed amount was measured using in situ null-ellipsometry [14]. The thickness and refractive index of the adsorbed layer were calculated from the ellipsometric angles Δ and ψ. These calculations were performed using a three-phase model (surface, adsorbate and ambient) [15], utilizing the measured pseudorefractive index for the silicon with the oxide layer. The amount adsorbed was calculated according to Cuypers et al. [16] using 0.75 ml/g and 4.08 g/ml as the values for the partial specific volume and the molar refractivity, respectively.

The instrument used was a Rudolph thin film ellipsometer type 43603-200E. It had been automated by adding stepping motors to the polariser and analyser which are controlled by a personal computer. The ellipsometric angles (Δ and ψ) were determined from the settings of polariser and analyser that gave a minimum in light intensity. The minima is obtained by using the

Progr Colloid Polym Sci (1998) 108:161–165
© Steinkopff Verlag 1998

method of swing. The experiment was performed in a cuvette using 4.5 ml of 10 mM sodium potassium phosphate with a buffer pH 7.5, to which stock solution of the sample solution was added. The final concentration was 17 or 1.7 μM. The temperature was $25 \pm 0.1\,°C$. The system was stirred by a magnetic stirrer at a rate of 325 rpm. After the pseudo-refractive index of the bare surface had been measured, the experiment was started by addition of the sample. The adsorption was carried out for 60 min after which the cuvette was rinsed with saline solution for 5 min at a flow rate of 20 ml/min. Desorption was then monitored for an additional 25 min.

Results and discussion

The adsorbed amount after 60 min of adsorption and after rinsing is presented in Table 1. The adsorbed amount for trunc 5 mutant was higher than for HCAII$_{pwt}$. The adsorbed amount for trunc 5 at 17 μM was in the region of a close packed layer of proteins, adsorbed end-on, while the value for HCAII$_{pwt}$ at this concentration is in the range for a side-on monolayer. Figure 1 shows representative adsorption curves for both variants. It can be seen that both proteins reach a semi-plateau value in the adsorbed amount before 60 min of adsorption, for both concentrations tested. However, the plateau value is reached after a longer time for the lower concentration. The amount removed upon rinsing is a larger for the HCAII$_{pwt}$ than for the trunc 5 mutant. Even though the desorption is larger for the HCAII$_{pwt}$, the initial adsorption kinetics show a similar pattern. This might indicate that a change to a less desorbable state happens after the initial adsorption.

In these experiments, the different versions of carbonic anhydrase adsorbed quite fast and in amounts that are in the region for close-packed monolayers of the proteins. One possible reason for the favourable adsorption is that the proteins are close to their isoelectric point (i.p). HCA II has an i.p of 7.5. Proteins are often seen to adsorb to the

highest degree close to their ip. This is due to a minimum in the electrostatic intra- and inter-molecular repulsive forces [17]. Often proteins form monolayers when they adsorb to interfaces, although proteins that form oligomers in solution have been seen to adsorb in bi- or multi-layers [18–20]. There have been no reports that the mutants used here of HCA II form oligomers in solution, and thus it is plausible that a monolayer of adsorbed proteins is formed. Unfortunately, we have not been able to investigate protein concentrations higher than those used in this paper and thus can not know for sure whether higher concentrations would lead to a higher amount adsorbed protein. However, the adsorbed amount is in the vicinity of close-packed monolayer. The values for a close-packed layer of Carbonic anhydrase II are calculated as squares with the axes 39 by 42 Å (side-on orientation) or 42 by 55 Å (end-on orientation) are 2.97 and 2.11 mg/m², respectively. The fact that the adsorbed amount is close to this value means that it has to be possible for the protein molecules to move after they have adsorbed, otherwise the adsorbed amount could only reach the value for random packing. This value is about 55% of the one for close packing. The rearrangement of protein molecules could be due to diffusion in the surface plane [21], or desorption and readsorption. As can be seen in Fig. 2, a portion of the adsorption proteins is desorbed upon rinsing with buffer. The observation that only part of an adsorbed layer of protein can be removed is quite common. The fact that only a fraction of the adsorbed proteins desorb has been taken as an evidence for a heterogeneous population of adsorbed proteins [22]. The adsorbed proteins could, e.g. differ in orientation and conformation.

The expected difference between the adsorption of the two mutants was that the truncated species would adsorb

Table 1 The adsorbed amount of the mutants after 60 min of adsorption and after rinsing with buffer solution

Mutant		Adsorbed amount after 60 min (mg/m²)	Adsorbed amount after rinsing (mg/m²)
HCAII$_{pwt}$	1.7 μM	1.24 ± 0.12	0.80 ± 0.12
HCAII$_{pwt}$	17 μM	1.84 ± 0.05	1.27 ± 0.03
Trunc 5	1.7 μM	1.80 ± 0.23	1.53 ± 0.31
Trunc 5	17 μM	2.53 ± 0.02	2.29 ± 0.07

Fig. 2 The adsorption of HCAII$_{pwt}$ (circles) and Trunc 5 mutant (squares) onto silicon oxide surfaces. The concentrations were 17 μM (filled symbols) and 1.7 μM (open symbols). The experiment was performed at 25 °C in a 0.01 M phosphate buffer at pH 7.5. After 60 min of adsorption the system was rinsed for 5 min with buffer

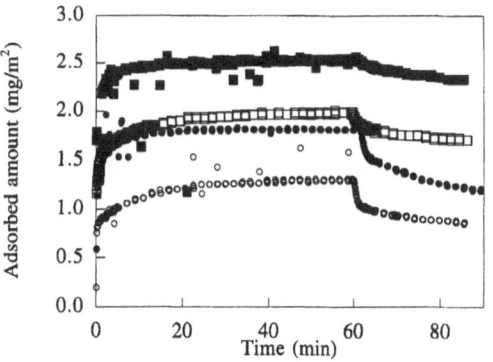

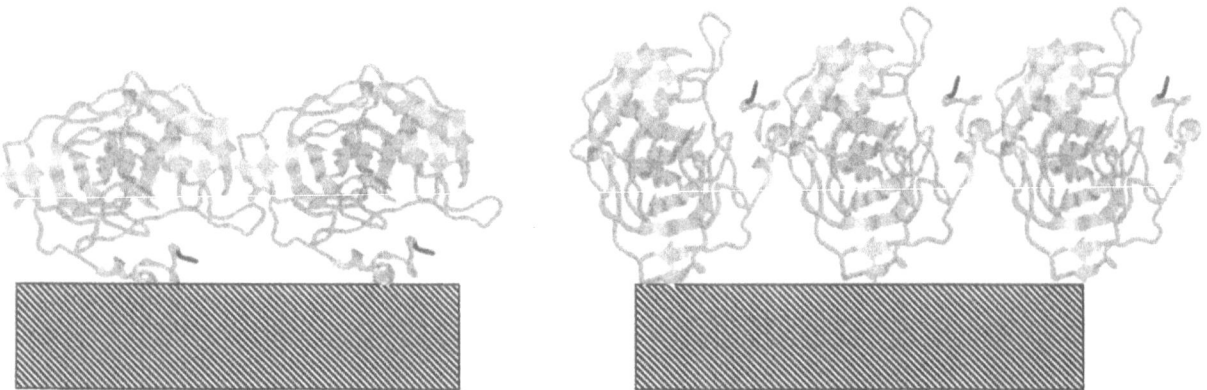

Fig. 3 Tentative suggestions for the orientation of HCAII$_{pwt}$ (left panel) and Trunc 5 (right panel) adsorbed to the silicon oxide surfaces

to a lesser extent, but more irreversibly to the silica surface than the HCAII$_{pwt}$. The reason for this expectation was that trunc 5 has a greater tendency to alter its conformation upon adsorption. Circular dichroism and fluorescence have previously been used to demonstrate that trunc 5, upon adsorption, obtains a structure similar to that of a folding intermediate, referred to as a "molten globule", and that the structural changes of HCAII$_{pwt}$ were less profound [1]. When a protein loses its native conformation, it expands, and fewer molecules would be expected to have enough space to adsorb.

The experimental data however was rather the opposite, as shown in Fig. 2; the amount of adsorbed trunc 5 exceeds that of HCAII$_{pwt}$ for both concentrations used. One reason for the low protein expansion could be that the experiments were performed close to the protein's i.p.

The adsorbed amount of trunc 5 is above what is expected for side-on adsorption, while the amount for HCAII$_{pwt}$ is slightly below this value. Thus, the difference in adsorbed amount could indicate a difference in the orientation of the proteins to the interface. Two plausible orientations are shown in Fig. 3. It is likely, however, that adsorbed proteins can adopt several orientations. The side-on orientation is probably dominant for the HCAII$_{pwt}$, while end-on dominates for the trunc 5 mutant. A tentative explanation for this adsorption behaviour for the truncated species is that the N-terminal region of the

trunc 5 mutant is in an unfolded conformation, leading to higher mobility in this region.

Our hypothesis is that the flexible N-terminal arm of the truncated version which is unfolded [2] functions to inhibit adsorption in the side-on fashion, due to steric repulsion caused by the high mobility in this region. It could be that the high adsorption of trunc 5 is due to that the protein is not allowed to alter its conformation on the surface but desorption from the surface, measured by rinsing with buffer, is lesser for the truncated version than for HCAII$_{pwt}$. Since the inability to desorb from the surface is an indirect indication of conformational rearrangement at the surface, it appears that trunc 5 is allowed to alter its conformation despite the lack of space.

As pointed out above, the two bulk concentrations investigated gave quite large difference in the adsorbed amount. Thus, it is not possible to determine if the higher value is in the plateau region of the adsorption isotherm. As can be seen from Fig. 2 and Table 1, the effect of an increase in protein concentration affects both the final adsorbed amount and the adsorption kinetics in a similar way for both mutants.

Acknowledgement The authors would like to express their gratitude towards Bengt Harlad Johnsson and Göran Aronsson for providing the bacteria strains used for production of the mutated proteins. This work was supported by the Swedish Research Council for Engineering Science (TFR).

References

1. Billsten P, Freskgård P-O, Carlsson U, Jonsson B-H, Elwing H (1997) FEBS Lett 402:67
2. Aronsson G, Mårtensson L-G, Carlsson, U, Jonsson B-H (1995) Biochemistry 34:2153
3. Carlsson U, Jonsson, B-H (1995) Curr Opin Struct Biol 5:482
4. Lindgren M, Svensson M, Freskgård P-O, Mårtensson L-G, Johnsson B-H, Carlsson U (1995) Biophys J 69:202
5. Mårtensson L-G, Jonsson B-H, Freskgård P-O, Kihlgren A, Svensson M, Carlsson U (1993) Biochemistry 32:224
6. Svensson M, Jonasson P, Freskgård P-O, Jonsson B-H, Lindgren M, Mårtensson L-G, Gentile M, Boren K, Carlsson U (1995) Biochemistry 34:8606
7. Andersson D, Freskgård P-O, Jonsson B-H, Carlsson U (1997) Biochemistry 36:4623

8. Håkansson K, Carlsson M, Svensson L-A, Liljas A (1992) J Mol Biol 227: 1192
9. Wahlgren M, Arnebrant T (1989) In Kessler HG, Lund DB (eds) Fouling and Cleaning in Food Processing. Prein, Federal Republic of Germany, p 200
10. Nair SK, Calderone TL, Christianson DW, Fierke CA (1991) J Biol Chem 266:17 320
11. Studier FW, Moffatt BA (1986) J Mol Biol 1986:113
12. Khalifah RG, Strader DJ, Bryant SH, Gibson SM (1977) Biochemistry 16:2241
13. Gill SC, von Hippel PH (1989) Anal Biochem 182:19
14. Azzam RMA, Bashara NM (1977) Ellipsometry and Polarized Light. North-Holland, Amsterdam
15. McCrackin FL (1969) A Fortran program for analysis of ellipsometer measurements. Nat Bur Stand, Tech Note 478, USA
16. Cuypers PA, Corsel JW, Janssen MP, Kop JMM, Hermens WT, Hemker HC (1983) J Biol Chem 258:2426
17. Norde W (1986) Adv in Colloid Interface Sci 25:267
18. Elofsson U (1996) Protein adsorption in relation to bulk phase properties: β-lactoglobulins in solution and at the solid/liquid interface. Thesis, University of Lund, Sweden
19. Tilton RD, Blomberg E, Claesson PM (1993) Langmuir 9:2102
20. Wahlgren M, Arnebrant T, Lundström I (1995) J Colloid Interface Sci 175: 506
21. Tilton RD, Robertson RC, Gast AP (1990) J Colloid Interface Sci 137: 192
22. Horbett TA, Brash JL (1987) In Brash JL, Horbett TA (eds) Proteins at Interfaces – Physicochemical and Biochemical Studies. American Chemical Society, Washington DC, p 1

Progr Colloid Polym Sci (1998) 108:166–174
© Steinkopff Verlag 1998

S.B. Engelsen
E. Mikkelsen
L. Munck

New approaches to rapid spectroscopic evaluation of properties in pectic polymers

S.B. Engelsen (✉)
E. Mikkelsen · L. Munck
The Royal Veterinary
and Agricultural University
Food Technology
Department of Dairy
and Food Science
Rolighedsvej 30
DK-1958 Frederiksberg C
Denmark

Abstract Pectins are highly desired carbohydrate polymers in the food industry due to their capabilities as gelling agents. As pectin molecules are very complex macromolecules extracted from plant cell walls, a 100% purification method for the pectin gelling principle has not yet been devised. For this reason pectins need to be carefully characterized for parameters such as degree of esterification, molecular weight, degree of polymerization, amount of plant residual materials, pH, calcium sensitivity, total amount of galacturonic acids, etc., as these factors will determine the final properties of the pectin product.

The purpose of this paper is to demonstrate the advantages of employing two different and highly complementary spectroscopic techniques, fluorescence and near infrared, for the determination of quality parameters in pectin powders. For this purpose three sample sets of pectinic powders are investigated: (I) citrus fruit raw materials, (II) standard extracted high methoxy pectins and (III) a pectin model set with large variations in the degree of esterification. The spectroscopic ensembles of these three pectin sets are subjected to chemometric analysis for the evaluation of spectroscopic correlations with reference physico-chemical measurements such as pectin yield, gel strength, simple sugar content and degree of esterification.

Key words Pectin – citrus – esterification – fluorescence – NIR – PCA – PLS

Introduction

Pectins are natural cell-wall polymers found in most flowering plants [1, 2] and is a most important carbohydrate-based gelling principle in food. Their non-toxicity and their ability to provide a gel with a very tender, short texture with excellent clarity and outstanding flavor release properties has made pectins popular as sugar-based hydrocolloids in the confectionery industry [3]. More recently pectins have attracted much attention as components in functional foods.

Two quite different types of pectin molecules are found (often simultaneously) in the plant cell walls: rhamno-galacturonan I (RG-I) [4] and rhamnogalacturonan II (RG-II) [5]. While the latter have a highly well-defined but very complex structure, the RG-I pectins are some of the most complex and variable polymers known. Their properties are also quite different; RG-II pectins in the form of borate-ester dimers may have interesting potential as heavy metal ion complexators [6], whereas the RG-I pectins have very desirable gelling properties. The most characteristic feature of pectin molecules is a partially methyl esterified $\alpha(1 \rightarrow 4)$ linked polygalacturonic backbone ("smooth region") which is responsible for the gelling properties. In the RG-I pectins this backbone is interrupted by $(1 \rightarrow 2)$-linked α-D-rhamnopyranose which serves as anchoring points for the attachment of lateral neutral oligosaccharides containing mainly D-galactopyranoses and L-arabinofuranose ("hairy region") [7].

For commercial applications RG-I pectins are usually extracted from citrus peel, apple pomace and sugar beet and processed to yield high and low methoxy pectins which gel differently. High methoxy (HM) pectins are capable of forming gel networks at acid pH in the presence of soluble solids and the degree of esterification (DE) controls their relative speed of gelation. The gelling properties of low methoxy (LM) pectins are dependent on the presence of divalent cations such as calcium and are much less pH-dependent than for the high methoxy pectins. In LM pectins the DE is the primary parameter in the control of their calcium reactivity. Because of the importance of DE in relation to pectin gelation properties, it is desirable to obtain a fast, non-invasive and robust method to predict (determine) the DE in pectin powders. We have previously with success developed spectroscopic methods to determine DE and degree of amidation (DA) for a range of amidated pectin powders [8]. The addition of amide groups to LM pectins strongly affects the calcium reactivity and the gelation properties, adding a new dimension to the design of pectin gels.

The present work represents a top-down spectroscopic/chemometric investigation of the quantitative performance of two different spectral techniques on a coherent set of pectin raw material and their standard extracted HM pectins. The investigation emphasizes the development of spectroscopic/chemometric methods for the prediction of pectin quality parameters in the raw material from fluorescence spectra and for the prediction of DE in the extracted pectin from NIR spectra. The main difficulties in developing such spectral methods is the complexity inherent to the pectin raw material and to the extracted pectin. Even the extracted pectin yields micro-heterogeneous pectin polymers, as a commercial purification method for the pectin gelling principle has not yet been devised. The samples and therefore the spectra will thus contain interferences from variations in: degree of polymerization, hairy regions, particle size, residual plant materials, water, geographic origin, etc. The relevant spectral information may be buried in such interferences and has to be unravelled by applying modern multivariate data analytical techniques such as principal component analysis (PCA) and partial least squares regression (PLS).

Experimental

The samples

The material examined consisted of three sets of pectinic powders: 71 samples of dried citrus fruit raw material (set I), ground to a fine powder, 71 standard extracted HM pectins (set II), extracted from the raw material (set I), and a set of five pectins with large variation in degree of esterification (DE) (set III). All sample material was kindly provided by Copenhagen Pectin A/S (Hercules Inc). The citrus fruit raw material were ground into powder. No other pre-treatment of the samples was performed.

Chemical reference data

Copenhagen Pectin A/S provided the physico-chemical reference data. The amount of simple sugars (SUGAR) (mono- and di-saccharides) were analyzed for the 71 raw material samples. Values ranged between 10.5–18.8%, with a mean of 15%. The yield of HM pectin from a standard extraction was measured in percent, ranged between 21.4–29.2%, with a mean of 25.5%. SAG is a gel strength measurement which was measured on all 71 HM pectins, ranged between 122–184 with a mean of 145. Methoxylation or DE was determined by titration on 25 of the 71 HM pectin samples, ranged between 60.1–68.8% with a mean of 65.2%. For the set of five pectin samples the DE were analyzed to be 7%, 22%, 45%, 60% and 75%.

Spectroscopic data

To resemble on-line measurements as closely as possible all spectroscopic evaluations of the samples were performed without any pre-treatment, except for the raw material which was ground. All spectral measurements were made at room temperature.

Fluorescence data were collected on a Perkin Elmer LS-50B spectrometer. The spectra were acquired by using a solid sample holder with a quartz window in reflectance mode. A pulsed xenon lamp excites the sample, and the fluorescence signal is registered with a photomultiplier. Complete excitation–emission fluorescence landscapes were collected for two raw materials and pectins that varied most in physico-chemical data in order to determine optimal excitation wavelengths. Nine excitation wavelengths: 240, 305, 330, 380, 415, 440, 470, 505 and 620 nm were selected from the landscapes. Emission spectra were recorded from the excitation wavelength plus 30 nm, to avoid the Rayleigh scattering, and up to 750 nm.

168
S.B. Engelsen et al.
New approaches to rapid spectroscopic evaluation of pectins

Table 1 The spectroscopic data

	Fluorescence	NIR
Instrument	Dispersive	Dispersive
Sampling technique	Reflectance	Reflectance rotating cup
Reference	None	Ceramic
X-variables		1050
X-units	nm	nm
X-min	270	400
X-max	750	2500
X-sampling	0.5	2
Scans	1	16

In subsequent data treatment the nine emission spectra for each sample were appended and stored as one spectrum (see Table 1).

Dispersive near infrared (NIR/VIS) data (including the visual region) were collected using a NIRSystems Inc. (model 6500) spectrophotometer. The spectrophotometer uses a split detector system with a Silicon (Si) detector between 400 and 1100 nm and a Lead Sulfide (PbS) detector from 1100 to 2500 nm. Angle of incident light was 180° and reflectance was measured at a 45° angle. The NIR/VIS reflection spectra were recorded using a rotating sample cup with a quartz window, and spectral data were converted to $\log(1/R)$ units.

Chemometrics

Chemometrics is the science of relating measurements made on a chemical system or process to the state of the system via application of mathematical or statistical methods (definition by the International Chemometrics Society). It is characteristic of chemometrics that it has been inspired by social sciences such as psychometrics and econometrics, which deal with real world multivariate data that are not suited for experiments. In this context, the advantage of chemometrics is that it is able to deal with spectral information containing multivariate co-linear data.

Principal component analysis (PCA) is a powerful technique for compression of large multivariate data sets, such as spectral information [9]. The multidimensional data set is resolved into orthogonal components whose linear combinations approximate the original data set in a least-squares sense. In PCA, the original data matrix ($\mathbf{X}$) is decomposed into a score matrix ($\mathbf{T}$) and a loading matrix ($\mathbf{P}$), and the residuals are collected in a matrix ($\mathbf{E}$): $\mathbf{X} = \mathbf{TP}^T + \mathbf{E}$. Only a limited number of principal components (PCs) equal to the chemical rank of the $\mathbf{X}$-matrix are relevant in describing the systematic information in $\mathbf{X}$. The

loading vectors for the principal components can be considered as pure hidden spectra that are common to all the measured spectra. What makes the individual raw spectra different are the amounts (scores) of hidden spectra. The scores contain information about samples and the loadings about the variables [8].

Partial least squares regression (PLS) is a predictive two-block regression method based on estimated latent variables and applies to the simultaneous analysis of two data sets (e.g. spectra and physical/chemical tests) on the same objects (e.g. pectin samples) [9]. The purpose of the PLS regression is to build a linear model enabling prediction of a desired characteristic ($\mathbf{y}$) from a measured spectrum ($\mathbf{x}$). In matrix notation we have the linear model $\mathbf{y} = \mathbf{Xb}$ where $\mathbf{b}$ contains the regression coefficients that are determined during the calibration step. PLS was first applied to evaluate NIR spectra by Martens and Jensen in 1983 [10] and is now used routinely to correlate spectroscopic data (rapid measurements) with related chemical/physical data (slow chemical/physical measurements).

Interval PLS (iPLS) is an in-house extension to PLS, which is designed to develop local PLS models on equidistant subintervals of the full-spectrum region. Its main force is to provide an overall picture of the relevant information in different spectral subdivisions and thereby narrow in the important spectral regions. The sensitivity of the PLS algorithm to noisy variables and its tendency to become saturated when applied to large data sets are highlighted by the informative iPLS plots.

Principal variables (PV) is a method for selection of a limited number of original variables (wavelengths) that describe as much as possible of the variance in the data matrix (spectra) or, alternatively, in a vector with a desired characteristic (chemical/physical measurement) [11]. Besides being important when developing robust PLS models and when optimizing filter instruments, PV selection is also very helpful in the interpretation of the models [8]. The PV method is initiated by finding the variable (wavelength) that covaries most with the y-vector (physical/chemical measurement). This variable is the first principal variable. The original data matrix is then reduced (orthogonalized) with respect to the first principal variable. Then new covariant variables in the reduced data matrix are selected iteratively. The result of the PV selection is a limited number of the original variables (e.g. wavelengths), while PCA/PLS selects latent factors based on information from all original variables (vectors). If the objective is to build predictive models, it is subsequently necessary to apply multiple linear regression (MLR) or PLS to the reduced matrix of principal variables.

Unless otherwise noted, results are all based on segmented cross-validation [12, 13]. In cases with the full sample material (71 samples), we used 7 (123123....) segments for cross-validation; in cases with a more limited sample material (DE and SUGAR), we used full (leave one out) cross-validation. Only validated results are reported.

Programs

Chemometric calculations were performed using Matlab ver. 4.2c 1 (Mathworks, Inc.) installed with the PLS – Toolbox ver. 1.5 (Wise & Gallagher; Eigenvector Technologies) and Unscrambler ver. 6.11a (CAMO A/S).

Results and discussion

NIR spectroscopy and pectins

Near infrared spectroscopy measures over- and combination-tones of the fundamental molecular vibrations found in the IR region. NIR represents an extremely reproducible and robust spectroscopic technique which in abundance has demonstrated its use as a rapid non-invasive spectroscopic method in the food and agrochemical sectors. It has already been shown that NIR has capabilities for detecting pectins and certain pectin quality parameters [8, 14]. Figure 1 illustrates the difference between the NIR spectra of the citrus raw material and the extracted HM pectin, respectively. In the visual section (400 to 900 nm) the raw material sample absorbs more strongly and has a characteristic peak at 668 nm from the chlorophyll component. In the NIR section from about 1400 nm the pectin sample absorbs more strongly and contains more details and sharper peaks, i.e. it is less amorphous or more crystalline.

Chemometric analysis of NIR spectra

A PCA was performed to investigate the variance structure in the raw physico-chemical data and in the spectral ensembles. Figure 2A shows a score plot of the NIR spectra of the raw material (I). The score plot readily reveals that (all) the samples from producer G are outliers (deviating). Figure 2B is a similar score plot, but based on the corresponding extracted HM pectins (II) according to the standard method. Although less clear, we still observe that all samples from producer G are outliers. This particular producer has left a fingerprint on the raw material which the extraction process involving solvation in hot acidified demineralized water, filtration, concentration, isopropanol precipitation, pH-adjustment and drying [15] was not able to remove completely. The fingerprint is still

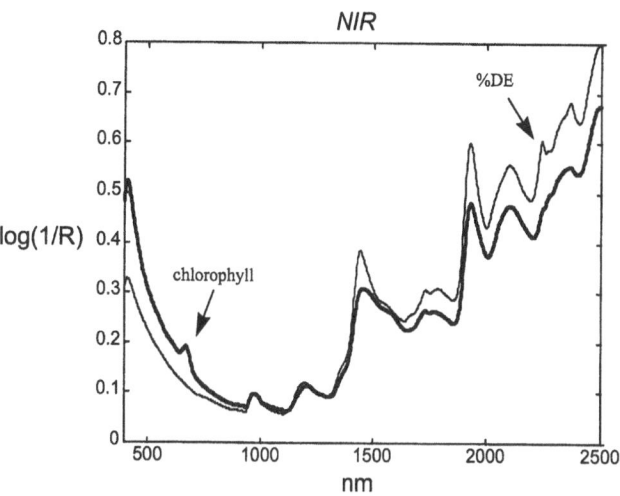

Fig. 1 NIR spectra of the same sample (a) as citrus raw material and (b) as standard extracted HM pectin

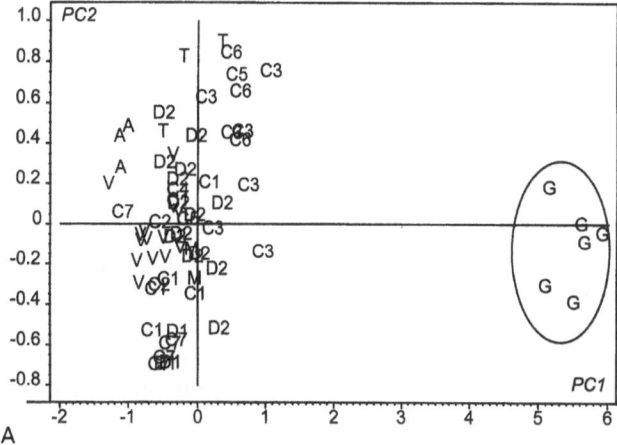

Fig. 2 PCA score plots based on NIR spectra of the (A) citrus raw material and of the (B) extracted HM pectins. The samples from producer G are enclosed in a circle

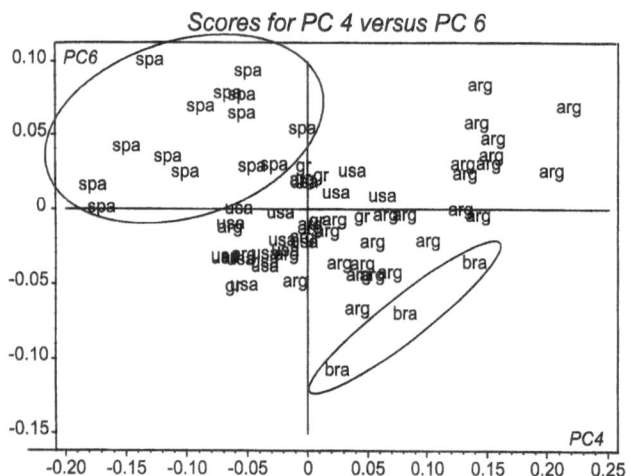

Fig. 3 PCA score plot based on the NIR spectra of the citrus raw material. The samples from Spain (spa) and Brazil (bra) are enclosed in a circle. Other producer countries included in the plot are Greece (gr), Argentina (arg) and USA (usa)

Table 2 PLS correlations between NIR spectra of HM pectins and pectin quality attributes

HM-pectins	NIR		
	PC	R	RMSECV
SAG	7	0.81	5.50
DE	3	0.82	1.40
DE (extended)	4	0.98	2.83

PC is the number of principal components or latent variables. R is the correlation coefficient and RMSECV is the root mean square error of cross-validation.

good precision. If the sample set with large variation in DE is included in the PLS modelling (extended), the correlation increases and so does the RMSECV. This is normal when the variation in the sample set is strongly increased. The fact that good PLS models to DE can be made without a large variation in the data set supports the conclusion that even small changes in DE give rise to detectable spectral changes.

As pectins usually are classified according to their degree of esterification, a rapid spectral model for the prediction of DE in the extracted HM pectins has great potential as a tool for internal quality control. In the following section we will focus on the NIR spectral PLS model to DE and attempt to refine the model by applying more refined chemometric procedures.

present in the standard extracted pectin samples. Perhaps it should be noted that the pectin raw material from producer G is only exceptional in the spectral characteristics, as it was not possible to separate these samples based on the physico-chemical data.

A closer look at the PCA based on the spectral NIR ensemble of the raw material (I) revealed that it was possible to separate samples originating from different countries. Figure 3 illustrates how a PCA score plot using PC 4 and PC 6 is clearly able to distinguish samples originating from Spain and Brazil. Again this type of classification was not possible using the chemical-based (Y-data) knowledge. Other classifications may exist in the spectroscopic data, but can only be extracted by accumulating more knowledge on the sample histories. This knowledge can be provided by, for example, dialogue with fruit growers, suppliers and process engineers. We call this process of accumulating knowledge "interview validation" and consider it to be a powerful tool for batch/process monitoring and quality control.

In this study we will only consider possible NIR spectral correlations to physico-chemical parameters measured on the extracted HM pectin samples. We set up the investigation as a screening experiment, measuring all the samples with VIS/NIR and then subsequently attempted to build PLS models for all available physico-chemical parameters. This procedure resulted in the two spectral PLS models listed in Table 2. Visual (VIS) information is treated together with the NIR information, due to instrumental considerations. In the case of the extracted pectins, NIR spectroscopy is capable of predicting the degree of esterification, DE, as well as the gel strength, SAG, with

Zooming in on degree of esterification of HM pectins

In Fig. 4 the spectra of the five extreme pectin samples (III) are displayed together with the normalized covariance spectrum with respect to DE. As indicated by the figure, by far the strongest co-variance is found in the small peak at 2244 nm. This small combination band (carbonyl C=O stretch + methylene C–H stretch) is found to co-vary almost perfectly with the DE. In a previous study [8] on amidated pectins with degrees of esterification between 20% and 55% we have indicated that this band should be located around 2228 nm. This information was based on another wavelength method, namely 1st derivative spectra, and thus represents the left sidelope of the peak. Other minor covariant bands are located at 1996 and 1436 nm. As practically all the information about DE is located in a very small spectral region, it might be relevant to reduce the spectral information before building correlation models. Figure 5 shows an iPLS plot of the NIR spectral information on the extracted pectin samples (II and III). In this figure the horizontal line represents the performance of a 4 PC full-spectrum PLS model (RMSECV = 2.7) and the small vertical bars indicate the performance of

Covarygram (DE)

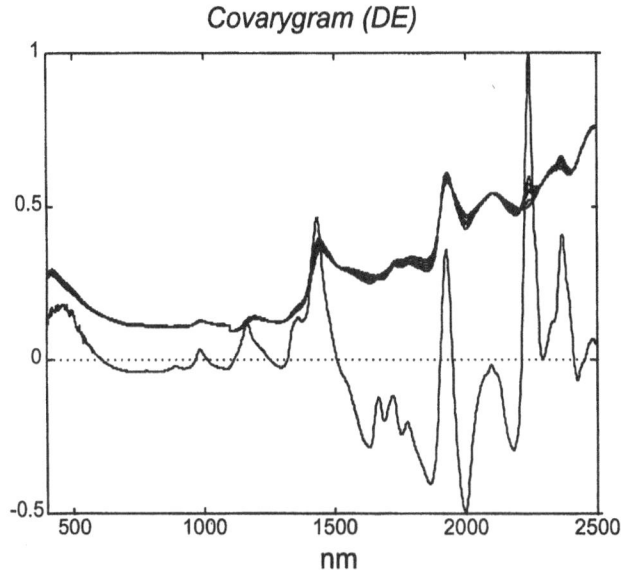

Fig. 4 Degree of esterification covarygram based on the NIR spectra of the 5 extreme samples (III). The covariation is max-normalized and the five extreme spectra superimposed

F.S. model: 4 LV's (line), INT. models: 2 LV's (bars)

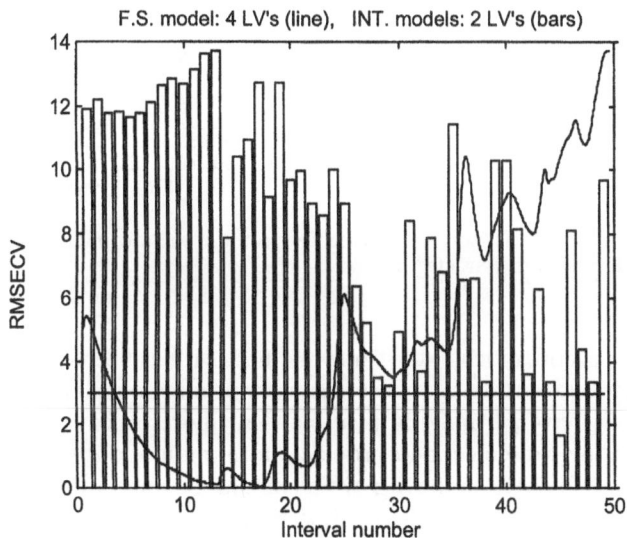

Fig. 5 iPLS plot of RMSECV (root mean square error of cross-validation) of DE based on the NIR spectra of the extracted HM pectin samples. The full-spectrum (F.S.) model (horizontal line) is based on 4 latent variables (LVs) and the subinterval iPLS models (bars) are based on 2 latent variables (LVs). The RMSECV bars are superimposed on the mean NIR spectrum

2 PC sub-interval PLS models (in this case 49 intervals). Indeed, we observe that a PLS on interval number 45 (including most of the peak at 2244 nm) performs much better than the global calibration (RMSECV = 1.7) using only 2 principal components. Obviously, the full-spectrum

Degree of esterification

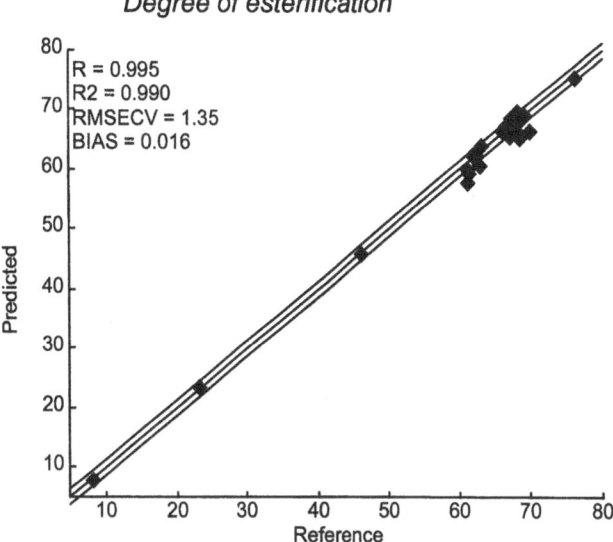

Fig. 6 Predicted vs. measured plot for sample set II + III based on the PV selected model for DE. R2 is the squared correlation coefficient providing the explained y-variance and BIAS is equal to $\Sigma(y_{pre} - y_{ref})/n$ where n is the number of samples

model needs more PCs to remove interferences, whereas the subinterval PLS models require fewer PCs due to less spectral interferences. The iPLS plot in Figure 6 also reveals that the visual region, as expected, contains no information about the DE. The iPLS model could only be slightly improved (RMSECV = 1.55) by translating and altering the size of the interval window (See Table 3). The optimal interval from 2206 to 2316 nm now includes the entire peak centered around 2244 nm (Table 3). The next step in our refinement procedure was to establish whether or not a filter instrument which only measures a few wavelengths can replace a full-spectrum instrument for this purpose. We used the selection method: principal variables to select a few wavelength variables which describe as much of the (DE) co-variant information as possible. By selecting only 5 variables (Table 3) we obtained an MLR model which was superior (RMSECV = 1.35) to the optimal iPLS model. This model illustrated in Fig. 6 is a so-called predicted vs. measured plot. As is apparent, this quantitative model is able to account for the five extreme samples ranging from 7% to 75% DE as well as for the 25 samples for which DE has been measured in the range from 60% to 69%. However, the 25 samples fall into two clusters in the ranges 60% to 62% DE and 65% to 69% DE which both have a standard deviation of less than 1% DE (0.8 and 0.9, respectively). The cluster with the lower values of DE are all from the same producer. As we estimate the standard deviation on the reference measurement to be approximately 1.5% DE,

172

S.B. Engelsen et al.
New approaches to rapid spectroscopic evaluation of pectins

Table 3 Zoom table with results from iPLS and PV on the NIR spectral model for the degree of esterification, DE

NIR DE	PLS	iPLS	opt-iPLS	PV
Variables	400 → 2498	2248 → 2288	2206 → 2316	2494, 2244, 1854, 1096, 1942
PC	4	2	2	–
RMSECV	2.71	1.68	1.55	1.35
R	0.98	0.99	0.99	1.00

Variables are the used (selected) wavelengths in nanometer. PC is the number of principal components or latent variables. RMSECV is the root mean error of cross-validation and R is the correlation coefficient.

we have no possibility of attempting to improve the model further by, for instance, making local models of the clusters.

Fluorescence spectroscopy and pectins

While NIR represents a direct and robust spectroscopic method which has already been adapted to the industrial environment, fluorescence spectroscopy represents an indirect and highly overseen spectroscopy which is only in its infancy with respect to industrial applications. Fluorescence spectroscopy is an extremely sensitive spectral method capable of detecting fluorescent trace substances such as flavonoids, amino acids, phenolic substances and NADH. Although citrus pectins are not fluorescent, the presence of fluorescent trace substances contributes to rather significant fluorescence spectra of the pectin samples. Figure 7 displays the fluorescence spectra at excitation wavelengths 330 and 414 nm for both raw material (I) and standard extracted pectin (II) samples. At excitation 330 nm an increase in fluorescence was observed due to up-concentration of fluorophores during the extraction process. At excitation wavelength 415 nm the opposite phenomenon is seen. In this case, the fluorophore with peak emission at 680 nm is completely removed during extraction. This peak originates from plant pigments (chlorophyll) and demonstrates clearly that the extraction process is efficiently removing all the chlorophyll.

Fluorescence spectroscopy might not be a suitable technique to monitor and predict physico-chemical parameters in pectin, as correlations have to be indirect via fluorophores such as secondary metabolites in plant cells. However, it has a great untapped potential for process and quality control of plant materials, as it has the potential to provide a fingerprint of the individual raw material batches, the process and the final product. By combining this rapid, non-destructive method with modern chemometric principles as demonstrated above, fluorescence spectroscopy can be used as a tool for process and quality control from the citrus fruit to the production

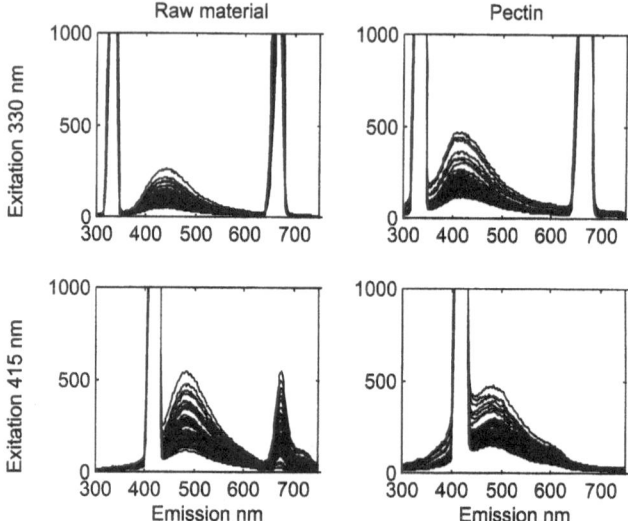

Fig. 7 Comparison of fluorescence spectra of the raw material and the standard extracted pectins

of pectin, much in the same manner as it has been demonstrated for the sugar beet to sucrose production chain [16]. Moreover, it may be possible that fluorescence spectroscopy can detect and provide information about samples that deviate in flavor, which might be relevant in the quality control of the final pectin powder.

Chemometric analysis of fluorescence spectra

When a PCA was performed on the fluorescence spectra, it was possible to perform classification quite similar to that done on the basis of the NIR spectra (Figs. 2 and 3). However, the ability to classify producers and countries was most prominent in the NIR PCA models.

As in the case of NIR, we performed a screening experiment measuring all the samples with fluorescence and subsequently attempted to build PLS models for all the physico-chemical parameters. In contrast to the

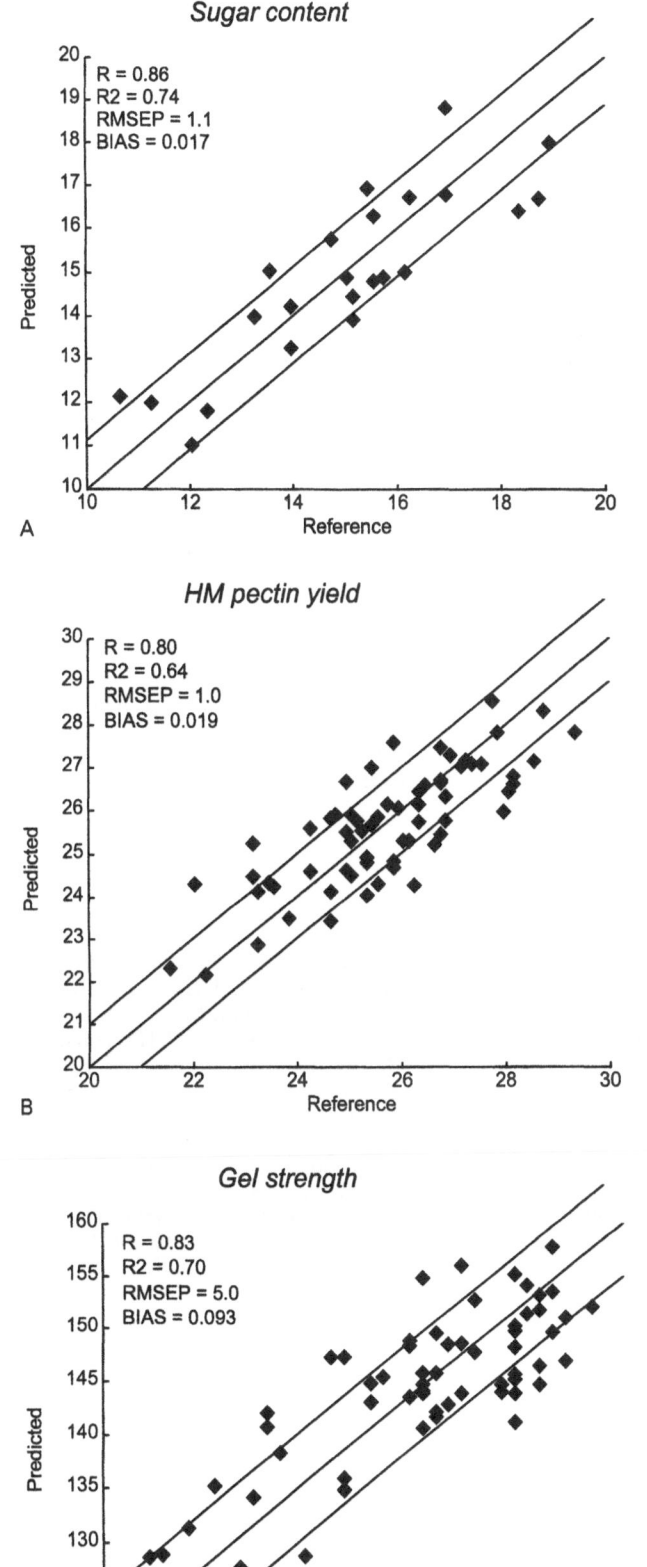

corresponding NIR analysis, PLS models using the fluorescence data on the extracted pectin did not lead to any useful correlation information. Apparently, the extraction process has removed the fluorescent correlation information and quite possibly transferred the information to the filter precipitate or the soluble phase. However, in the case of classification of the pectinic raw material samples (I), fluorescence spectroscopy was capable of providing reasonable models for sugar content, HM pectin yield and HM pectin gel strength.

Figure 8A shows a predicted vs. measured plot of a 2 principal component PLS model for sugar content in the citrus raw material. This simple model, which has a strong influence from the chlorophyll peak (negative correlated) in the emission spectrum using excitation radiation with 505 nm wavelength, has a correlation coefficient of 0.86 and a RMSECV of 1.1%. In our opinion there is no doubt that such a model can be realized and further refined for routine quality control of the raw material. However, for any further validation and optimization it will be necessary to include a larger sample set and to optimize the spectral repeatability by, e.g., improving the sampling procedure and by including true replicates. This also holds for the 8 component PLS model for the HM pectin yield displayed in a predicted vs. measured plot in Fig. 8B. As a rule of thumb PLS calibration usually require 5–10 samples per principal component (the more components, the more samples are required). For this reason about 50–80 samples are required for the calibration procedure, leaving no further samples available for test set evaluation. It is intriguing, however, that it is apparently possible to predict the pectin yield from fluorescence spectra measured on the citrus raw material. The reason has to be found in indirect information from, e.g., fluorescent secondary metabolites detected in the cell walls. However, it should be born in mind that also a direct spectroscopic method based on NIR spectroscopy will have to unravel the relevant information from multiple interferences such as varying amounts of plant materials such as cellulose, varying amounts of plant pigments such as chlorophyll, varying amount of water, etc. In addition, the model has to be able to differentiate between extractable and non-extractable pectin. The model with a cross-validated correlation coefficient of 0.8 and a RMSECV of 1% for sample set I ranging from 21% to 29% can be classified as good but preliminary. There still remains a

Fig. 8 Predicted vs. measured plot for the cross-validated PLS models based on fluorescence spectra of sample set I: (A) the sugar content; (B) HM pectin yield and (C) the SAG gel strength. R2 is the squared correlation coefficient providing the explained y-variance and BIAS is equal to $\Sigma(y_{pre} - y_{ref})/n$ where n is the number of samples

challenging future chemometric task to further validate and implement the model. Obviously the amount of extractable HM pectin in the citrus raw material (YIELD) is an important factor from the point of view of raw material supply and purchase, since a successful predictive model can be used for introducing payment for the active compound.

Finally, it was possible to develop a PLS model for predicting the HM pectin gel strength from the fluorescence (Fig. 8C). This cross-validated model, which is remarkably simple using only 4 PCs, has a correlation coefficient of 0.83 and is able to predict the gel strength ranging from 122 to 160 (the only sample with a value higher, namely 184, was a strong outlier in the model, indicating perhaps that the reference value was erroneous) with an RMSEP of 5. Together with the two abovementioned PLS models this model underlines the great untapped potential of using fluorescence spectroscopy in the pectin and related industries, especially as spectrophotometers become faster and more robust.

Conclusions

We have spectrally analyzed a unique set of HM pectin samples and compared it to related raw materials. At first we investigated whether non-invasive spectroscopic methods have the potential to characterize some of the many possible variations which raw materials employed in pectin production may possess owing to production site, weather conditions, pre-processing and storage. It was found that NIR and fluorescence spectroscopy have possibilities of characterizing pectinic raw materials in a complementary way.

An accurate quantitative model for predicting the degree of esterification in extracted HM pectins from NIR spectra covering the entire possible range was developed and analyzed. During this process several new chemometric tools were used to illustrate and to focus on the relevant spectral regions for information. Finally, preliminary and indirect spectral models for the prediction of yield and sugar content based on fluorescence spectra of the citrus raw material are proposed.

Acknowledgements We are grateful to Jan Staunstrup and Jørgen Søderberg, Copenhagen Pectin A/S (Hercules Inc.) who provided us with the pectin samples and related physico-chemical data. This investigation was made possible by funds from the Føtek-2 programme to professor Lars Munck and by funds from the Danish Research Councils and the Department of Education to the Centre for Advanced Food Studies (LMC).

References

1. McNeil M, Darvill AG, Fry SC, Albersheim P (1984) Ann Rev Biochem 53:625
2. Carpita NC, Gilbeaut DM (1993) The Plant Journal 3:1–30
3. Carr JM, Sufferling K, Poppe J (1995) Food Technology 7:41
4. O'Neill M, Albersheim P, Darvill A (1990). In: Dey PM (ed) Methods in Plant Biochemistry, Vol 2, Carbohydrates. Academic Press, England, pp 415–441
5. Darvill AG, McNeill M, Albersheim P (1978) Plant Physiol 62:418
6. O'Neill MA, Warrenfeltz D, Kates K, Pellerin P, Doco T, Darvill AG, Albersheim P (1996) J Biol Chem 271:22 923
7. Engelsen SB, Cros S, Mackie W, Pérez S (1996) Biopolymers 39:417
8. Engelsen SB, Nørgaard L (1996) Carbohydrate Polymers 30:9
9. Geladi P, Kowalski BR (1987) Analytica Chimica Acta 185:1
10. Martens H, Jensen SA (1983) In: Holas J, Kratochvil J (eds) Progress in Cereal Chemistry and Technology, Vol 5a, Elsevier, Amsterdam, pp 607–647
11. Höskuldsson A (1994) Chemometr Intell Lab Syst 23:1
12. Wold S (1978) Technometrics 20:397
13. Martens H, Næs T (1993) Multivariate Calibration, Wiley, New York
14. Horváth L, Norris KH, Horvárth-Mosonyi M, Rigó M, Hegedüs-Völgyesi (1984) Acta Alimentaria 13:355
15. Rolin C (1993) In: Whistler RL, BeMiller JN (eds) Industrial Gums, Polysaccharides and Their Derivatives, Academic Press, San Diego, CA, pp 257–294
16. Nørgaard L (1995) Zuckerind 120:970
17. Wold S, Esbensen K, Geladi P (1987) Chemometr. Intell Lab Syst 2:37

Progr Colloid Polym Sci (1998) 108:175–191
© Steinkopff Verlag 1998

O. Midttun
J. Sjöblom
O.M. Kvalheim

A multivariate study of diffuse reflectance infrared profiles of resin fractions from crude oils

O. Midttun (✉) · J. Sjöblom
O.M. Kvalheim
Department of Chemistry
University of Bergen
Allegaten 41
N-5007 Bergen
Norway

Abstract Interfacially active fractions were separated from deasphalte

ned crude oils using adsorption onto different solid particle surfaces followed by desorption using combinations of different solvents.
The fractions were studied using diffuse reflectance infrared fourier transform spectroscopy (DRIFT) and the resulting matrices of spectra analysed using principal component analysis (PCA). The analysis of the infrared spectra show that different functionalities can be separated using this adsorption/desorption procedure, and that the functionalities separated depend on the crude, adsorption surface and solvents used. A two step desorption procedure gave more variation in the spectra of the isolated resin fractions than a single desorption step.

Key words Crude oil – asphaltenes – resins – adsorption – infrared spectroscopy – principal component analysis

Introduction

The emulsification of crude oil and water in oil production and transport causes problems because of the high stability and viscosity of the corresponding emulsions. These emulsions are stabilized by solid asphaltene particles that precipitate as a result of changes in pressure and temperature. In the laboratory it is common to precipitate the heaviest fractions from crudes by mixing the crude with an alkane. Several similar procedures are in use, combining different *n*-alkanes, time scales and precipitation conditions [1–4]. These precipitated fractions are also called asphaltenes, but cannot be expected to be identical to the asphaltenes formed in the production. The asphaltene molecules are considered to be sheets of polyaromatics also containing alkyl chains and heteroatoms [1, 5]. It has been shown that the precipitated fractions vary with changes in the precipitation method used [3, 6]. Førdedal et al. [7] have shown that the interaction between the asphaltene particles and monomeric resins are important in determining the stability of model emulsions, and Schildberg et al. [3] found evidence of aggregation/complexation between asphaltenes and resins. Resins are usually separated from the crude using chromatographic methods [1, 3, 8]. The resins are found to be structurally similar to, but smaller than, the asphaltenes [1, 3, 9]. They are polar compounds not precipitated in the solvent precipitation. Since the definitions of both asphaltenes and resins are more practical than strictly scientific, the characteristics of both fractions are method dependent. Possible models for the interplay between the resins and asphaltene particles in emulsion stabilization are given by Førdedal et al. [7] and McLean and Kilpatric [9].

Using infrared spectroscopy the contents of the ir-active bonds in a sample can be examined. Several authors [8, 10–13] have shown infrared spectra of crude oils and some fractions separated from them. Differences in aromatic and oxygen compound contents between the fractions studied are observed.

When large numbers of data are to be analysed, principal component analysis (PCA) [14, 15] can be used to reveal information that would otherwise be difficult to extract from such large sets of data. PCA sequentially extracts the linear combinations (called principal components, PC) of the original variables in the data matrix that explains the largest fraction of the variation in the original matrix under the contstraint of orthogonality between the extracted PC. Using this procedure one can often explain the majority of the variation in the original datamatrix with just a few principal components. When dealing with spectroscopic data the benefit is evident since hundreds or even thousands of variables describe a large number of often overlapping adsorption bands. The data reduction achieved by PCA often facilitates the interpretation of the data, both with regard to similarities and differences between the samples studied, and when assessing the importance of and correlations between variables. The value of a sample on a given PC is called its score on that particular PC, and the weight of a variable on a PC is called the variable loading. PCA makes it possible to display graphically the major trends of dataset in just a few plots, revealing sample similarity in so-called score plots. Correlations between variables ad the weight of each variable on the principal components can be seen in the corresponding loading plots. PCA has previously been used to examine variation in asphaltene composition as studied by infrared microspectroscopy [13].

Pretreatment of the spectra in order to correct for baseline differences and differences in sample concentrations and amounts are often necessary to make them suitable for PCA. First, it is common to correct for baseline drift [16]. In infrared spectroscopy baselines can often be described by a first or second order polynomial. After baseline correction common pretreatment methods used to make spectra comparable are normalization by dividing each datapoint in a spctrum by a chosen reference datapoint serving as internal standard, or normalization to constant sum by dividing each point in the spectrum by the total spectrum to give a total absorbance of 1 [17].

By chromtographic separation of resins from crudes one benefits from the fact that some interfacially active molecules in deasphal, adsorb at the surface of silica particles and can be desorbed using appropriate solvents. In this work we report the results of a study of crude fractions first adsorbed on and then desorbed from several different solid surfaces using different solvents. The idea is that by changing the hydrophilic/hydrophobic surface characteristics of the particles one will adsorb fractions from the crude that show varying interfacial activity. Geise et al. [17] have characterized a variety of particles and surfaces with regard to hydrophobic/hydrophilic character. Combining different particle surfaces with sol-

vents showing varying intermolecular interactions to desorb adsorbed fractions, further increases the separation into resin fractions showing different interfacial activity in different media. The classification of solvents in solvent classes by solvent power is given by Davison [18].

Methods

Chemicals

The crudes where supplied by Elf Aquitane (crude F and crude N) and Statoil (crude V). Crude V is from a production field in Venezuela, crude F from a field in France and N from the North Sea. Some characteristics of the crudes are given in Table 1. n-Pentane, dichloromethane (DCM), methanol (MeOH) and benzene were all of p.a. quality from Merck Darmstadt, Germany. Silanol was obtained from Waters. Talc, CaO and KBr were from Merck Darmstadt, Germany. The siloxane was obtained from Elkem. Qualitative filter papers used were of type 1 from Whatman, England.

Separation of asphaltenes and resins

The asphaltenes were separated at room temperature by stirring 0.90 g of crude with 5.0 ml of n-pentane followed by centrifugation of the mixture at 2400 rpm for 10 min. To obtain the resin fractions, the supernatant from the asphaltene precipitation was mixed with the particles to adsorb interfacially active fractions that did not precipitate. The amount of particles added is given in Table 2 together with some characteristics of the particle surfaces. The silanol particles are the most hydrophilic and most

Table 1 Some characteristics of the crude oils investigated

Crude oil	Viscosity	Asphaltene content (% by weight)
France (F)	Medium	9.5
North Sea (N)	Low	1.5
Venezuela (V)	Very high	18

Table 2 Some characteristics of particles used and amounts (in grams) of particles added to preprecipitated crude to adsorb resins

Particle	Characteristics	Amount (g)
CaO	Basic	4
Silanol	Hydrophilic, Acidic	2
Siloxane	Hydrophobic	2
Talc	Hydrophobic, Basic	6

Progr Colloid Polym Sci (1998) 108:175–191
© Steinkopff Verlag 1998

acidic particles, the talc is the most hydrophobic and CaO the most basic. The mixture of deasphaltened crude/*n*-pentane and particles was stirred, centrifuged for 2 min at 1000 rpm, and the supernatant (if any) removed. The particles with the adsorbed crude oil molecules were then mixed with 50 ml of one of the solvents (benzene, DCM or 7% MeOH in DCM), stirred and centrifuged for 2 min at 1000 rpm. This was followed by decantation and filtration of the supernatant. This desorption process was repeated until the supernatant was colorless after centrifugation. The desorption process was repeated using the same procedure with another of the three solvents. The solvent was evaporated from the solvent/resin mixture at 90 °C in a rotary evaporator and then under a $N_2(g)$ blanket to ensure that all the solvent was removed. In this way the resin fraction coded ABCD was obtained. The first letter gives the crude, the second the particle type, the third letter

Table 3 Abbreviations used in coding the resin fractions ABCD. For example, the FSBD fraction were obtained from deasphaltened crude F fractions which were adsorbed on Silanol particles that were not desorbed by benzene but by dichloromethane to obtain the resin

Crude (A)	Particle (B)	Solvent (C and D)
F = France	C = CaO	B = Benzene
N = North Sea	S = Silanol	D = DCM
V = Venezuela	T = Talc	M = 7% MeOH in DCM
	X = Siloxane	

gives solvent 1 (in which the resin fraction was *not* desorbed), and the last character gives the solvent used in the second desorption step, in which the resin fraction was desorbed from the surface. The abbreviations used in naming the resin fractions ABCD are given in Table 3. Figure 1 shows the separation process.

Fig. 1 The extraction procedure

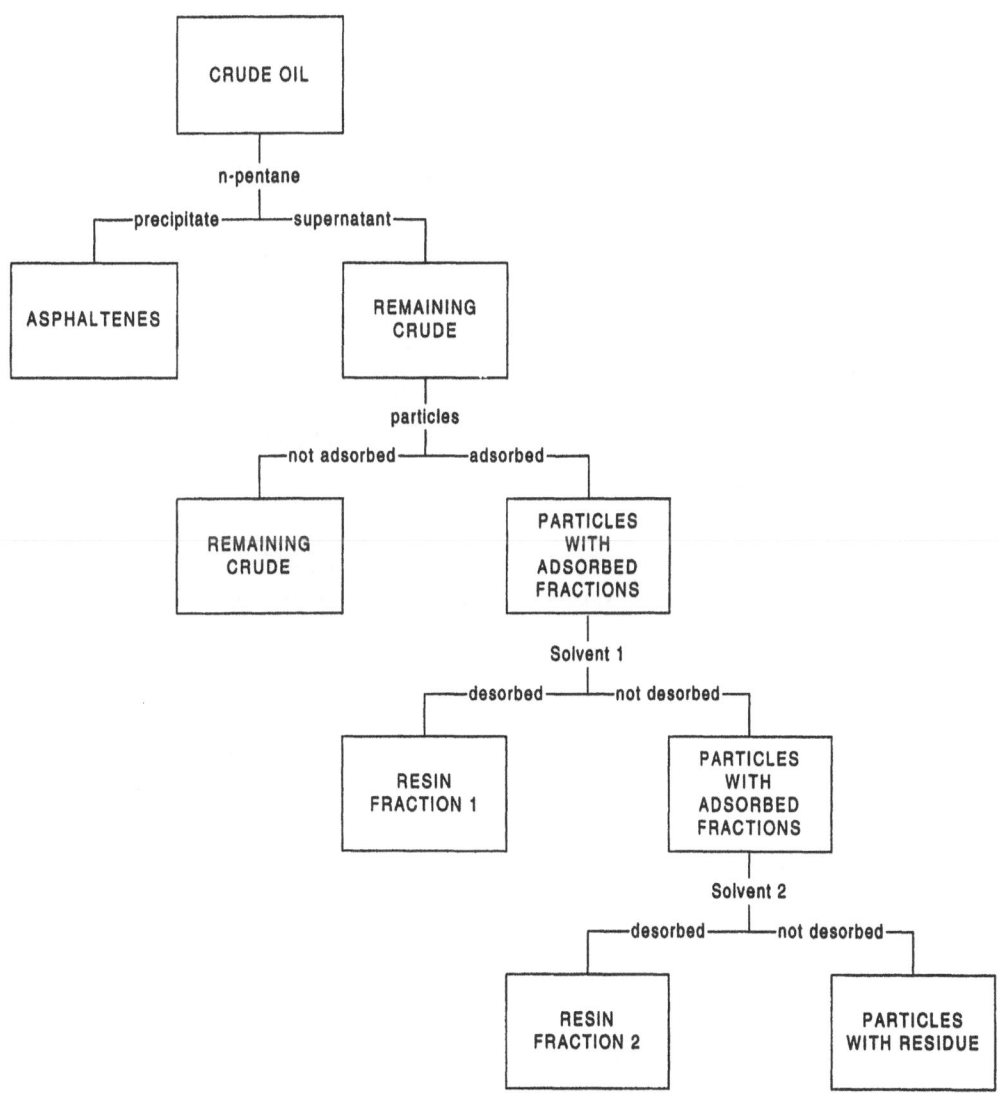

178
O. Midttun et al.
Diffuse reflectance infrared profiles of resin fractions

Diffuse reflectance infrared fourier spectroscopy (DRIFT)

The DRIFT spectra were obtained using a Perkin-Elmer 1720X spectrophotometer equipped with a DTGS detector. Spectra were taken in the wavenumber region 4000–6000 cm^{-1} using 40 scans at a resolution of 4 cm^{-1}. Cylinders of KBr pressed at 2 bars for 5 min were used as background for the spectra. Sample preparation before the aquisition of the infrared spectra consisted of making 1% solutions of the crude fractions in DCM and dripping 4 drops of this solution from a pipette onto the same cylinder of KBr powder that was used as background for that sample. The DCM were then evaporated by heating the sample to 550 °C for 5 min at before the spectrum of the sample was obtained [19].

Data treatment and analysis

The reflectance spectra were first baseline corrected using a first-order baseline correction in two parts: 4000–1800 cm^{-1} and 1800–600 cm^{-1}. Only the latter part of the spectra were used in the following PCA. The part of the spectra from 1800 to 600 cm^{-1} were normalized and then transformed using the Kubelka-Munk transformation [20]. The resulting matrices (one matrix of spectra for each crude) of second step desorption resin, crude and asphaltene spectra were then analysed using PCA. The crude and asphlatene spectra were included in the PCA for reference. Data treatment and multivariate analysis were performed using Matlab version 4.2c on a personal computer.

Results and discussion

Amounts

When using silanol particles the supernatant was light yellow after stirring. For the siloxane particles the supernatant was light brown. When using CaO and talc there were no supernatant after centrifugation. This can be explained by the fact that the volume of CaO and talc particles added were larger than the volume of deasphaltened crude because of their smaller specific surface area. The first solvent step (which gave the fraction called resin fraction 1 in Fig. 1) gave in all cases fraction with large yields (>50%) and spectra similar to the respective deasphaltened crudes. These fractions are therefore believed to consist mainly of deasphaltened crude trapped in the volume between the particles and not adsorbed on them. Because of this, these spectra were not analysed any further. For all three crudes the resins desorbed from siloxane and silanol using the MeOH/DCM as solvent in solvent

Table 4 Second step solvent treatments that did not result in any desorbed resins

Crude F	Crude N	Crude V
FSMB	NCDB	VCDB
FSMD	NCMB	VCMB
FXMB	NCMD	VCMD
FXMD	NSMB	VSMB
	NSMD	VSMD
	NXMB	VXMB
	NXMD	VXMD

step 2 gave the largest resin yield. For crude F the yield in these fractions was ~1.5% of the crude. From crude N the same fractions gave about 1% yield and from crude V 4–5%. The remaining resins all yielded less than 0.5%. Some of the second step solvent treatments gave no resins. This was the case for all the crudes when MeOH/DCM were used as solvent 1 and silanol or siloxan particles were used to adsorb the resins. For crude N and V the use of CaO and MeOH/DCM as solvent 1, or DCM as solvent 1 and benzene as solvent 2, also gave no resins. The particle/solvent combinations that gave no resins are given in Table 4. It then remained spectra from 22 crude F fractions and 19 crude N and V fractions to be analysed.

Inspection of the spectra

The baseline corrected, normalized and Kubelka-Munk transformed DRIFT spectra of the crudes are given in Fig. 2, and the asphaltene spectra are shown in Fig. 3. The aliphatic CH$_2$ and CH$_3$ stretching bands are seen at around 2900 cm^{-1}, and their deformation bands are at 1465 and 1377 cm^{-1}. Comparing the different crude spectra, the largest intensity in the band at 1600 cm^{-1} (which is assigned to aromatic and conjugated C=C bonds) is found for crude V and the lowest intensity for crude N. In the asphaltene spectra the highest intensity in this band is found in the asphaltene from crude V and the lowest in crude N asphaltene. It is also seen that there is a relative increase in the absorption in the region 1500–1100 cm^{-1} in the asphaltene spectra compared to the spectra of the crudes. Two sharp peaks at 730 and 720 cm^{-1} are seen in the spectrum of the asphaltene from crude N.

The baseline corrected, normalized and Kubelka-Munk transformed spectra of the region 1800–600 cm^{-1} for the crude, asphaltene and second step desorption resins are shown in Figs. 4–6. We observe significant differences both internally in the matrices from each crude and between the matrices from different crudes. Band assignments for DRIFT spectra are shown in Table 5.

Progr Colloid Polym Sci (1998) 108:175–191
© Steinkopff Verlag 1998

Fig. 2 DRIFT spectra of the three crudes

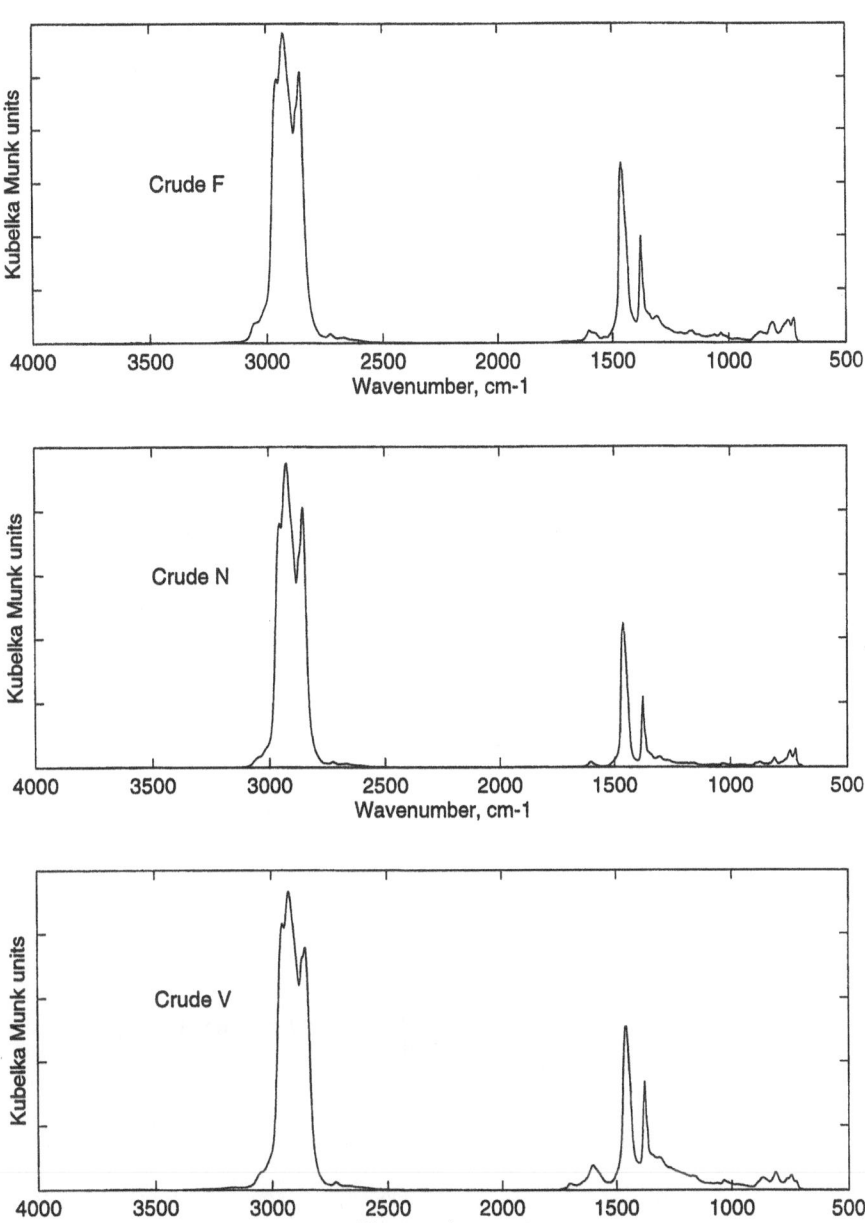

Intercrude similarities and differences

It is seen that few resin fractions are obtained after using the MeOH/DCM mixture as solvent 1. Only talc gives detectable amounts of second step resins from all three crudes after first using this solvent. This may be explained by the hydrophobic nature of the talc surface and the polar/hydrogen bonding nature of the MeOH/DCM solvent. Crude F also gives resins after using this solvent in solvent step 1 when using CaO as adsorbent. The three crudes give matrices containing mostly the same bands.

Aliphatic and aromatic bands are found in all the spectra, while ether/sulfoxide, ester and different carbonyl bands are found in one or more spectra from all crudes. It is seen that the North Sea crude (N) has the largest variation in the peaks around 700 cm^{-1} and the lowest relative intensity in the region 1750–1500 cm^{-1}. The crude from Venezuela (V) has no fraction with a large peak around 1030 cm^{-1}, and is the only crude to have fractions with a large peak centered at about 1570 cm^{-1}. Crude F is the only crude having fractions with an intense, narrow peak at 1650 cm^{-1}.

Fig. 3 DRIFT spectra for the
three asphaltene fractions

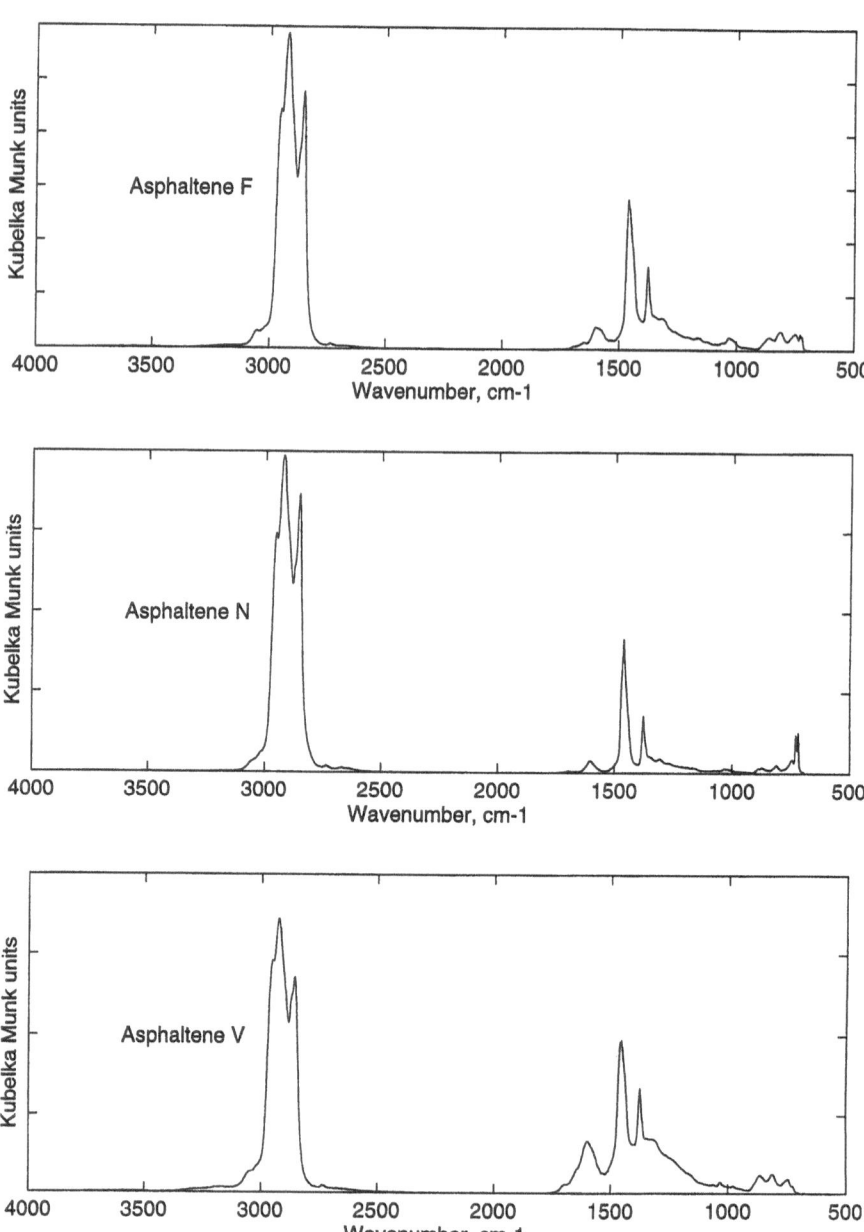

Crude F

This crude (spectra shown in Fig. 4) gives resin fractions that have the highest intensity in the peak at 1030 cm^{-1} when siloxane particles are used. Table 5 assigns the peak centered around 1030 cm^{-1} to ether and/or sulfoxide functionalities. Figure 4 further shows that all the second step solvent treatments that gave a detectable amount of resins from the siloxane adsorbed fractions have approximately the same relative intensity in this peak. At the same time this particle type gives low intensities in the region above 1500 cm^{-1}, except for the FXBD fraction.

Some of the resins desorbed from the very hydrophobic talc particles show a good separation of a peak at 1730 cm^{-1} from the carbonyl groups absorbing at lower wavenumbers. The intensity of this band is particularly high when using the MeOH/DCM mixture as solvent in desorption step 1 and DCM in step 2, indicating that molecules containing the responsible functionality alone are not desorbed by the MeOH/DCM solvent. This is also seen from the spectra of fractions were MeOH/DCM is used as solvent 2; they do contain this peak, but it is accompanied by an amide carbonyl peak at 1650 cm^{-1}. The peak at 1730 cm^{-1} is also seen in the FTBD, FTBM

Progr Colloid Polym Sci (1998) 108:175–191
© Steinkopff Verlag 1998

Fig. 4 DRIFT spectra (in Kubelka–Munk units) for the studied region (1800–600 cm^{-1}) for all the studied fractions from crude F

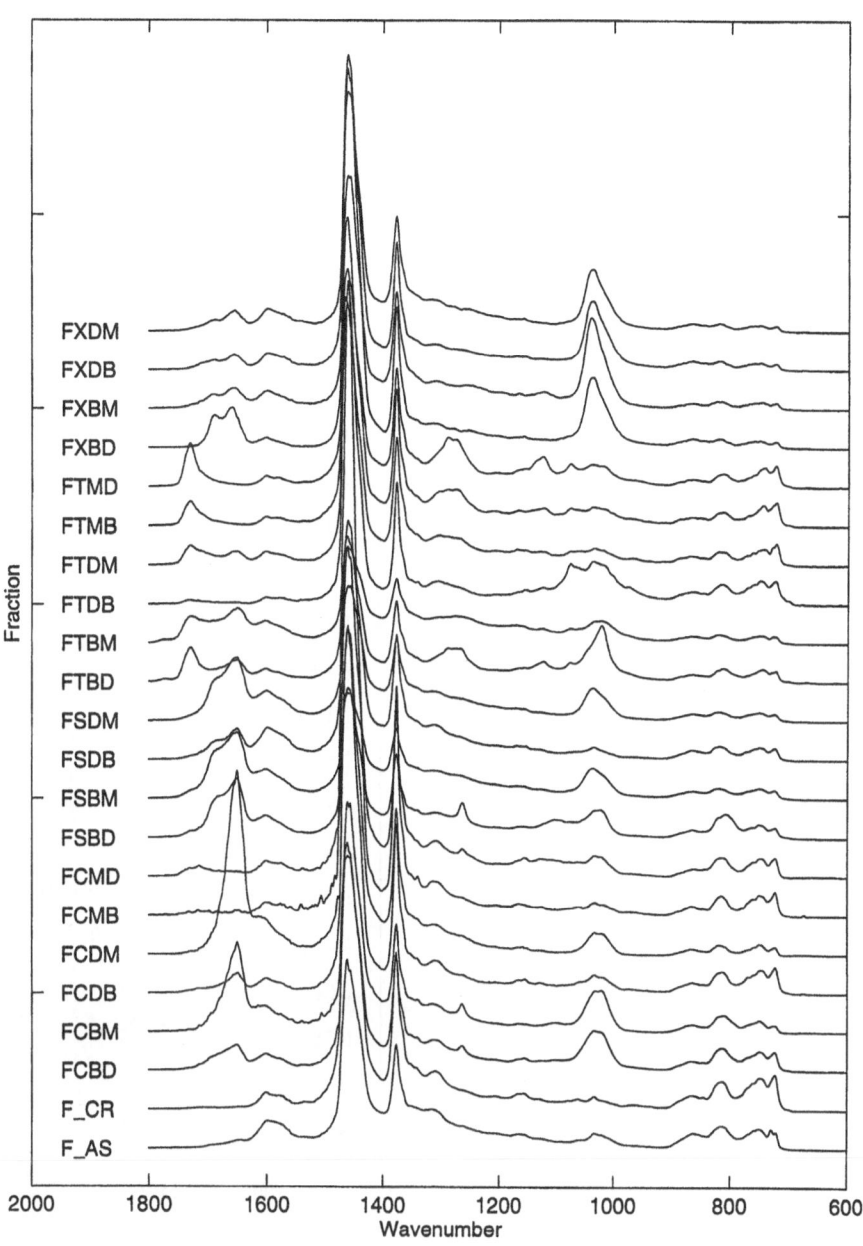

FTDM and FTMB resins. It is not seen in any CaO, silanol or siloxane resins. The fact that this peak is in all cases accompanied by a band at 1280 cm^{-1} indicates the presence of ester functionalities, see Table 5. Using silanol as adsorbent it is seen that all these resin fractions have an amide band at 1650 cm^{-1} with a shoulder on the higher wavenumber side, indicating that aldehyde and/or ketone functionalities are included in the desorbed fractions. This group of peaks is most intense when MeOH/DCM or pure DCM are used as second step solvent, except when DCM is used after MeOH/DCM when no resins were obtained. It is also seen that the FCDM fraction has a large, sharp

peak centered at 1650 cm^{-1}. This sharp amide peak is also observed at a lower intensity in the FCBM resin spectrum and at low intensities in the other resins desorbed from CaO, except when MeOH/DCM is used in solvent step 1. It is concluded that the MeOH/DCM solvent removes all molecules containing this functionality which can be desorbed by DCM or benzene from the CaO surface.

From Fig. 4 it is seen that it is in fact possible to separate these different functionalities from one another. The amide functionality is well separated in the FCDM resin fraction, the ether/sulfoxide functionality, in several

Fig. 5 DRIFT spectra (in Kubelka–Munk units) for the studied region (1800–600 cm^{-1}) for all the studied fraction from crude N

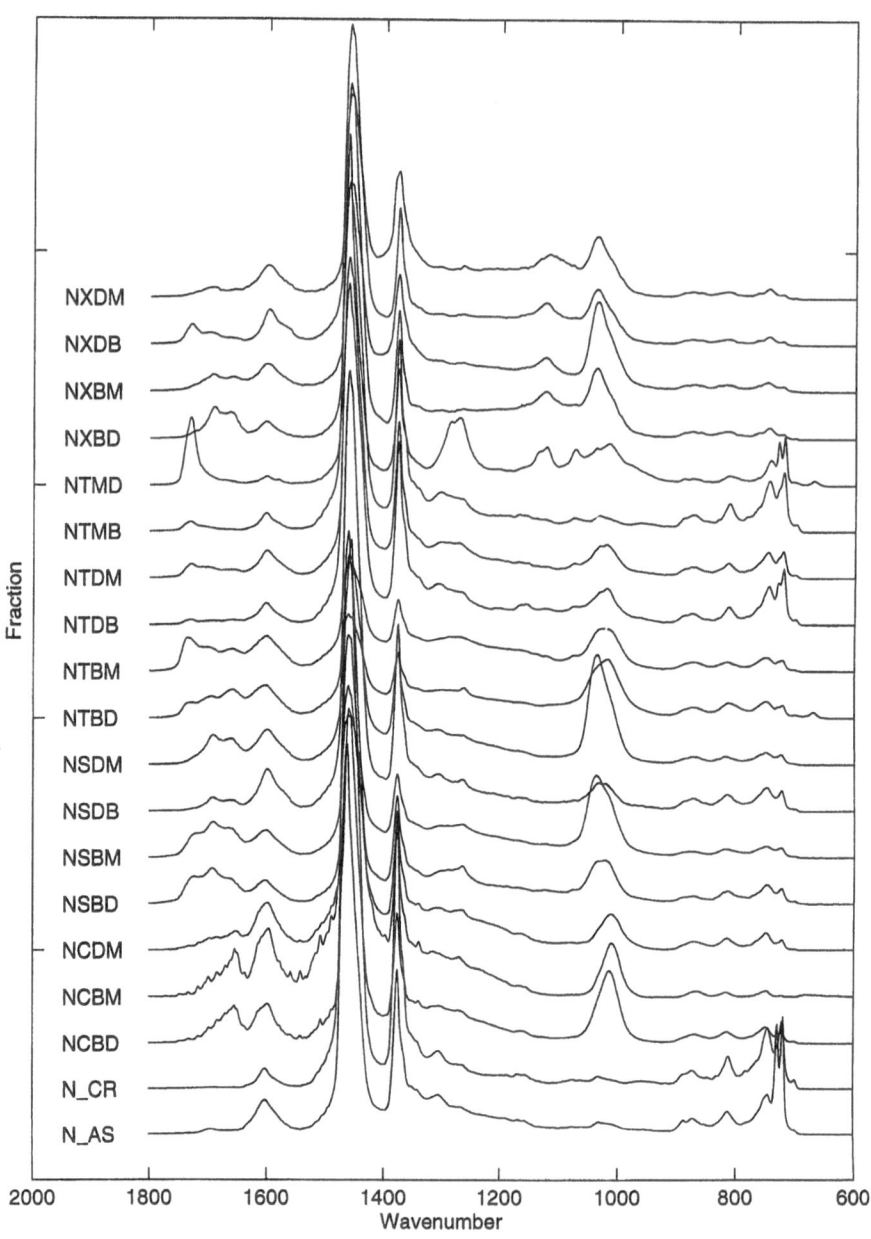

siloxane resins and the ester functionality is well separated in FTMD.

Crude N

This light crude from the North Sea has the lowest asphaltene content of the three crudes (Table 1). Figure 5 shows that ether/sulfoxide groups (the peak around 1030 cm^{-1}) are present with a relatively high intensity in several of the resin fractions. It has a large relative intensity in one or more fractions desorbed from all particles except talc.

As for crude F, all the siloxane resins contain the ether/sulfoxide peak. These siloxane resins also include a peak at 1120 cm^{-1}. Assigment of this peak is not certain, but several sulfur functionalities absorb in this region [20]. The talc resin fraction NTMD contains strong ester bands, and they are also observed in other talc, siloxane and silanol spectra, but at much lower intensities. The talc fractions have relatively low intensity in the ether/sulfoxide peak, and large variation in the aromatic CH deformation peaks from 900 to 700 cm^{-1}, with largest intensities in the fractions obtained when benzene is used as solvent 2 and in the NTMD fraction. These molecules then have

Progr Colloid Polym Sci (1998) 108:175–191
© Steinkopff Verlag 1998

Fig. 6 DRIFT spectra (in Kubelka–Munk units) for the studied region (1800–600 cm^{-1}) for all the studies fractions from crude V

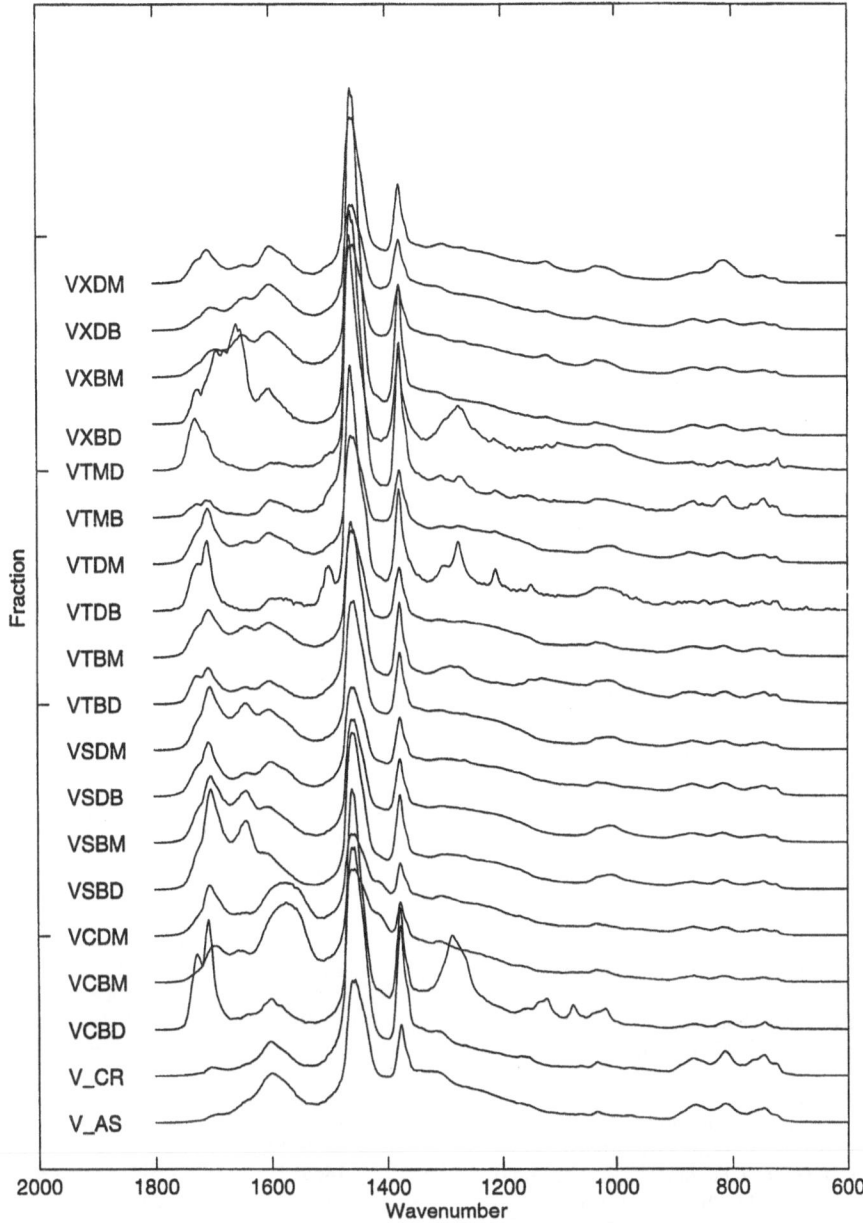

a low solubility (rather than being adsorbed on the surface of the talc particles) in MeOH/DCM and are best desorbed by benzene, since the intensity in the NTBD and NTBM spectra is low. The two sharp peaks in the aromatic C–H deformation region around 700 cm^{-1} are unique to this crude with respect to the other two studied in this work. According to Lin-Vien et al. [21] these peaks indicate the presence of mono- or disubstituted-benzene rings. The talc fractions desorbed when using benzene as solvent 1, contain some peaks in the carbonyl region 1730–1650 cm^{-1}. These peaks are also present in all the resins desorbed from silanol and the ester bands are included in the NSBD and NSBM spectra. All the silanol

resins also contain the ether/sulfoxide peak at 1030 cm^{-1}. The sharp amide carbonyl peak centered at 1650 cm^{-1} is present in those CaO resins which were not desorbed by benzene, but with a lower intensity than in the resins from crude F that contains this sharp band. None of the CaO resins contain significant carbonyl bands at wavenumbers higher than the amide carbonyl band.

Crude V

This is the heaviest of the three crudes analysed, being almost bitumenic. No large ether/sulfoxide peak is observed in any of the fractions studied. Large differences are seen in the region 1650–1730 cm^{-1} between the siloxane resins. The amide band is most intense in the VXBD fraction and least intense in VXDM. The ester bands are seen in the VTBD, VTDB and VTMD spectra, but their band shapes are different from each other and from the ester bands seen in the spectra of resin fractions from crude F and N. The amide band at 1650 cm^{-1} is of low intensity in the talc resins. The carbonyl bands at ~1700 cm^{-1} are stronger than the amide band in all silanol fractions, showing that other carbonyl groups than amide carbonyl are more abundant in these resins. The VCBM and VCDM resins contain a broad band centered at 1580 cm^{-1} which distinguish the spectra of these two fractions from the other obtained in this work. It is reasonable to assume that this band consists of the aromatic band centered at 1600 cm^{-1} and at least one more band. The most likely candidate for the second band being the N=O symmetric stretching mode in nitro compounds, which, according to Lin-Vien et al. [21], have an absorption band in the 1600–1480 cm^{-1} region. This assignment is strengthened by the presence of a band at 1400 cm^{-1}

where the asymmetric stretching mode of N=O can be found (Table 5). The ester bands are most pronounced in the VCBD resins. This is the only of the three crudes to give large bands from ester in resins desorbed from other particles than talc.

Principal component analysis

The principal component analysis describes and quantifies the variation present in the data sets. The amount of variation described by the first 4 principal components are given for each crude in Table 6. PC1 explains approximately 79% of the variation in the spectra for crudes F and N, while the first PC explains 57% for crude V. The first four PCs together explain 98, 95 and 94% of the total variation for crudes F, N and V, respectively. From Table 7 it is seen that the first PCs extracted from each dataset are quite similar, having cosines of 0.97 (cosine to the angle

Table 5 Band assignments for DRIFT spectra. The assignments are taken from Refs. [20, 21]

Functional groups	Absorption bands (cm^{-1})
OH (stretch, free)	3650–3600
O–H, N–H (stretch, H-bonded)	3500–3200
C–H (stretch, C=C and aromatics)	3050–3000
CH$_3$, CH$_2$ (stretch in aliphatics)	2950–2850
Ester (C=O stretch)	1750–1710
Ketones (C=O stretch)	1730–1700
Aldehydes (C=O stretch)	1730–1690
Amide (C=O stretch)	1670–1620
C=C (conjugated and aromatic)	1600
Nitro (N=O symmetric stretch)	1600–1480
C–CH$_3$, C–CH$_2$ (asymmetric bending)	1465
Nitro (N=O antisymmetric stretch)	1400–1300
C–CH$_3$ (symmetric bending)	1377
Ester (C–O stretch)	1320–1100
Aromatic ether (C–O stretch)	1300–1200
ROH (C–O stretch or O–H def. (coupled))	1210–1000
Aliphatic ether (C–O stretch)	1150–1050
Sulfoxide (C$_2$S=O)	1060–1020
Aromatic C–H def.	900–700

Table 7 Cosines between the principal components

Crude F	Crude N			
	PC1	PC2	PC3	PC4
PC1	0.97	0.03	−0.19	0.05
PC2	0.05	0.50	0.19	0.12
PC3	0.20	−0.34	0.73	−0.10
PC4	−0.04	−0.45	−0.17	0.78

Crude F	Crude V			
	PC1	PC2	PC3	PC4
PC1	0.94	−0.06	0.23	0.16
PC2	−0.05	0.73	0.52	0.02
PC3	−0.12	−0.18	0.16	0.09
PC4	0.16	0.41	−0.59	−0.03

Crude N	Crude V			
	PC1	PC2	PC3	PC4
PC1	0.88	−0.08	0.31	0.18
PC2	−0.09	0.26	0.45	0.34
PC3	−0.35	−0.08	0.26	0.17
PC4	0.12	0.38	−0.42	0.28

Table 6 Variation of data sets described by each of the first four principal components for the three crudes

PC	Crude F		Crude N		Crude V	
	% variation	cum. variation	% variation	cum. variation	% variation	cum. variation
1	78.7	78.7	79.3	79.3	57.1	57.1
2	12.1	90.8	7.9	87.2	18.2	75.2
3	4.9	95.7	4.9	92.2	13.2	88.4
4	1.9	97.6	2.8	95.0	5.5	93.9

Progr Colloid Polym Sci (1998) 108:175–191
© Steinkopff Verlag 1998

Fig. 7 Loadings for the first four principal components from PCA of the matrix of spectra of crude F fractions

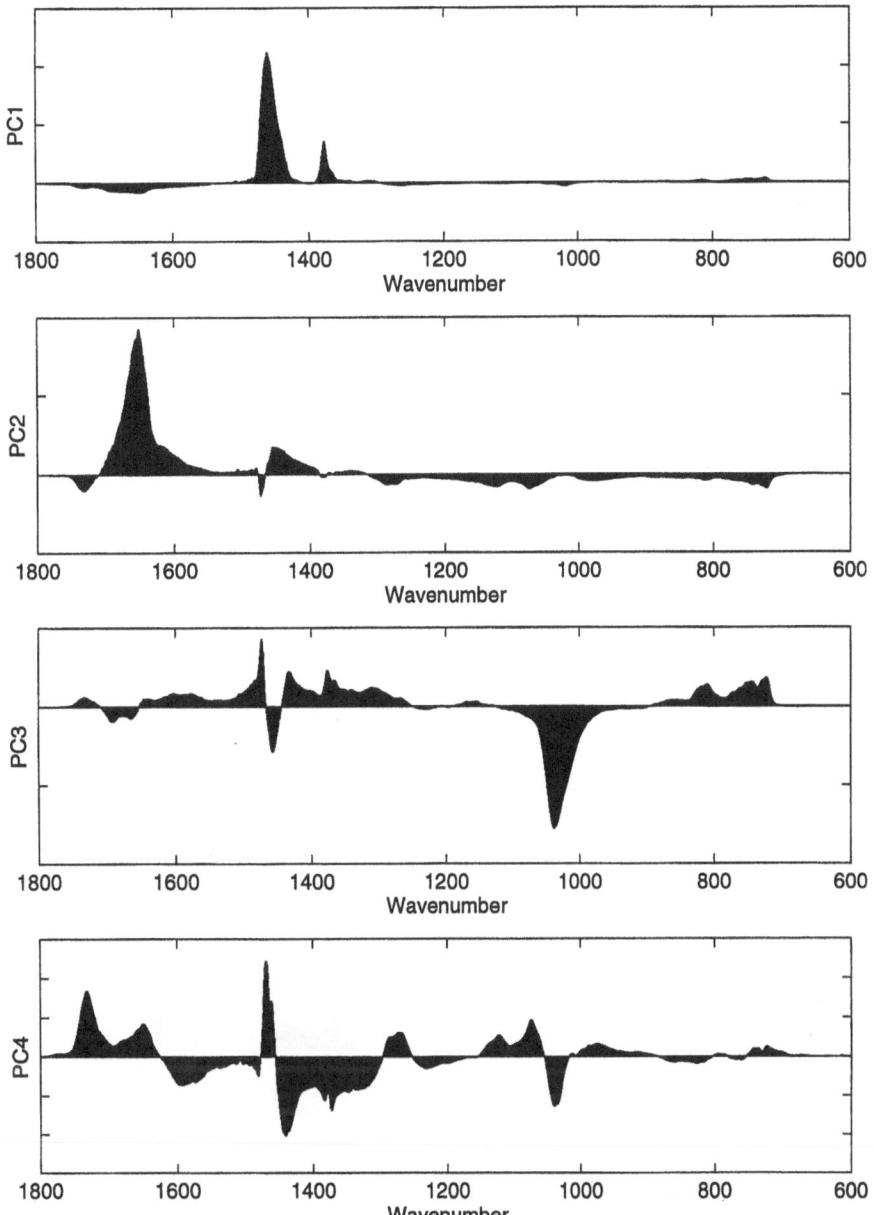

between PC1 from crude F and PC1 from crude N), 0.94 (PC1 from crude F vs. PC1 from crude V) and 0.88 (PC1 from crude N vs. PC1 from crude V). The most disimilar being PC1 from crude N (the lightest crude) and PC1 from crude V (the heaviest crude). The remaining principal components show to significant similarity, all the other cosines have absolute values lower than 0.79. The loading plots from the principal component analysis are shown for the three crudes in Figs. 7–9, and the score plots are in Figs. 10–12. From the loading plots PC1 is seen to account mainly for the aliphatic CH_2 and CH_3 bending vibrations centered around 1377 and 1465 cm^{-1}, indepen-

dent of which crude is analysed. This explains the similarity of the first principal components described by the cosines between their loading vectors. It is also seen from Figs. 7–9 that the loading patterns for PC2, PC3 and PC4 of the crudes are quite different. Since the variation described by PC1 from crude V is much lower than for crudes F and N, and observing that the loadings from the aliphatic bands are not large on any other PCs for any of the crudes, it is demonstrated that the studied fractions from crude V contains a larger fraction of variation in the intensity in other bands than the aliphatic compared to crude N and F.

Fig. 8 Loading for the first four principal components from PCA of the matrix of spectra of crude N fractions

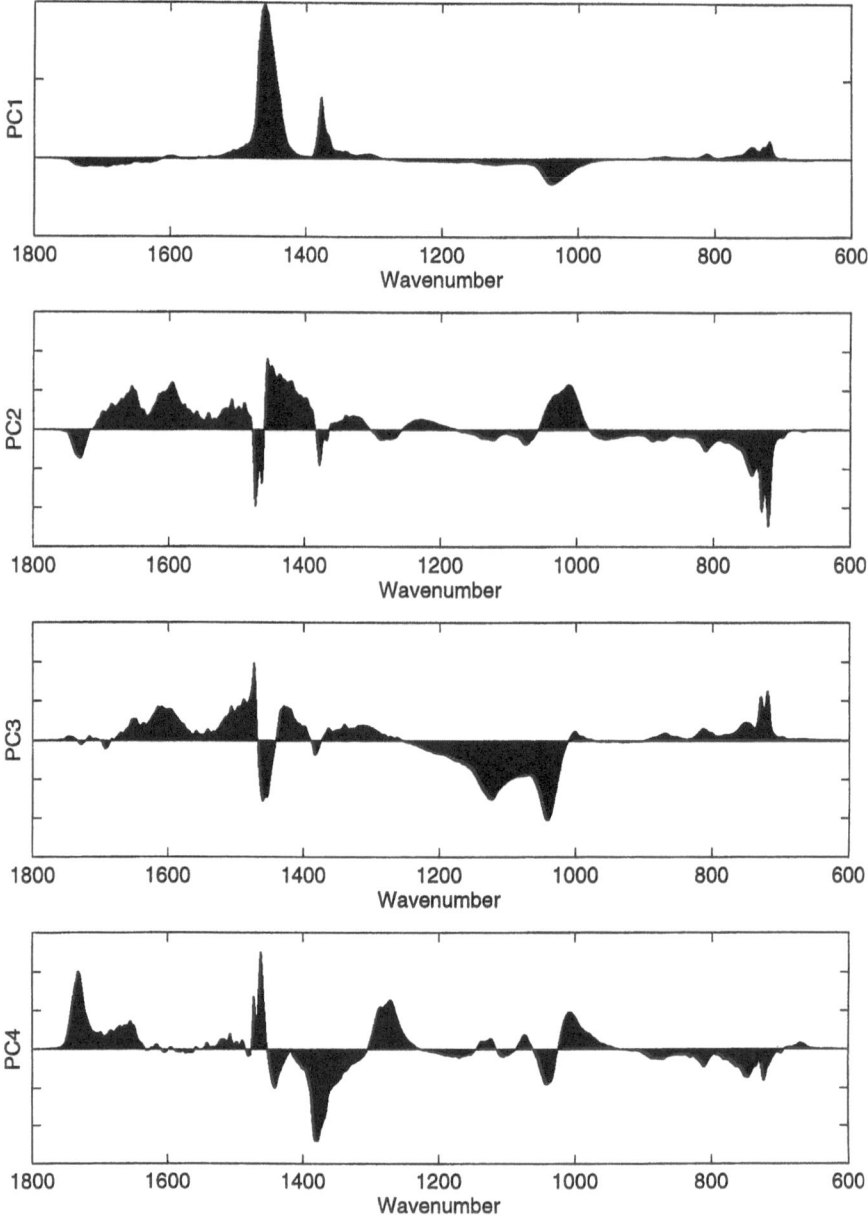

Crude F

For the scores of crude F (Fig. 10), it is observed that the resin fraction FCMB has the highest score on PC1, while the resins desorbed from silanol using DCM or MeOH/DCM in the second desorption step and the talc resins FTBD and FTBM have the lowest score in this PC. Fractions with high scores on this PC are seen from the loading plot (Fig. 7) to have large relative intensities in the aliphatic bonds. Those fractions with a low score on this PC have a relatively large amount of their absorption in other bands than the aliphatic. FCDM is the fraction that has the highest score on PC2 while all the talc resins have

a negative score, FTMD and FTMB having the largest negative values. From the loading plots it is seen that this PC is dominated by the amide carbonyl peak at 1650 cm^{-1}. Spectra with relatively large intensities in this peak are then expected to score high on this PC, and this is indeed the case for FCDM. The ester bands are seen to have negative loadings on this PC, contributing to the negative scores of the talc resins. All the siloxane resins have a negative score on PC3. The loadings on this PC are strongly negative for the ether/sulfoxide band, and this band is seen in Fig. 4 to be strong in all the siloxane resins, and less strong (but not absent) in the other fractions. The crude and asphaltene spectra are seen to have the highest

Progr Colloid Polym Sci (1998) 108:175–191
© Steinkopff Verlag 1998

Fig. 9 Loadings for the first four principal components from PCA of the matrix of spectra of crude V fractions

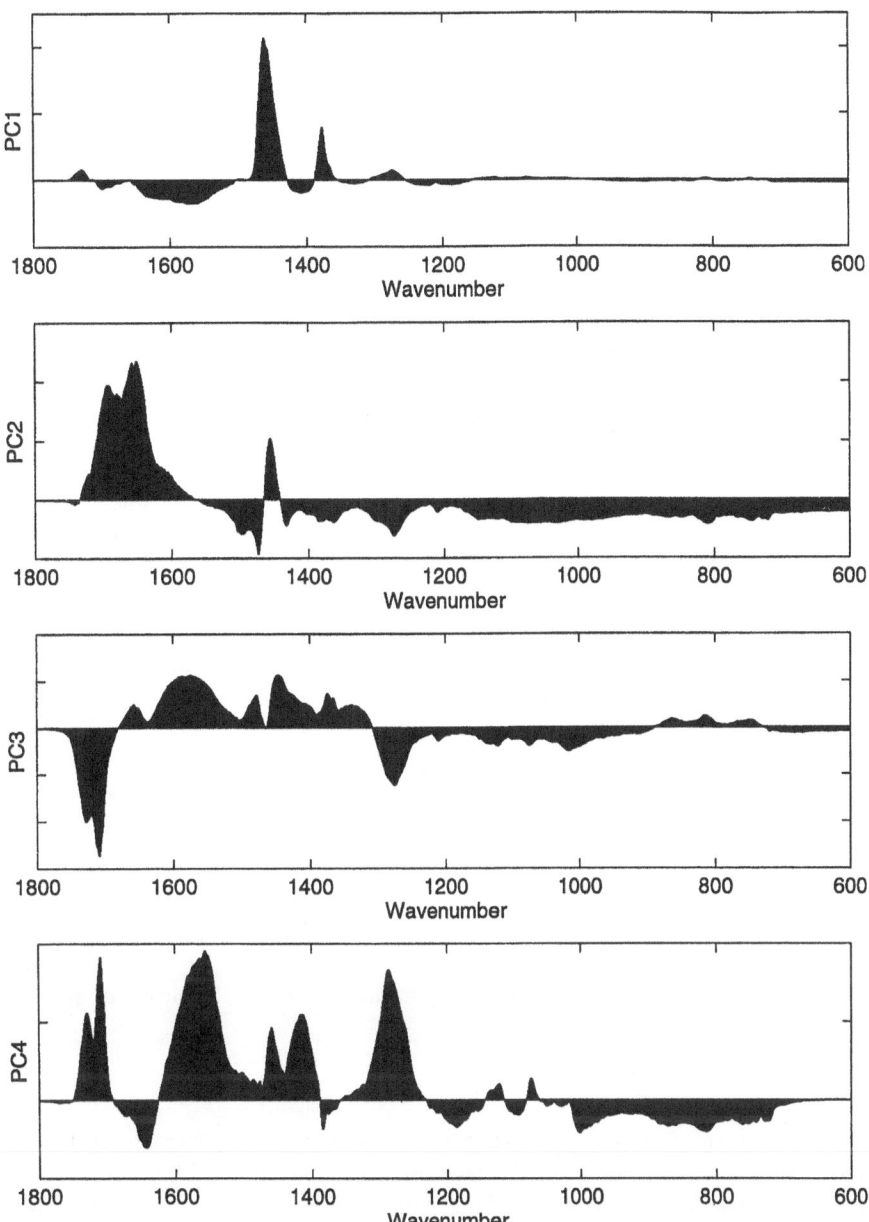

scores on this PC. On PC4 all the talc resins have positive scores. For most of them the explanation is seen from the loadings to be in the ester peaks, which are seen to have positive loadings from Fig. 7. The asphaltene and crude spectra have the largest negative scores on PC4. The variables that cause the asphaltene and crude spectra to have high scores on PC3 and low scores on PC4 seem to be found in the region $1450–1300\ cm^{-1}$, since these region has opposite loadings to the ether/sulfoxide and ester bands on PC3 and PC4, respectively.

It is interesting to note that PC2 are dominated by the amide band and PC3 by the ether/sulfoxide band. The loadings on PC4 demonstrates the correlation be-

tween the two ester bands at 1730 and $1280\ cm^{-1}$. It should also be noted that the amide band have large loading only on PC2, the ether/sulfoxide band only on PC3 and the ester bands only on PC4. This demonstrates that the variation in these bands are largely independent in the studied fractions from this crude.

Crude N

Turning to the scores on PC1 from the PCA of crude N, it is seen that the crude spectrum has the highest score and that NCBM and NSDB also have high scores on this PC.

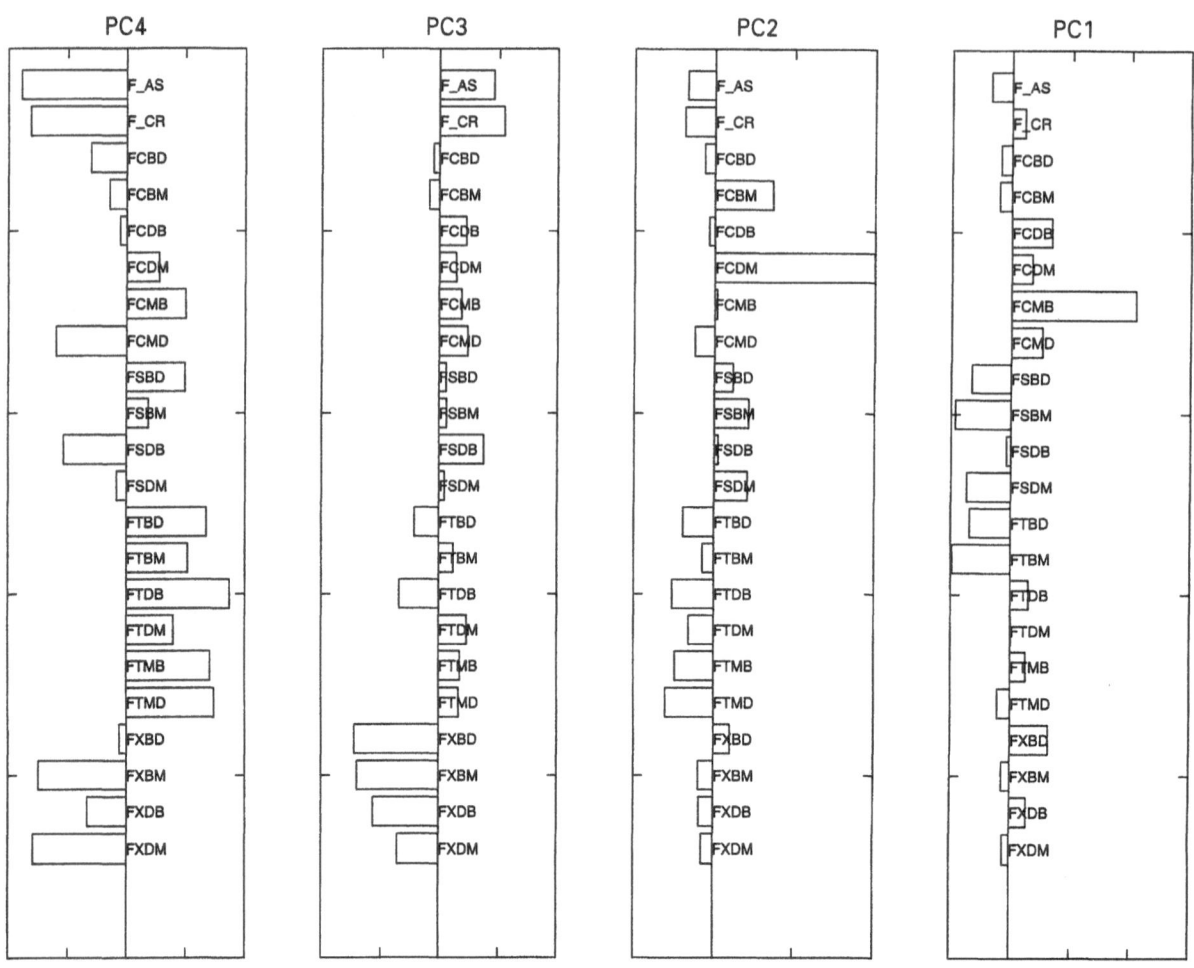

Fig. 10 Scores on the first four principal components from PCA of the matrix of spectra of crude F fractions

The loading plot in Fig. 8 shows that this PC separates the more aliphatic fractions from the less aliphatic ones, the most aliphatic fraction being the crude itself. In the opposite direction, NTBD and NTBM, together with NSBM have the largest negative score. These spectra are seen from Fig. 5 to have a relatively large part of their absorption in the regions 1700–1550 cm^{-1} and ∼ 1030 cm^{-1}. Both these regions are seen from Fig. 8 to have negative loading on PC1. The loadings on PC1 shows that the ether/sulfoxide band at 1030 cm^{-1} has the largest total variation (negatively) correlated to the variation in the aliphatic bands. The score on PC2 is highest for NCBM while the asphaltene and NTMD spectra have the largest negative values. The large negative loadings of the bands around 700 cm^{-1} explain the negative score of the asphaltene spectrum. Comparing this with the positive loading of the ether/sulfoxide, aromatic and amide carbonyl bands, the high score of the NCBM spectrum can be

explained. PC3 give low scores for all the siloxane resins. This can be explained by the negative loadings in the region 1300–1000 cm^{-1}. On PC4 NTMD has the highest score. This is the only fraction from this crude with a large part of its intensity in the ester bands, and from the loadings it can be seen that this explains the large score on this PC.

The loadings of the functional group absorptions are not separated by PCA on this crude in the same manner as for crude F. Only PC1 are dominated by one or two bands, the rest are complex combinations from the whole spectral region used in the PCA. It is, however, seen from Fig. 5 that the ester functionality is well separated in the NTMD resin. The best separation of the ether/sulfoxide band is achieved in the NXBM resin. No fractions separates the amide functionality well from this crude, the amide band is sharp in two CaO fractions, but they also include the ether/sulfoxide band.

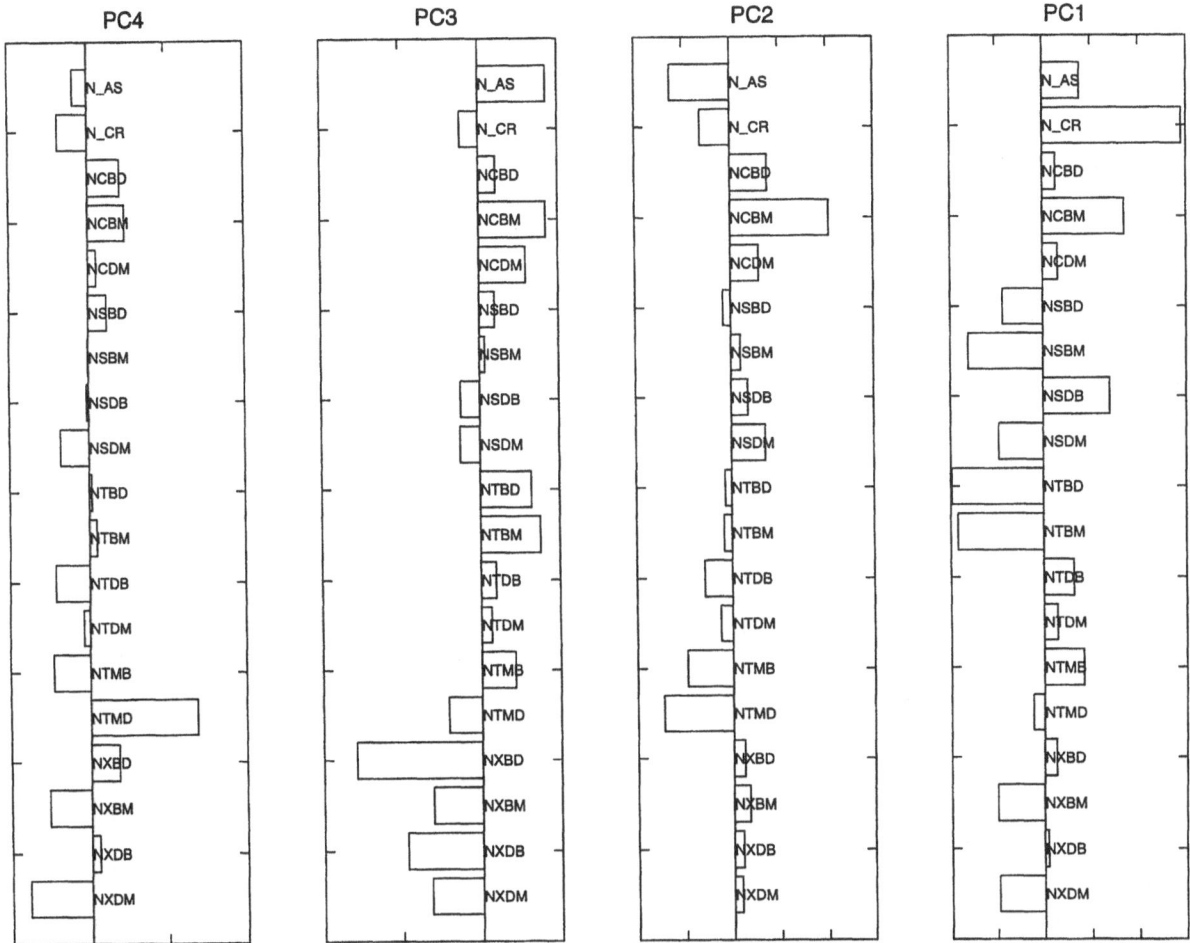

Fig. 11 Scores on the first four principal components from PCA of the matrix of spectra of crude N fractions

Crude V

The fractions with the largest positive scores on PC1 are VTMB and VXBD. The largest negative scores are seen to be from VCBM and VCDM. The loadings in Fig. 9 show that PC1 give high scores to spectra with large aliphatic bands, and low scores for those with large absorption between 1500 and 1650 cm^{-1}. On PC2 VXBD is seen to have the largest positive score, followed by VSBD, while the asphaltene and the crude spectra have large negative scores. It is seen from the loading that the amide band at 1650 cm^{-1} and the carbonyl band at 1700 cm^{-1} have large positive loadings on this PC, causing the positive loadings of VSBD, VSBM, VSDM and VXBD. On PC3 the loadings of the ester bands are seen to cause VCBD, VTDB and VTMD to have low scores. The asphaltene and crude spectra have large positive scores, together with VCBM and VXBD on this PC. From the loadings it is seen that this may in part be explained by their missing ester bands. VCBD and VCBM are the fractions that have the largest scores on PC4. Loadings on this PC are dominated by the ester bands and the nitro bands. The ester bands seem to explain the large score of VCBD and the nitro bands explains the score of VCBM. The higher score of VCBM than VCDM on PC3 and PC4 are explained by the larger relative intensity in the nitro bands in the VCBM fraction.

The only functionality that is well separated from other functionalities in any of these spectra is the ester group. It is well separated in VTMD and VCBD.

Conclusions

It is clear from the results that the presented separation procedure gives large variations in the spectra of the resin fractions. In general, the distribution of functional groups in the spectra of the fraction obtained seem to vary with all three variables: crude oil, adsorption surface and extraction solvent. The only adsorption–desorption procedure

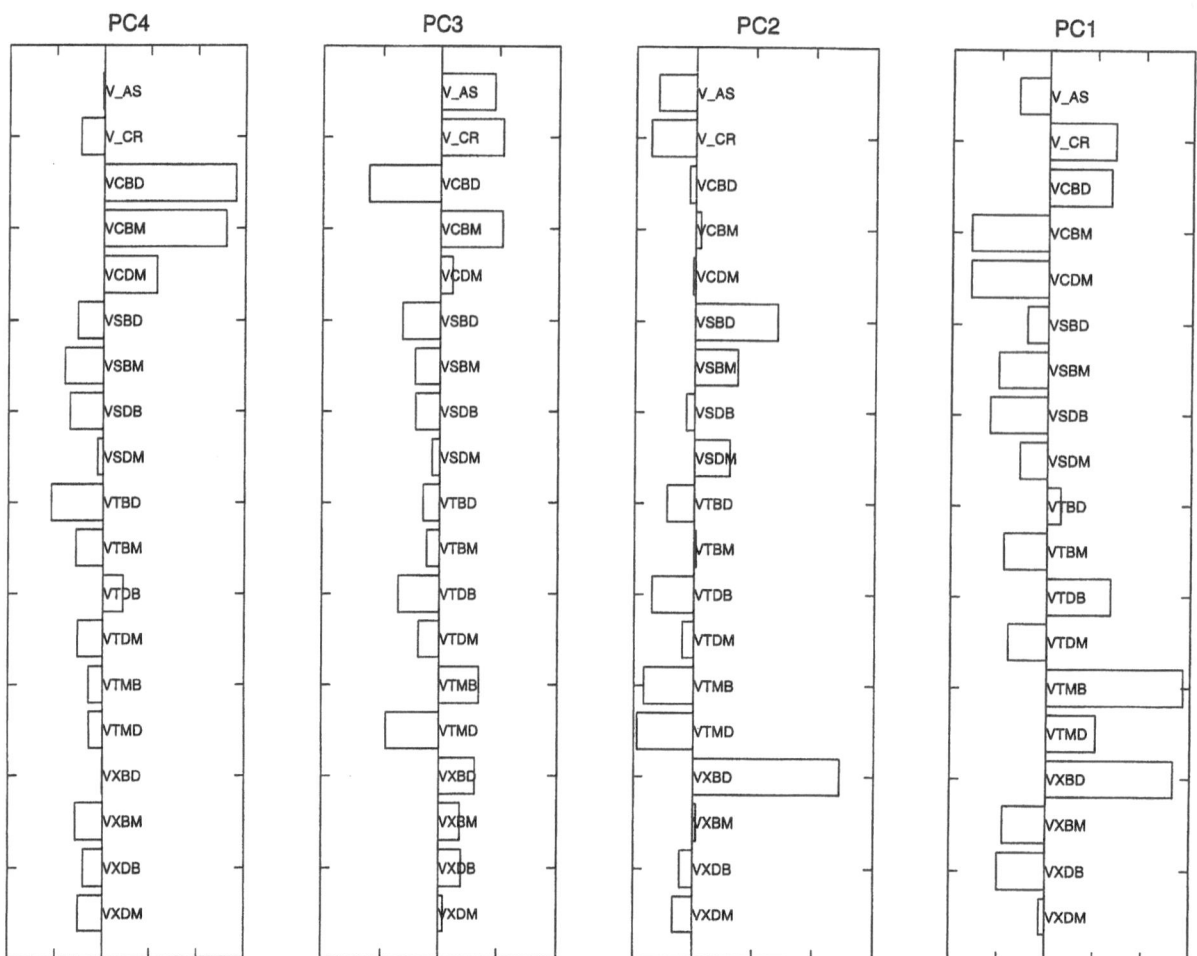

Fig. 12 Scores on the first four principal components from PCA of the matrix of spectra of crude V fractions

which gave similar spectra of the obtained fraction from all three crudes is talc–MeOH/DCM–DCM, this separates the ether/sulfoxide band well from the other oxygen containing functionalities observed in this work. All the talk resins from all crudes seem to have some ester bands present, but they are not always separated from other functionalities. All resins using the siloxane–benzene–DCM procedure and all resins desorbed from silanol using DCM or MeOH/DCM in the second desorption step include the amide band and a carbonyl band at $1700 \, \text{cm}^{-1}$. Amide and ether/sulfoxide functionalities are well separated in some crude F resins. From crude N and crude V a good separation of these functionalities was not achieved.

In the present study we have undertaken a thorough mapping of the properties of isolated resin based fractions from different non-related crude oils. This paper deals only with analytical and spectroscopic properties of the fractions. In a subsequent paper we will report on the emulsion stabilizing properties of these fractions alone and in combination with asphaltene particles.

Acknowledgements Øivind Midttun acknowledges the Norwegian Research Council for a Ph.D. grant. Elf Aquitaine and Statoil are thanked for providing the crudes. The technology programme Flucha financed by NFR and oil industry is also thanked for financial support.

References

1. Speight JG (1991) The Chemistry and Technology of Petroleum, Marcel Dekker, New York

2. Førdedal H, Midttun Ø, Sjöblom J, Kvalheim OM, Schildberg Y, Volle J-L (1996) J Colloid Interface Sci 182: 117–125

3. Schildberg Y, Sjöblom J, Christy AA (1995) J Dispersion Sci Technol 16(7): 575–605

Progr Colloid Polym Sci (1998) 108:175–191
© Steinkopff Verlag 1998

4. Sjöblom J, Mingyang L, Christy AA, Rønningsen HP (1995) Colloids Surfaces 96:261–272
5. Andersen SI, Birdi KS (1990) Fuel Sci Technol Int 8(6):593–615
6. Corbett LW, Petrossi U (1978) Ind Eng Chem Prod Res Dev 17(4):342–346
7. Førdedal H, Schildberg Y, Sjöblom J, Volle J-L (1996) Colloids Surfaces 106: 33–47
8. Koots JA, Speight JG (1975) Fuel 54: 179–184
9. McLean JD, Kilpatric PK, to be published
10. Mingyuan L, Christy AA, Sjöblom J (1992) In Sjöblom J (ed) Emulsions – a fundamental and practical approach. Kluwer, NATO ASI series, C 363: 157–172

11. Christy AA, Dahl B, Kvalheim OM (1989) Fuel 68:430–435
12. Christy AA, Kvalheim OM, Høiland H (1994) Chemometrics and Intelligent Laboratory Systems 23:197–204
13. Yen TF, Wu WH, Chilingar GV (1984) Energy Sources 7:203–235
14. Johnson RA, Wichern DW (1992) Applied Multivariate Statistical Analysis. Prentice-Hall, Englewood Cliffs, NJ, pp 356–395
15. Wold S, Esbensen K, Geladi P (1987) Chemometrics and Intelligent Laboratory Systems 2:37–52
16. Karstang TV (1996) In Nordtvedt R, Brakstad F, Kvalheim OM, Lundstedt T (eds) Anvendelse av kjemometri innen industri og forskning. Tidsskriftforlaget kjemi AS, Oslo, pp 129–145

17. Geise RF Jr, Wu W, van Oss CJ (1996) J Dispersion Sci Technol 17(5): 527–547
18. Davison P (1996) In: Whim BP, Johnson PG (eds) Directory of Solvents. Blackie Academic & Professional, London, p 4
19. Christy AA (1990) Sampling techniques and data processing in diffuse reflectance FT-IR spectroscopy. Thesis, University of Bergen, Norway
20. Kemp W (1987) Organic Spectroscopy. Macmillan, Hampshire, England
21. Lin-Vien D, Colthup NB, Fateley WG, Grasselli JG (1991) The Handbook of Infrared and raman Characteristic Frequencies of Organic Molecules. Academic Press, New York

Progr Colloid Polym Sci (1998) 108:192–198
© Steinkopff Verlag 1998

Detection of phase and structural changes in a 7-aminocoumarin monolayer by dynamic and optical methods

J.B. Rosenholm
A. Alekseev
A. Nikitenko
J. Peltonen

J.B. Rosenholm (✉) · J. Peltonen
Department of physical Chemistry
Åbo Akademi University
Porthansgatan 3-5
FIN-20500 Turku
Finland

A. Alekseev · A. Nikitenko
General Physics Institute
Russian Academy of Sciences
Vavilov street 38
117942 Moscow
Russian Federation

Abstract The aggregation induced by the increased surface pressure of aliphatic coumarin dye molecules in a Langmuir monolayer has been investigated with the aid of surface pressure, UV-Vis 45° angle and Brewster angle reflectometry as a function of the total surface area. The dynamic properties of the structural changes were recorded as compression speed and temperature-dependent shear viscosity and dynamic elasticity of the coumarin monolayer. A sequence of compression-induced phase and structural changes were found as a function of the compression. These changes are related to the formation of H-aggregates which was dependent on the duration and number of the compression–expansion cycles.
A simple model is offered to explain the main aggregation equilibrium of the 7-aminocoumarin molecules. The compression-induced structure was dependent on the compression speed, the temperature and the dilution in a stearic acid matrix.

Key words Coumarin dye – LB-film – rheology – spectroscopy – aggregation

Introduction

The majority of studies on aggregated monolayers have been carried out on spiropyrans, cyanine and azo dyes [1–4]. The existence of H-aggregates in LB films of a rigidified aliphatic 7-aminocoumarin derivative was suggested in a previous report where the absorption and emission properties as well as fluorescence kinetics of the LB films were studied [5]. The kinetics of H-aggregation on a coumarin dye Langmuir monolayer was demonstrated as a compression-induced blueshift of the absorption maximum [6]. Also the nonlinearily and discontinuously changing elastic modulus indicated that much more structural changes took place than expected from the shape of the compression and surface potential isotherms [6].

In this work the shift of the wavelength is evaluated using a simple model for the monomeric and aggregated state, respectively. Moreover, the p-polarized Brewster angle reflected light was recorded in order to further investigate the structural features of the thin film. The viscoelastic properties of the coumarin monolayer was investigated by measuring the elasticity and viscosity of the film as well as the hysteresis of the surface pressure–area cycles at different initial compressions.

Experimental details

A coumarin 102 derivative, 7-aminocoumarin, was provided by Professor M.A. Kirpichonok of Timiriazev Agricultural Academy, Moscow. All monolayer studies were performed on a KSV 5000 Langmuir trough (KSV Instruments, Helsinki, Finland). The coumarin molecules in chloroform with a concentration of 1 mg/ml were spread on the water surface. The subphase water of pH 5.6–5.8

was purified with a Millipore Milli-Q filtering system (Millipore Corp., USA). The subphase temperature was maintained at 20.3 or 21.3 $\pm$ 0.2 °C in order to enable the comparison of the data with previous results [6].

A Photal MCPD-100 spectrophotometer (Otsuka, Osaka, Japan) was applied to carry out the *in situ* spectroscopic measurements with the experimental details reported elsewhere [6]. The Brewster angle laser light reflectometry method used is the same as that used by Hönig et al. [7]. In our experiment the subphase surface was illuminated at the Brewster angle by an Ar-laser (514.5 nm, 10 mW) through a P-polarizer. The intensity of the reflected light was detected by a photomultiplier and recorded during the monolayer compression.

The dynamic Youngs modulus of elasticity was determined as the area corrected ratio of the change in stress over the change in strain, expressed as

$$E^s = A \, d\pi/dA \, . \tag{1}$$

The modulus of elasticity thus corresponds to the negative inverse of the isothermal dynamic surface compressibility. The barrier oscillating at 40 mHz and with an amplitude of 0.5–1% of the surface area was used for the dynamic elasticity measurements [6, 8]. Using these parameters, no viscous flow was observed to occur.

In order to determine the in-plane shear viscosity of the floating monolayer at the air–water interface, the surface canal viscosimeter technique was used [9]. In this technique, the monolayer is forced to flow through a deep canal under the action of a constant surface pressure while the monolayer flow rate, Q, is measured. No anomalous flow could be detected under the experimental conditions selected. The monolayer viscosity values, η_s, were calculated using the two-dimensional Poiseuille equation [10]:

$$\eta_s \approx \frac{\Delta\pi w^3}{12LQ} - \frac{w\eta}{\pi} \tag{2}$$

where $\Delta\pi$ is the surface pressure gradient between the ends of the canal, w the canal width, L the canal length and η the shear viscosity of the subphase. The monolayer flow rate (in m²/s) was measured about 1–2 min after the opening of the canal.

Results and discussion

Optical properties

The characteristics of the compression-induced blue shift of the UV–Vis surface reflection spectrum has been presented previously [6]. Maximum intensity was found at

45° from the plane of the surface, indicating a preferred orientation of the H-aggregates. An equilibrium was found between the monomeric state and the surface-pressure-induced aggregate as evidenced by an almost common intersection point of the spectra. We may thus assume that the spectra recorded at different surface pressures may be used to quantify the equilibrium. Since the aggregation induces changes in the state of the chromophore (variable extinction coefficient) an evaluation of the average energy of the unresolved peak seems more appealing than the intensity analysis. Furthermore, because of the alkyl chain rendering the molecule amphiphilic we assume that 7-aminocoumarin is fully insoluble in subphase at all surface pressures.

The population (surface concentration) of coumarin may be determined as $p_{tot} = N_{tot}/A$, where N_{tot} is the constant total number of molecules per unit area A. The total fraction of molecules may then be subdivided into the fraction of monomers and molecules in the aggregates, respectively,

$$p_{tot} = p_{mon} + p_{aggr} \, . \tag{3}$$

Denoting the average wavelength for each state (i) as $\langle \lambda \rangle_i$ the equilibrium may be described as

$$p_{tot}\langle \lambda \rangle_{obs} = p_{mon}\langle \lambda \rangle_{mon} + p_{aggr}\langle \lambda \rangle_{aggr} \, . \tag{4}$$

Equation (4) may be written in the form

$$\langle \lambda \rangle_{obs} = \langle \lambda \rangle_{mon} - (p_{aggr}/p_{tot})(\langle \lambda \rangle_{mon} - \langle \lambda \rangle_{aggr}) \, . \tag{5}$$

In order for this simple two-state model to apply a plot of $\langle \lambda \rangle_{obs}$ vs. $1/p_{tot}$ should give two straight lines. If the line corresponding to a large surface area is extrapolated to $A = \infty$ it should give $\langle \lambda \rangle_{mon}$ as a limiting value. If the line corresponding to a small surface area is extrapolated to $1/p_{tot} = 0$ the intercept should represent $\langle \lambda \rangle_{aggr}$. If, however, the monomers remain in some kind of oligomeric aggregates upon dilution the slope is non-zero and the intercept at $1/p_{tot} = 0$ ($\langle \lambda \rangle_{oligo}$) may be used as a reference state for the oligomer-aggregate equilibrium instead. A curvature in the region of the intersection of the lines is typical for a comparable population of both states.

The fraction of coumarin molecules in each state may now be estimated from the population of each state divided by the total population:

$$X_{aggr} = p_{aggr}/p_{tot} = (\langle \lambda \rangle_{mon} - \langle \lambda \rangle_{obs})/(\langle \lambda \rangle_{mon} - \langle \lambda \rangle_{aggr}) \, , \tag{6a}$$

$$X_{mon} = 1 - X_{aggr} \, . \tag{6b}$$

Plotting $\langle \lambda \rangle_{obs}$ of the absorbance maximum against A (proportional to $1/p_{tot}$) two roughly straight lines are indeed obtained (Fig. 1) with a transition at the point

194
J.B. Rosenholm et al.
Phase and structural changes in a 7-aminocoumarin monolayer

Fig. 1 The shift of the absorbance maximum as a function of compression of a 7-aminocoumarin monolayer as measured using two different compression speeds. The boundary values for the monomeric ($\langle\lambda\rangle_{\text{mon}}$) and aggregated states ($\langle\lambda\rangle_{\text{aggr}}$) as extracted from the plot are represented by the dashed lines

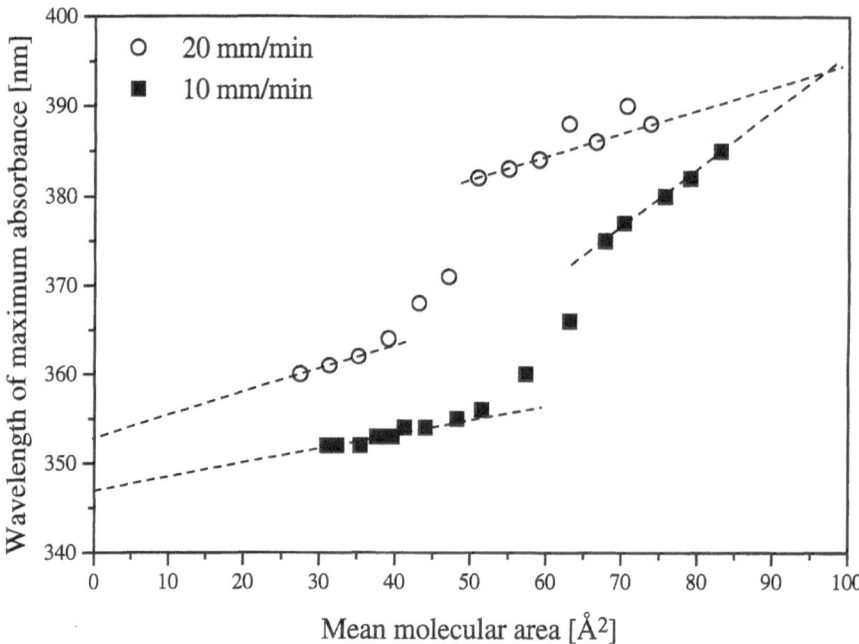

where the surface pressure rises towards its solid condensed limit.

As seen the slopes of the lines are dependent on the compression speed. The following limiting values were extracted from the plots:

$$\lim\langle\lambda\rangle_{\text{obs}}(A = 97 \text{ Å}^2) \equiv \langle\lambda\rangle_{\text{mon}} = 394 \text{ nm} \approx \lambda_{\text{soln}},$$

$$\lim\langle\lambda\rangle_{\text{obs}}(A = 0 \text{ Å}^2, 20 \text{ mm/min}) \equiv \langle\lambda\rangle_{\text{aggr}} = 353 \text{ nm},$$

$$\lim\langle\lambda\rangle_{\text{obs}}(A = 0 \text{ Å}^2, 10 \text{ mm/min}) \equiv \langle\lambda\rangle_{\text{aggr}} = 347 \text{ nm}.$$

The limiting value chosen for the monomeric state 394 nm was found as an intersection point for the largest expansions used in the experiments (97 Å^2) and it is nearly equal to the wavelength found for 7-aminocoumarin dissolved in chloroform (390 nm). With these limiting values inserted in Eq. (5) we obtain the monomer-aggregate fraction distribution shown in Fig. 2, both for the 10 mm/min (a) and for the 20 mm/min (b) compression speed. There is, indeed, a very rapid formation of aggregates at areas below the compression limit where the aggregate formation should in ideal case be negligible. This explains the nonzero slope in Fig. 1 and indicates that preaggregation (oligomerization) may have occurred prior to compression.

The association of the monomers to aggregates is particularly strong when compression exceeds the critical compression limit found at 50–68 Å^2 (0.015–0.020/Å^2). The equilibrium is dependent on the compression speed. However, the main difference is that an increased compression speed only reduces the degree of aggregation.

The contribution of the compression procedure was further investigated by applying compression–expansion cycles to the coumarin monolayer. The π–A isotherm hysteresis curves of the dye molecules measured at the temperature of 21.3 °C are shown in Fig. 3a. Some hysteresis was observed already at small surface pressures such as 5 mN/m. The repeated compression–expansion cycle caused the displacement of the isotherm to smaller molecular areas demonstrating the reorganization of the molecules within the monolayer. The simultaneously measured absorption spectra showed that the blue shift of the absorption maximum induced by the surface pressure increase [6] was irreversible within the time frame of the experiment, i.e. the absorption maximum remained at the same spectral position when the surface pressure was returned to a low value (Fig. 3b). If the compression speed was increased from 10 to 20 mm/min, the pressure increase took place at larger mean molecular areas. The hysteresis was smaller and the absorption maximum shifted partly back when decreasing the pressure. The kinetics of the aggregation process could thus be controlled by the compression speed. More stable and probably also larger and polydisperse aggregates being represented by a wider band width were formed when the film was compressed carefully with low speed (Fig. 2).

In order to investigate the complex aggregation features detected by the elasticity measurements previously reported [6], the reflection of the p-polarized Brewster angle laser light at the air–water interface was recorded.

Progr Colloid Polym Sci (1998) 108: 192–198
© Steinkopff Verlag 1998

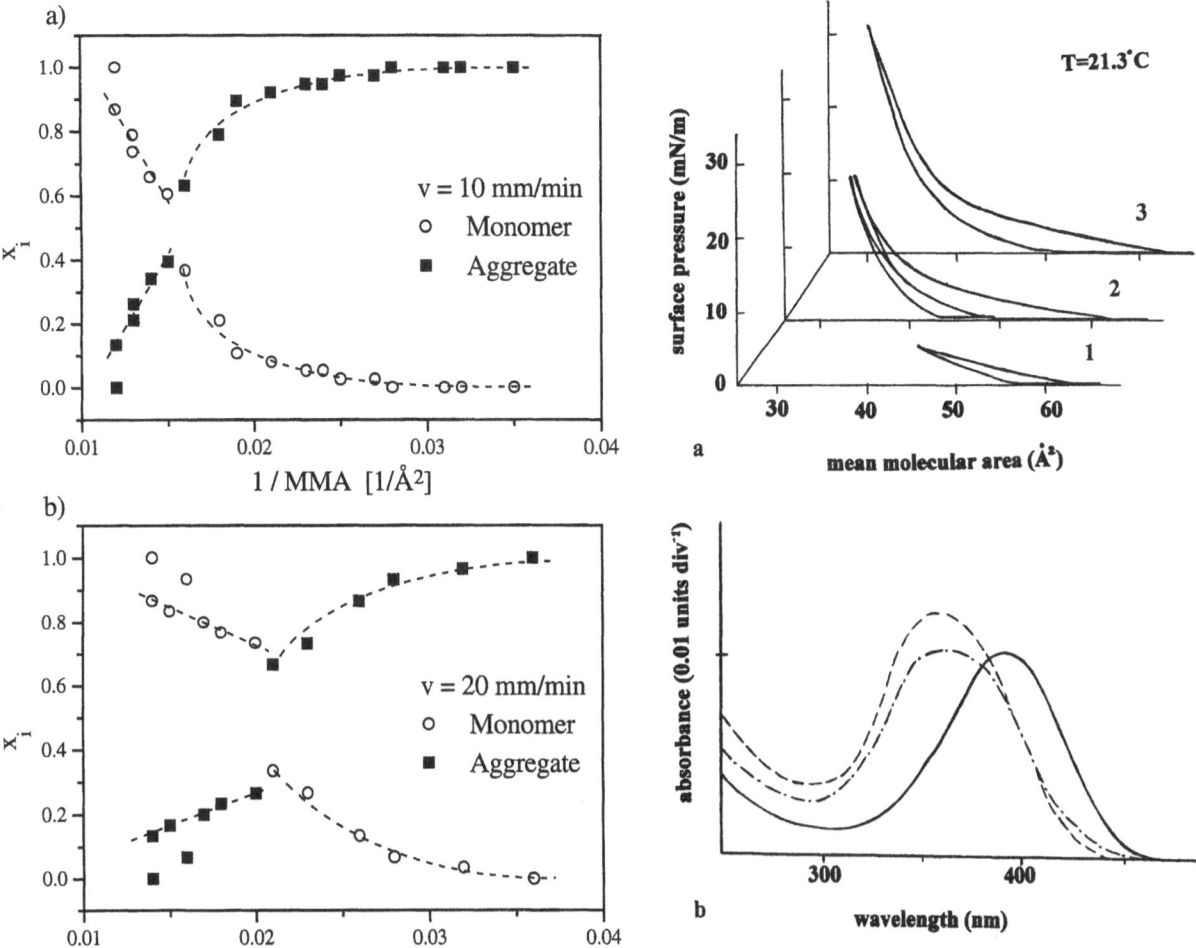

Fig. 2 The fraction distribution of the monomeric and aggregated states vs. inverse of the mean molecular area (MMA) for two different compression speeds v

Fig. 3 (a) Surface pressure-mean molecular area isotherm hysteresis (compression-decompression) of a monolayer of 7-aminocoumarin for maximum surface pressures 5 (curve 1), 20 (curve 2) and 32 mN/m (curve 3). (b) Surface reflection spectra measured during the hysteresis measurement at the following surface pressures (mN/m): —— 0.2 (compressed), — — 18.0 (compressed); —●— 2.0 (decompressed)

The experiments have shown that the deposition of a monolayer changes the Brewster angle value negligibly but the intensity of the reflected light changes significantly [11]. It is known that the reflected p-polarized light never vanishes at the Brewster angle (contrary to the Fresnel formals). In conformity with the theories explaining this effect [12, 13], the reason is in the existence of a very thin transition layer on a surface of water or a solid substrate. The characteristics of this layer are different from those of a bulk substance and they determine the intensity of the reflected p-polarized light. This interpretation is slightly different from the view of Hönig and Möbius who suggested that the practically zero reflection from pure substrates becomes observable only when a monolayer is introduced modifying the Brewster angle condition [7]. The deposition of an organic monolayer on the transition layer can

instead be considered as a formation of a composite surface layer producing changes in the original p-polarized reflected light. It is thus expected that the coumarin layer changes the optical properties of this composite surface layer and influences the intensity.

The dependence of the intensity of the reflected light at the p-polarized Brewster angle on the compression is presented in Fig. 4. For comparison, the modulus of the dynamic elasticity recorded under equal conditions is shown. It is interesting to note that the local maxima of both curves coincide at the same surface pressure values. Hence, the reflectance data confirms the existence of the successive subtle structural changes of the monolayer during compression. However, there are also some characteristic differences from the elasticity curve demonstrating the different mechanisms responsible for the changes,

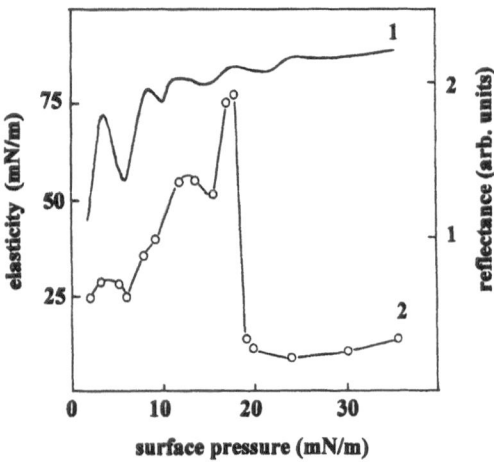

Fig. 4 The intensity of the laser light reflected at the Brewster angle (curve 1) and elasticity (curve 2) as a function of surface pressure for a 7-aminocoumarin monolayer. The subphase temperature was 20.3 °C

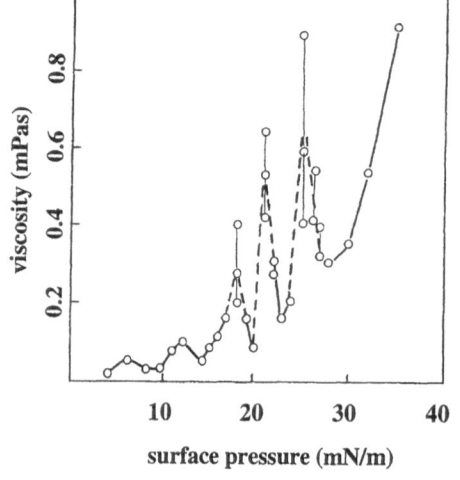

Fig. 5 The surface viscosity as a function of surface pressure of a 7-aminocoumarin monolayer. The subphase temperature was 21.3 °C

respectively. First, the largest intensities found upon the initial aggregation, reflected as the main effect on the surface potential [6]. Second, an additional peak maximum was found when the elastic modulus was dramatically reduced at ca. 20 mN/m surface pressure. The transitions were found to be highly temperature dependent, a reduction in temperature producing less noise and smaller fluctuations.

Rheological properties

In a recent investigation of Peng et al. [14] of the viscoelastic properties of, e.g. octadecanol and behenic acid it was shown that the shear modulus of behenic acid exhibited a number of dramatic structural transitions in the 10–25 mN/m range while only one loop with a maximum at 26 mN/m was found for the octadecanol monolayers. The results were recorded with torsion pendulum (rotor) in contact with the spread film and the response were typical for films of low stiffness and damping. The shear viscosity was found to be in the range of 0.06–0.20 mN/m for the octadecanol and it followed closely the relative changes of the shear modulus. For behenic acid the viscosity was found to be in the range of 0.016–0.052 mN/m and had an overall shape similar to the octadecanol, but did not exhibit the salient phase changes similar to the shear modulus.

According to synchroton X-ray scattering investigations reported by Kenn et al. [15] the monolayer of behenic acid at 25 °C and low surface pressures is in the liquid condensed state with distorted hexagonal packing

of the molecules and the chains tilted towards the nearest neighbour and the tilt angle decreasing during the compression. This resembles roughly the formation of H-aggregates. The sudden and steep increase in both shear modulus and shear viscosity of behenic acid through an apparently homogeneous liquid condensed phase region was interpreted to be due to an enhanced interaction between the domains in the film or a continuous change in the angle of the molecular tilt. However, the sudden extensive reduction of the modulus and viscosity at approximately 25 mN/m was related to an increased disorder in the monolayer induced by a changed tilt direction of chains towards the nearest neighbours.

Since the rheological properties seem to provide valuable structural information it was logical to study the rheological behaviour of the monolayer in more detail (Fig. 5). The nonlinear increase of the viscosity was disrupted by several local fluctuations in the curve, again being an indication of some structural changes taking place in the monolayer. Compared with the elasticity curve Fig. 4), it is quite obvious that the local maxima of the viscosity curve correspond to local minima of the elasticity curve.

An exception is found at compressions exceeding 20 mN/m. The exceedingly low surface elasticity then erases the sensitivity towards the changes in the molecular structure of this viscous phase. However, the low elasticity allows for a complex flow behaviour which is dependent on the molecular interaction and orientation, respectively. The viscosity minima correspond to the Brewster angle reflection maxima found in this compression range (Fig. 4) indicating a condensed structure of this transition state.

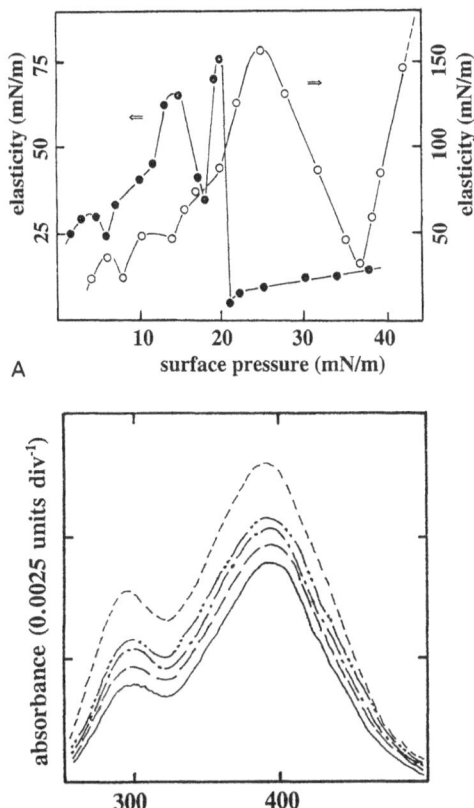

Fig. 6 (A) The surface elasticity as a function of surface pressure of a pure 7-aminocoumarin monolayer (•) and a steraic acid:7-aminocoumarin (70:30) mixed film (○). (B) Surface reflection spectra of the mixed monolayer as measured at the following surface pressures (mN/m): —— 0.3; — — 6.0; —•— 17; —••— 25.0; – – – – 25.0 (after 2 min UV-irradiation). The subphase temperature was 21.3 °C

The absence of the fluctuations in the shear viscosity representing these intermediate states seems to be due to the much lower sensitivity of the torsion pendulum technique (ca 10^{-3}–10^{-4} mPa s) as compared to the flow technique (ca 10^{-6} mPa s) [16].

The aggregation tendency of the coumarin molecules was also investigated by diluting them in a stearic acid monolayer (Fig. 6A). It was then found that the corresponding transition states were substantially enhanced and shifted to higher surface pressures. Moreover, it was found that the sudden fluctuating changes are indeed continuous and represented by a large number of experimental points.

In the least compressed state the transition seems moreover to be split into two subsequent, but well-resolved transitions. The presence of stearic acid seems thus to stabilize the monolayer structure to such extent that the final transition from the disordered to the ordered

viscous state has not been reached at 42 mN/m where the elasticity is 150 mN/m. Despite these remarkable rheological effects the spectra remained unchanged at all surface pressures. This indicates that the aggregation process is markedly inhibited by the stearic acid matrix (Fig. 6B). On the other hand, the reduction of the temperature by only 1° to 20.3 °C shifted the main transition to lower surface pressures by 2.5 mN/m and reduced considerably the development of the viscous structures above ca. 18–22 mN/m (cf. Fig. 4 (curve 2) and Fig. 6a). The high temperature dependence was also found in the Brewster angle reflection study.

Conclusions

Brewster angle reflectiometry and rheology have been proven to be very sensitive to salient structural changes occurring in 7-aminocoumarin monolayers as evidenced by X-ray and neutron reflectometry on comparable systems. Only a fraction of the transitions can be detected by ordinary surface pressure and surface potential measurements.

The initial rise of the coumarin monomers to a more perpendicular orientation is readily detected by an enhanced surface potential value and it produces a very strong reflected intensity at Brewster angle. These oligomeric aggregates do, however, produce only small compression-dependent elastic and viscous responses.

As deduced from a simple association model, the extent of aggregation of the monomers becomes important over a very narrow range of surface area. Despite this strong effect only small changes are found for the surface pressure and the surface potential. The Brewster angle spectra as well as the viscoelasticity of the film indicate that a series of transitional structural changes are surpassed when the compression is increased.

Finally, the coumarin film reorganises into a viscous arrangement as evidenced by an exccedingly low elasticity. The new viscous film has a low elasticity but the flow behaviour indicates a continuous series of subsequent aggregated states appearing upon further compression. Some of these are detected with the Brewster angle spectroscopy.

The aggregation is dependent on the compression speed (the time allowed for aggregation), the temperature and the dilution in non-active matrix molecules, all evidencing the very rich and subtle association process occurring in the monolayers of surface active molecules.

Acknowledgments The work was supported in part by the Scientific-Technical Committee of the Academy of Finland and Technology Development Centre, TEKES, Finland.

References

1. Heesmann J (1980) J Am Chem Soc 102:2167–2173
2. Nakahara H, Uchimi H, Fukada K, Tamai N, Yamazaki I (1989) Thin Solid Films 178:549–553
3. Seki T, Ichimura K (1990) J Phys Chem 94:3769–3775
4. Hibino J, Moriyama K, Suzuki M, Kishimoto Y (1992) Thin Films 210/211:562–564
5. Alekseev A, Konforkina T, Savransky V, Kovalenko M, Jutila A, Lemmetyinen H (1993) Langmuir 9:376–380
6. Alekseev A, Peltonen J, Savransky V (1994) Thin Solid Films 247:226–229
7. Hönig D, Möbius D (1991) J Phys Chem 95:4590–4592
8. Blank M, Lucassen J v.d. Tempel M (1970) J. Colloid Interface Sci 33:94–100
9. Joly M (1972) In Matijevic E (ed) Surface and Colloid Science, Vol 5. Wiley-Interscience, New York
10. Naito K, Iwakiri T, Miura A, Azuma M (1990) Langmuir 6:1309–1315
11. Nikitenko A, Savransky V (1993) Opt Spectrosc 74:327–332 (in Russian)
12. Sivukhin D (1956) JETP (in Russian) 30:374–382
13. Drude P (1959) Theory of Optics. Dover, New York
14. Peng J, Barnes G, Abraham B (1993) Langmuir 9:3574–3579
15. Kenn R, Böhn C, Bibo A, Peterson I, Möhwald H, Als-Nielsen J, Kjaer K (1991) J Phys Chem 95:2092–2097
16. Kanner B, Glass J (1971) In Gushee E (ed) Chemistry and Physics of Interfaces II. ACS Publications, Washington, pp 51–61

Progr Colloid Polym Sci (1998) 108:199–202
© Steinkopff Verlag 1998

R. Sjövall
S. Lidin

Coordination in inorganics

I. Aspects of tetrahedral distortions

R. Sjövall* (✉) · S. Lidin
Inorganic Chemistry
Arrhenius Laboratory
Stockholm University
S-106 91 Stockholm
Sweden

*Present address
SAFT AB, Tech. Dept.
P.O. Box 709
S-572 28 Oskarshamn
Sweden

Abstract Tetrahedral distortion values are calculated as the mean quadratic elongation and the bond angle variance for 63 MX_4 tetrahedra (M = transition metal or thallium, X = Cl → I). The degree of distortion according to these two geometrical parameters are more or less linearly correlated. The magnitude of the distortion is affected by the ratio of the central to vertex atom radii, and by the character of the atoms surrounding the vertex atoms. The tendency is that the heavier the elements, the lower the degree of distortion. Simulated distortion modes could not be assigned to the observed ones. However, the presented results from the observed MX_4 tetrahedra suggest that the distortions are attributed to effects related to the crystal structure and chemistry rather than to any strictly mathematical phenomena.

Key words Tetrahedral distortion – simulated models

Introduction

In 1971, Robinson, Gibbs and Ribbe reported some novel ideas reflecting on the distortions of MO_4 tetrahedra and MO_6 octahedra in rock-forming minerals [1]. The authors introduced parameters, or quantitative measures, for the determination of the polyhedral distortions, e.g. the mean quadratic tetra-/octa-hedral elongation and the angle variance. The former parameter is dimensionless and often used in strain analysis showing the ratio between a strained and an unstrained state, whereas the second parameter represents the vertex–center–vertex angle variance of the polyhedron. As a result from implementation of these parameters, it was found that there exists a linear correlation between the elongation and the variance values.

One of us (R.S.) has worked with synthesis and crystal structure determination of cesium cadmium/mercury iodides [2]. The present investigation and the study concerning coordination number (Part II of this contribution) are intended to create trends in coordination. This paper aims

at providing some reflections on the distortion of MX_4 tetrahedra using the Robinson definitions, where M represents a transition metal or thallium and X represents chlorine, bromine or iodine atoms. These tetrahedra form building units in the following compounds, A_2MX_4, AMX_4, A_3MX_5, $A_xM_yX_z \cdot wH_2O$, where A represents an alkali/alkaline earth metal and M and X as previously defined. It is common for the tetrahedra to be quite regular in shape. The tetrahedra in the investigated non-hydrate compounds form isolated units, while those in the hydrates consist either of isolated or vertex-sharing units.

Finally, the observed tetrahedral distortions are compared to six simulated modes of distortion.

Definitions and calculations

The two most important parameters selected for the calculations were the mean tetrahedral quadratic elongation and the bond angle variances according to the Robinson definitions.

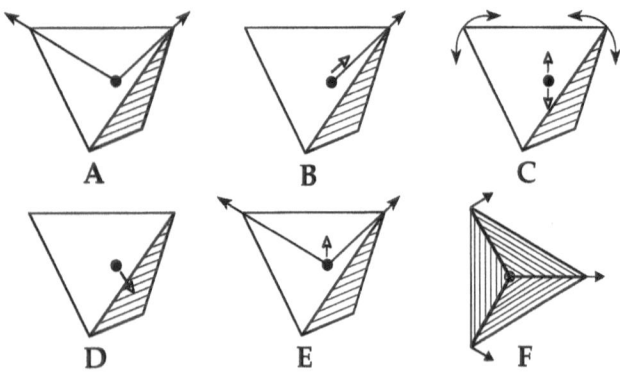

Fig. 1 Definitions of the simulated tetrahedral distortion modes: (A) Biaxial stretch, denoted by filled arrows, along two M–X directions while the central atom is stationary. (B) Uniaxial stretch along one M–X direction with central atom in mass center throughout the stretch (denoted by an open arrow). (C) Moving of two vertex atoms from a triangular towards a final square planar state with constant M–X distances. Central atom in mass center. (D) Central atom moving towards the mass center of one face. (E) Biaxial stretch along two M–X directions with central atom in mass center throughout the stretch. (F) Biaxial stretch in the X–X′–X plane, perpendicular to the M–X directions, combined with a uniaxial stretch along the M–X′ direction. The stretches start from the vertex positions of the regular tetrahedron. The magnitude of the biaxial stretch is twice that of the uniaxial one. The central atom is stationary. Note that stretches within a mode occurs at intervals of equal length, unless stated otherwise

Mean tetrahedral quadratic elongation (MTQE):

$$\langle \lambda_{\text{tet}} \rangle = \sum_{i=1}^{4} (l_i/l_c)^2/4$$

Tetrahedral angle (X–M–X) variance:

$$\sigma^2_{\theta(\text{X}-\text{M}-\text{X})} = \sum_{i=1}^{6} (\theta_i - 109.47°)^2/5$$

where l_i represents the observed center (M)–vertex (X) distance and l_c the calculated center-vertex distance for a regular tetrahedron with the same volume as that of the observed compound, and θ_i the calculated vertex–center–vertex angle [3].

The compounds investigated are divided into the following categories according to the compound composition:

– A$_2$MX$_4$. P2$_1$/m Sr$_2$GeS$_4$-type [4] or closely related P2$_1$, Pnma β-K$_2$SO$_4$-type [5] or closely related space groups, Pnma Olivine-type [6] or closely related space groups,

– A$_3$MX$_5$. Pnma (NH$_4$)$_3$ZnCl$_5$-type [7] or closely related Pbca Cs$_3$HgI$_5$-type [8], I4/mcm Cs$_3$CoCl$_5$-type [9],
  – AMX$_4$. No structure-type restrictions.
  – A$_x$M$_y$X$_z$·wH$_2$O. No structure-type restrictions.

The structural data and references to original papers were mainly obtained from the volumes of *Structure Reports* and *Chemical Abstracts*.[1]

The definitions of the selected distortion modes are summarized in Fig. 1. It is easy to realize that there is an unlimited number of ways to distort a tetrahedron. Hence, only six modes have been simulated in order to demonstrate a few types of distortions.

Results and discussion

The observed tetrahedra exhibit an almost linear dependence between the mean tetrahedral quadratic elongation (MTQE) and the tetrahedral angle (X–M–X) variance as displayed in Fig. 2a and b. This linearity is particularly pronounced for MTQE values below 1.015. The overall correlation coefficient, ρ_{corr}, for 63 tetrahedra is 0.992. When the categories are considered separately the following results are obtained.

A$_2$MX$_4$ (31):

$$\langle \lambda_{\text{tet}} \rangle = 0.999 + 0.000297\sigma^2_{\theta(\text{X}-\text{M}-\text{X})}, \quad \rho_{\text{corr}} = 0.998$$

A$_3$MX$_5$ (15):

$$\langle \lambda_{\text{tet}} \rangle = 1.000 + 0.000248\sigma^2_{\theta(\text{X}-\text{M}-\text{X})}, \quad \rho_{\text{corr}} = 0.999$$

AMX$_4$ (5):

$$\langle \lambda_{\text{tet}} \rangle = 1.001 + 0.000132\sigma^2_{\theta(\text{X}-\text{M}-\text{X})}, \quad \rho_{\text{corr}} = 0.890$$

A$_x$M$_y$X$_z$:wH$_2$O (12):

$$\langle \lambda_{\text{tet}} \rangle = 1.000 + 0.000252\sigma^2_{\theta(\text{X}-\text{M}-\text{X})}, \quad \rho_{\text{corr}} = 0.991$$

The number of tetrahedra in each category is given in parentheses, where $\langle \lambda_{\text{tet}} \rangle$ is the MTQE, and $\sigma^2_{\theta(\text{X}-\text{M}-\text{X})}$ is the tetrahedral angle (X–M–X) variance. The equations and correlation coefficients coincide for all except the AMX$_4$ category. Since this category only contains five tetrahedra, the regression analysis is greatly affected when one data point deviates from the curve (which is the case here). The obtained expressions above agree with the expression obtained by Robinson et al. ($\langle \lambda_{\text{tet}} \rangle = 1.00 + 0.00024\sigma^2_{\theta(\text{X}-\text{M}-\text{X})}$) for M′O$_4$ tetrahedra containing cations (M′) other than Si or Al in a variety of silicates and alumino-silicates [1].

[1] List of original structural papers will be available from the authors on request.

Progr Colloid Polym Sci (1998) 100: 199–202
© Steinkopff Verlag 1998

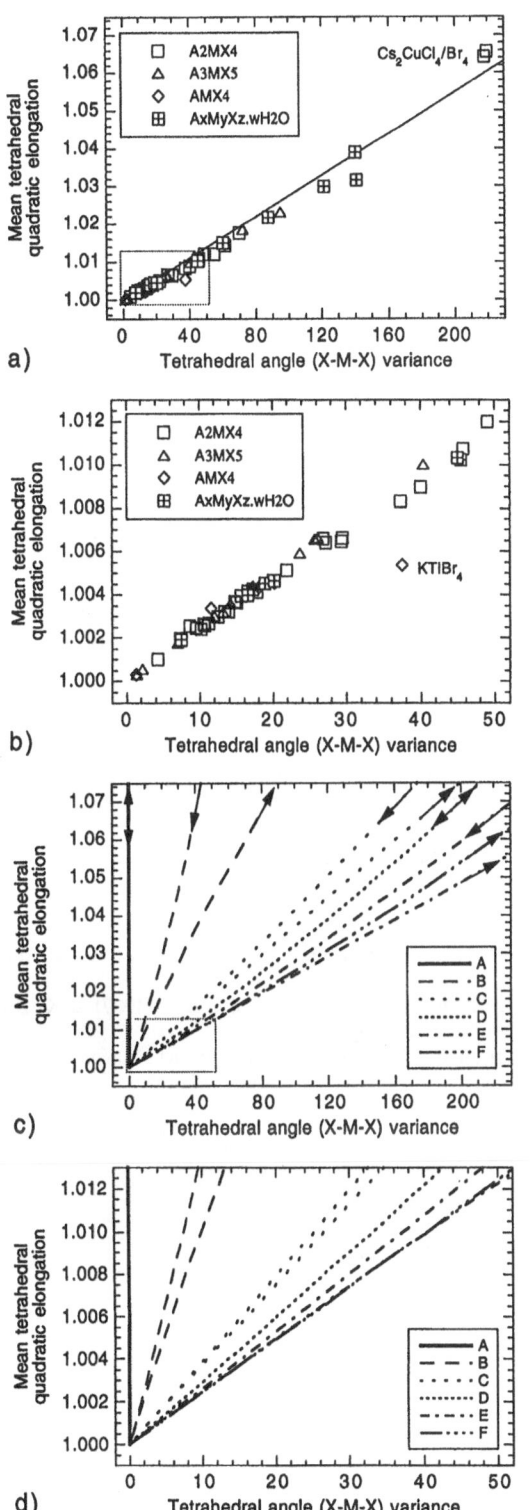

Fig. 2 Mean tetrahedral quadratic elongation versus the tetrahedral angle (X–M–X) variance for observed tetrahedra (a, b) and for simulated distortion modes (c, d). An arrow directed to the origin indicates a compression of the tetrahedron in that direction, while an extension is indicated by an arrow directed from the origin. The overall regression line is also displayed in (a). The enlargement of the areas within the broken lines in (a) and (c) are shown in (b) and (d)

The observed linearity is remarkable since this investigation covers a wide range of different structures and structure types. However, some values need to be commented on, e.g. the distortions of the MBr_4 tetrahedra in β-K_2SO_4-type Cs_2CuBr_4 (see Fig. 2a) and Cs_2ZnBr_4 (lower left corner of Fig. 2a). The copper compound tetrahedra is rather distorted ($\langle \lambda_{tet} \rangle = 1.0643$; $\sigma^2_{\theta(X-M-X)} = 218.1$) with vertex–center–vertex angles in the range 100.0–$132.3°$ and with center–vertex distances between 2.310 and $2.372\,\text{Å}$. The corresponding angles and bond lengths for the far more perfect $ZnBr_4$ tetrahedron ($\langle \lambda_{tet} \rangle = 1.0025$; $\sigma^2_{\theta(X-M-X)} = 9.5$) in Cs_2ZnBr_4 are 106.4–$114.9°$ and 2.377–$2.410\,\text{Å}$, respectively. The larger scatter of angles and distances in the copper compound is a result of the Jahn–Teller effect. Despite the observed distortion differences for these compounds, they are still strongly correlated as are the rest of the A_2MX_4 compounds.

A general trend among the compounds in the A_2MX_4 category is that there is a tendency towards a decreasing distortion in the series $A = Li \rightarrow Cs$, e.g. for the A_2ZnBr_4 compounds (Fig. 3). The same tendency of decreasing distortion is found with increasing size of the halogen atom, e.g. Cs_2HgX_4 compounds where $X = Cl \rightarrow I$ (Fig. 3). The variation of the tetrahedral distortions depends on the match between the central and vertex atoms, and between the vertex atoms and the surrounding alkali/alkaline earth metal as well.

Another trend observed, shown in Fig. 3, is that the distortion decreases in the sequence from chlorine to iodine for the homologous series of the closely related orthorhombic Cs_3HgX_5 compounds [Cs_3HgCl_5 (Pnma), Cs_3HgBr_5 (Pnma) and Cs_3HgI_5 (Pbca)]. This is in agreement with the observations made for the A_2MX_4 compounds. A couple of explanations for this decrease is that the Hg/I radius ratio is more favorable with respect to the

Fig. 3 Mean tetrahedral quadratic elongation versus the tetrahedral angle (X–M–X) variance for tetrahedra in three groups of compounds: Cs_2HgX_4, $X = Cl \rightarrow I$; A_2ZnBr_4, $A = Li \rightarrow Cs$; Cs_3HgX_5, $X = Cl \rightarrow I$. Within each group the distortion decreases for the series $X = Cl \rightarrow I$ or $A = Li \rightarrow Cs$

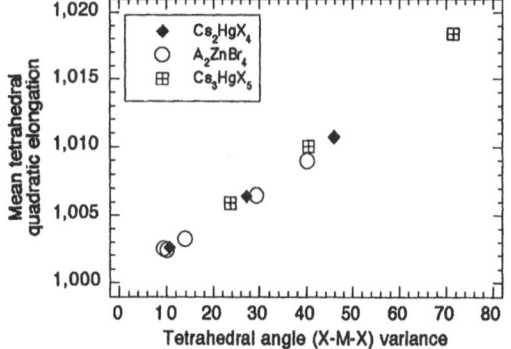

tetrahedral coordination than are the Hg/Cl and Hg/Br ratios, and/or that the vertex atoms are less affected by the cesium atoms in the higher homologues.

The inferior correlation coefficient obtained for the AMX_4 compounds depends on the shape of the $TlBr_4$ tetrahedron in $KTlBr_4$ (Fig. 2b), which have fairly equal center–vertex distances (2.549 to 2.557 Å) and vertex–center–vertex angles which range from 102.2 to 115.0°. Due to the definition of MTQE expression, the very small bond distance variations will result in a low elongation value. Thus, the actual elongation value ($\langle \lambda_{tet} \rangle = 1.0054$) deviates significantly from the predicted value ($\langle \lambda_{tet} \rangle = 1.0094$) at the same angle variance. Nevertheless, there is a distortion of the $TlBr_4$ tetrahedron confirmed by the scatter of angle variance values.

The distortion values displayed for the hydrates are significantly larger than the majority of values for the remaining categories (Fig. 2a and b). The reason for this is that many of the hydrates exhibit vertex sharing of tetrahedra. Also, the presence of water molecules will affect the tetrahedral shape and, consequently, the structure will become more strained.

The results from plotting MTQE versus the angle variance for the simulated distortion modes are shown in Fig. 2c and d. The slope varies greatly in magnitude depending on the applied mode of distortion. Note that arrows indicate the direction of compression or extension of the tetrahedron vertices, or motion of central atom to new mass center.

Although the observed tetrahedral values seem to coincide well with those obtained for modes D–F in Fig. 2c and d, it is not correct to associate them with any such mode. The reason is simply that the observed values can be obtained from linear combinations of a number of various distortion modes, while the simulated distortions are the result of a change of one applied mode at a time. In addition, each mode is accompanied with the restraints defined in the text of Fig. 1. Thus, the modes of deformation should not be considered as an attempt to rationalize the geometrical behavior of the compounds in question. They should only be considered to show that geometrical freedom is much greater than chemical freedom.

In conclusion, this investigation has shown that the tetrahedral distortions are the results of effects related to crystal structure and chemistry rather than to any strictly mathematical phenomena.

Acknowledgements B.S.E. Kerri Oikarinen is gratefully acknowledged for linguistic comments on the manuscript.

References

1. Robinson K, Gibbs GV, Ribbe PH (1971) Science 172:567
2. Sjövall R (1997) Thesis, Lund University, Sweden
3. Sjövall R (1996) TETRA. A Program for Calculation of Tetrahedral Distortions and Related Geometrical Parameters. Department of Inorganic Chemistry 2, Lund University, Sweden
4. Philippot E, Ribes M, Maurin, M (1971) Rev Chimie Minér 8:99
5. Lidin S, Larsson A-K (1991) Acta Chem Scand 45:856
6. Hyde BG, Andersson S (1989) Inorganic Crystal Structures. Wiley, New York
7. Klug HP, Alexander L (1944) J Amer Chem Soc 66:1056
8. Sjövall R, Svensson C (1997) Z Kristallogr 212:732
9. Figgis BN, Gerloch M, Mason R (1964) Acta Crystallogr 17:506

Progr Colloid Polym Sci (1998) 108:203–208
© Steinkopff Verlag 1998

R. Sjövall
S. Lidin

Coordination in inorganics

II. Coordination number – a geometrical consideration

R. Sjövall* (✉) · S. Lidin
Inorganic Chemistry
Arrhenius Laboratory
Stockholm University
S-106 91 Stockholm
Sweden

Present address
SAFT AB, Tech. Dept.
P.O. Box 709
S-572 28 Oskarshamn
Sweden

Abstract Coordination number (CN) is discussed from a geometrical point of view. Two models for the determination of CN 8 to 11 have been developed and tested with good results on 78 polyhedra consisting of alkali/alkaline earth metals as central atoms and halogen atoms/water oxygens as ligands. In these models the polyhedral volume (Vol_{poly}), the volume of a sphere – least-squares fitted to the observed polyhedron (Vol_{sph}), and the CN itself are used in the calculations in order to optimize and motivate the selected number of ligands. The first model shows that for a certain CN there exists a linear correlation between Vol_{poly} and Vol_{sph}. Since the values of the observed polyhedra lay close to the line obtained for ideal polyhedra, it was assumed that the CN is fairly well described. In the second model the Vol_{poly}/Vol_{sph} ratio is plotted against the coordination number for observed and ideal polyhedra. With increasing numbers of ligands the observed values exhibit a trend towards a closer adaption to the curve obtained from the ideal polyhedra. This adaption, or optimum CN, can be used as a criterion that motivates the choice of a certain coordination number. Finally, comparisons are made between the second model and the calculated bond distances in some observed polyhedra.

Key words Coordination number – polyhedral models

Introduction

Many investigations concerning the coordination around atoms have been reported in the literature over the years. One common method that is used to illustrate coordination is to define a central atom surrounded by a polyhedron (coordination polyhedron) with ligand atoms at its vertices. For coordination numbers (CN) up to six it is often rather easy to describe the most frequently occurring polyhedra, e.g. a tetrahedron is obtained for CN 4, a trigonal bipyramid for CN 5 and an octahedron or a trigonal prism for CN 6. From CN 7 and forth, however, the number of possible polyhedra available increases markedly. A simple measure used for these larger polyhedra is

to regard all atoms with an opposite charge from the central atom as ligands, as long as the bond distances are reasonable with respect to the sum of ionic radii, and there is not a counter-charge atom at a shorter distance. One could also calculate the ratio between potential central-vertex atom distances and the shortest central-vertex atom distance. A significant jump in the series of ratio values might, in combination with structural knowledge, indicate the limit for the choice of the proper coordination number.

In this paper the coordination concept for low symmetrical polyhedra is approached from a purely geometrical point of view. First, the emphasis is spent on the introduction of three polyhedral measures for higher order coordination polyhedra. These measures, the polyhedral volume (Vol_{poly}) and the volume of a fitted ellipsoid (Vol_{ell}) or sphere ($\mathrm{Vol_{sph}}$), are exemplified for CN 8 to 11 in order to motivate the selected CN for a large number of observed polyhedra of alkali/alkaline earth metal (A) – transition metal (M) – halides (X). The crystal structures for a majority of these halides have previously been extensively discussed [1]. Then a novel measure, based on the $\mathrm{Vol_{poly}}/\mathrm{Vol_{sph}}$ to CN ratio, is implemented to show the agreement between ideal and observed polyhedra. Finally, some reflections are made and they relate the geometrical concept and the information gained from the chemistry of the compounds studied.

Definitions and calculations

The polyhedron calculations were performed with the program IVTON, developed and supplied by Balić Žunić and Vicković [2]. This program defines, using least-squares refinement, the best central position of a polyhedron, the centroid, to its vertices [3]. Using the program a number of parameters can be derived. Those used in the present study are:

– average vertex-to-centroid distance, r, defined as the centroid sphere radius,
– volume of calculated sphere, $\mathrm{Vol_{sph}} = 4\pi r^3/3$, circumscribing the centroid,
– volume of the observed polyhedron of a compound, $\mathrm{Vol_{poly}}$, with vertices at the mass centers of the ligands,
– volume of calculated ellipsoid, $\mathrm{Vol_{ell}}$. The ellipsoid surface is fitted to the vertex positions of the centroid.

Calculations were performed on a multitude of polyhedra in various halide compounds.[1] In all calculations the alkali or alkaline earth metals represent the central atom of the coordination polyhedron, while the halogen atom or water oxygen represent the ligands. The selected compounds are, depending on their structural relationships, divided into the following categories:

– A_2MX_4; P – $P2_1/m$ Sr_2GeS_4-type [4] or closely related $P2_1$,
– A_2MX_4; P′ – Pnma β-K_2SO_4-type [5] or closely related space groups,
– A_3MX_5; P – Pnma $(NH_4)_3ZnCl_5$-type [6] or related Pbca Cs_3HgI_5-type [7]
– A_3MX_5; I – $I4/mcm$ Cs_3CoCl_5-type [8],
– $A_xM_yX_z\cdot wH_2O$, no unit cell restrictions,
– $CsMX_4$; P – Primitive unit cells,
– $CsHg_2Br_5$; P – Primitive unit cell.

The polyhedra studied were mainly those having 8, 9, 10 or 11 coordination. To each of these CN an ideal

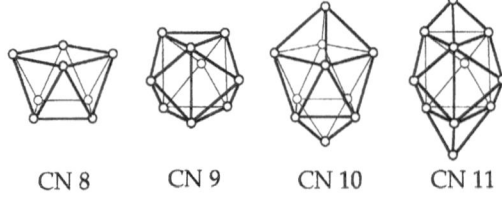

CN 8   CN 9   CN 10   CN 11

Fig. 1 Schematic drawing of the ideal polyhedra selected with CN ranging from 8 to 11. Note that the distance between adjacent atoms (circles) is equal, and that the central atoms are omitted

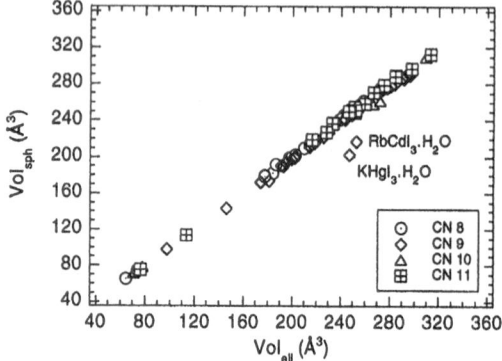

Fig. 2 $\mathrm{Vol_{sph}}$ with radius r plotted vs $\mathrm{Vol_{ell}}$ fitted to polyhedra with CN ranging from 8 to 11

polyhedron was matched. In order to show the agreement between the observed and the ideal polyhedron, the criterion for the latter was a polyhedron with all edges equal in length, neglecting the diagonal distances of square faces. In the case of an eight coordination a square antiprism was chosen, while for nine, ten and eleven coordination a tricapped trigonal prism, a bicapped square antiprism and a pentacapped trigonal prism were selected, respectively. The ideal polyhedra are displayed in Fig. 1.

Results and discussion

Preliminary analysis showed that the $\mathrm{Vol_{sph}}$ with radius r was strongly correlated to the $\mathrm{Vol_{ell}}$. This is in complete agreement with the results obtained for sulfides in a previous study [3]. In Fig. 2 this relationship is displayed for 78 polyhedra. A curve fit utilizing all data points and the linear regression method, gave the equation $\mathrm{Vol_{sph}} = 0.998 + 0.989\,\mathrm{Vol_{ell}}$ with a correlation coefficient $\rho_{corr} = 0.981$. Since no significant difference between these volumes was detected, excluding the two hydrates which lay slightly off the line, either the spherical or ellipsoidal

---

[1] List of original structural papers will be available from the authors on request.

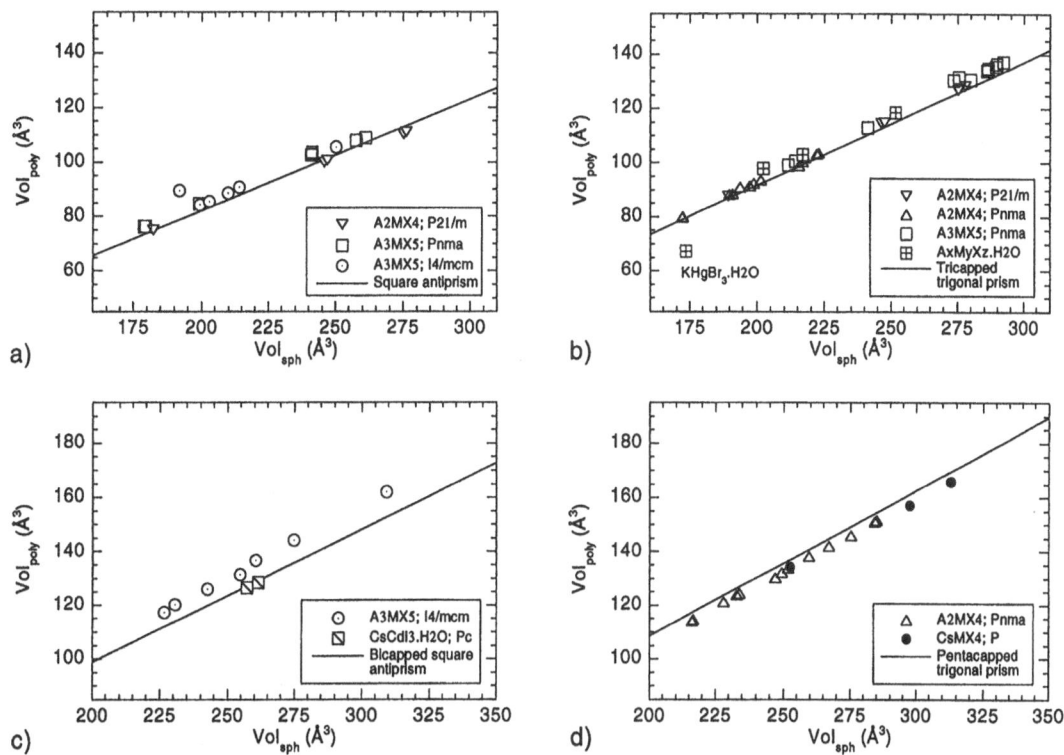

Fig. 3 Vol_{poly} vs Vol_{sph} of observed compound polyhedra. In each graph the volume of an ideal polyhedron is shown (line). The CN 8 to 11 corresponds to Figs. a–d, respectively

model could be used. Note that the sphere does not by necessity have the same extension as the ellipsoid although its volume is equal. The spherical model is perferred because of its simplicity.

An interesting feature observed is that this linear behavior appears to be independent of the chemistry of the central atom as well as that of the ligands. Hence, the linearity holds for isostructural compounds containing atoms which are completely different in size, e.g. potassium and chlorine in K_2CoCl_4, and cesium and iodine in the high temperature modification (β-) of Cs_2CdI_4. It should be noted that the polyhedra in some hydrates have two types of vertex atoms, X (halogen) and O (water oxygen), and thus different center-to-vertex distances. This may affect the calculated shape of the polyhedron, in combination with the sharing of vertices for these hydrates. Despite this, however, only two hydrates (outlined in Fig. 2) were found to significantly deviate from the fitted curve.

The calculated/observed data can be compared and plotted using many methods to give appropriate descriptions of complex polyhedra. One such method is to plot the volume of a compound polyhedron as a function of the volume of a sphere with radius r. This method has been used as a measure for the deviation of a polyhedron from

its ideal shape and to show the similarities between various polyhedra [9]. This method also shows that there is a relatively linear relationship between the Vol_{poly} and Vol_{sph} of the investigated polyhedra, see Fig. 3. An interesting and important observation to be made is the lack of atomic size dependence for isostructural compounds, and for compounds with the same CN. This is due to the fact that many of the plotted values lie on or very close to the curve of the ideal polyhedron selected. The conclusion is that the coordination could fairly well be described by the proposed model of the ideal polyhedron.

It is not clear how to assign the number of ligands around a particular atom. The question of which ligand should be included and which should not becomes apparent. So far in this text the criterion has simply been to choose the same CN as reported in the literature for the compounds studied. The next section of this paper will reflect on the coordination number from different angles of approach. The subject was discussed for binary compounds by Hansen [10], who plotted the ratio of the average center–vertex, $\langle d(A-X) \rangle$, to average vertex–vertex, $\langle d(X-X) \rangle$, distances against the coordination number for highly symmetric polyhedra with constant center–vertex and vertex–vertex distances. The graph obtained showed a strongly linear dependence, $\rho_{corr} = 0.991$.

Table 1 Definition of the types of ideal polyhedra used

CN	Polyhedron type
2	Line
3	Triangle
4	Tetrahedron*
5	Trigonal bipyramid
6	Octahedron*
7	Pentagonal bipyramid
8	Square antiprism
9	Tricapped trigonal prism
10	Bicapped square antiprism
11	Pentacapped trigonal prism
12	Cube octahedron
12	Icosahedron*
14	Rhombic dodecahedron
20	Regular pentagonal dodecahedron*
24	Truncated octahedron

* Denotes Platonic solids with constant center-to-vertex and vertex-to-vertex distances

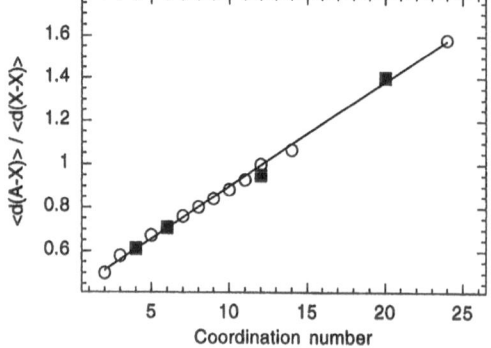

Fig. 4 $\langle d(A-X)\rangle/\langle d(X-X)\rangle$ vs CN according to the polyhedra in the table. Filled squares represent the Platonic solids

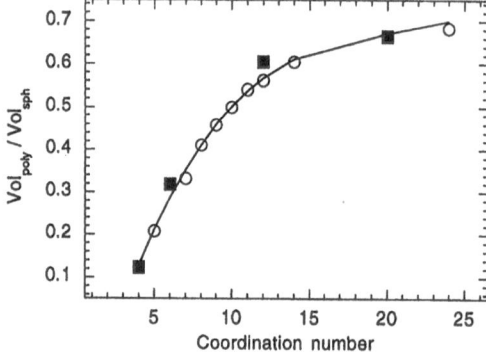

Fig. 5 Vol_{poly}/Vol_{sph} vs CN according to the polyhedra in the table. Filled squares correspond to the Platonic solids

The criterion used in this investigation was to have an ideal polyhedron with constant vertex-to-vertex distances, see the Table 1 for a list of selected polyhedra. Here, the number of polyhedra has been extended in comparison to the study above. The graph obtained is shown in Fig. 4. The correlation is extremely accurate, $\rho_{corr} = 0.999$, for the equation $\langle d(A-X)\rangle/\langle d(X-X)\rangle = 0.414 + 0.0484$ CN. For the four Platonic solids the correlation coefficient was 0.997. Since there might be some ambiguity of whether the diagonal of a square surface should or should not be included in the calculations (here, it is not), the aim was to find another measure which could relate a polyhedron to a certain coordination number. Therefore, the Vol_{poly}/Vol_{sph} ratio was employed and plotted as a function of CN for the polyhedra listed in Table 1. The results obtained are displayed in Fig. 5. The curve (smoothened) for the ideal polyhedra can be approximated by a polynomial of third order having a correlation coefficient of 0.998. The corresponding values for the Platonic solids deviate from the curve slightly.

The four graphs in Fig. 6 illustrate what occurs when the Vol_{poly}/Vol_{sph} ratio increases with respect to the number of ligands around a certain atom. In the respective graph, the calculated Vol_{poly}/Vol_{sph} values are plotted as a function of the CN for polyhedra with 8, 9, 10 and 11 vertices in compounds Cs_2CdI_4 [11], Cs_3CdI_5 [12], β-Cs_2CdI_4 [13], $CsTlI_4$ [14] and $CsCdI_3 \cdot H_2O$ [15]. These polyhedra consist of cesium (central atom) and iodine atoms (vertices), except for the hydrate which also includes two water–oxygen vertices. The solid curve outlined is that obtained in Fig. 5 for ideal polyhedra. In general, it is clear that for low CN the deviation from the curve is larger than for higher CN. As the number of ligands increases

there is a trend towards a closer adaption to the curve. This means that the ligands will adopt a more homogeneous distribution around the entire sphere. Addition of further ligands will improve the model, but consequently the polyhedron grows. This growth will result in an overcritical ligand representation and a gigantic coordination polyhedron not relevant for structural description. A subcritical representation produces a bad fit to the curve due to a poor and inhomogeneous distribution of ligands.

From a purely geometrical point of view the interpretation of the observed trend above is that the optimum number of ligands has been reached according to a specific central atom with a certain ligand distribution. This may then motivate the chosen coordination number. The choice is, however, made without taking any bond distance values into consideration at all. The last part of this section will reflect on aspects concerning bond distances for the cesium–iodine–(water–oxygen) polyhedra in Fig. 6. One

Progr Colloid Polym Sci (1998) 108:203–208
© Steinkopff Verlag 1998

Fig. 6 Graphs that illustrate the Vol_{poly}/Vol_{sph} ratio as a function of CN for some selected compounds having (a) 8, (b) 9, (c) 10 and (d) 11 ligands

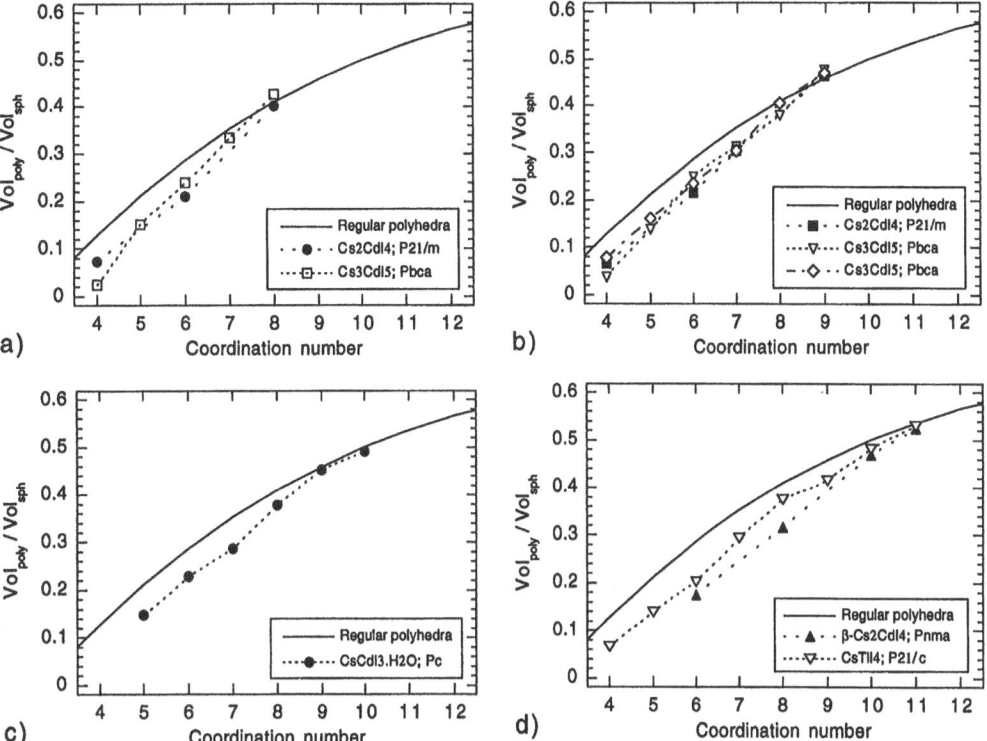

way to represent bond distances is shown in Fig. 7, where the ratio, $d(Cs–I)_n/d(Cs–I)_1$, $n = 1 \rightarrow 11$, vs. CN is outlined for the above-mentioned polyhedra. A significant jump in the series of ratio values, e.g. between 1.09 and 1.19 (corresponding to a bond distance change from 4.04 to 4.41 Å) in Fig. 7a for Cs_3CdI_5, may be indicative for the coordination number limit in combination with the exceeding of a certain ratio level. In the figure text, the bond distance ranges are also given. Note that the variations are large: 3.72–3.97 Å for the shortest Cs-I distances and is up to 4.02–4.90 Å for the longest one. As a parenthesis it should be mentioned that, according to Shannon [16], the sum of the ionic radii of cesium and iodine is 3.87 Å in polyhedra with six iodine vertices.

In the following text a comparison will be made between the chosen CN, based on the polyhedra in Fig. 6, and the results obtained from the representation of bond distance ratios of the same polyhedra in Fig. 7.

There is doubt as to whether the two polyhedra in Fig. 7a should be described with CN 8. Both curves exhibit significant jumps indicating that the CN with respect to the bond distances should be six (Cs_2CdI_4) and seven (Cs_3CdI_5) rather than eight. However, according to the geometrical model a CN of 6 is found for the polyhedron in Cs_2CdI_4, which is contradictory to the results obtained in Fig. 6a.

In the case of nine coordination, Fig. 7b, there are two polyhedra, Cs_2CdI_4 and Cs_3CdI_5 (squares), which exhibit large differences between the shortest and the longest distances, and one polyhedron, Cs_3CdI_5 (triangles), with a small scatter of distances. From a bond distance approach the proposed CN for the first two polyhedra tend to be seven, while the CN for the third one is nine. The geometrical model, Fig. 6b, suggests CN 9 for the first two polyhedra, and 8 or 9 for the last polyhedron.

For CN 10 some discrepancy could be observed. Fig. 7c indicates that a polyhedron with seven vertices may be chosen, while the geometrical interpretation in Fig. 6c shows that either CN 9 or 10 will suffice. For the structural description of $CsCdI_3 \cdot H_2O$ ten coordination was chosen [15].

The two graphs in Fig. 7d exhibit completely different behavior, with large distance variations for the polyhedron in β-Cs_2CdI_4 and small variations for that in $CsTlI_4$. Despite this, eleven coordination is motivated for both polyhedra, in Fig. 6d.

It is obvious that there exists a discrepancy concerning the interpretation of the results from the geometrical and the bond distance methods discussed previously. In conclusion, simply using one method is not beneficial in determining a coordination number. Therefore, the chemistry of the compound should be used in conjunction with the

Fig. 7 Ratio $d(\text{Cs–I})_n/d(\text{Cs–I})_1$, $n = 1 \rightarrow 11$, vs CN for some selected compounds. The bond distance ranges are: (a) CN 8: Cs_2CdI_4 3.85–4.28 Å; Cs_3CdI_5 3.72–4.41 Å; (b) CN 9: Cs_2CdI_4 3.77–4.39 Å; Cs_3CdI_5 3.90–4.19 Å; Cs_3CdI_5 3.82–4.41 Å; (c) CN 10: $CsCdI_3 \cdot H_2O$ 3.86–4.55 Å; (d) CN 11: $\beta\text{-}Cs_2CdI_4$ 3.76–4.90 Å; $CsTlI_4$ 3.97–4.46 Å. Note that the two Cs–water oxygen distances are omitted in (c)

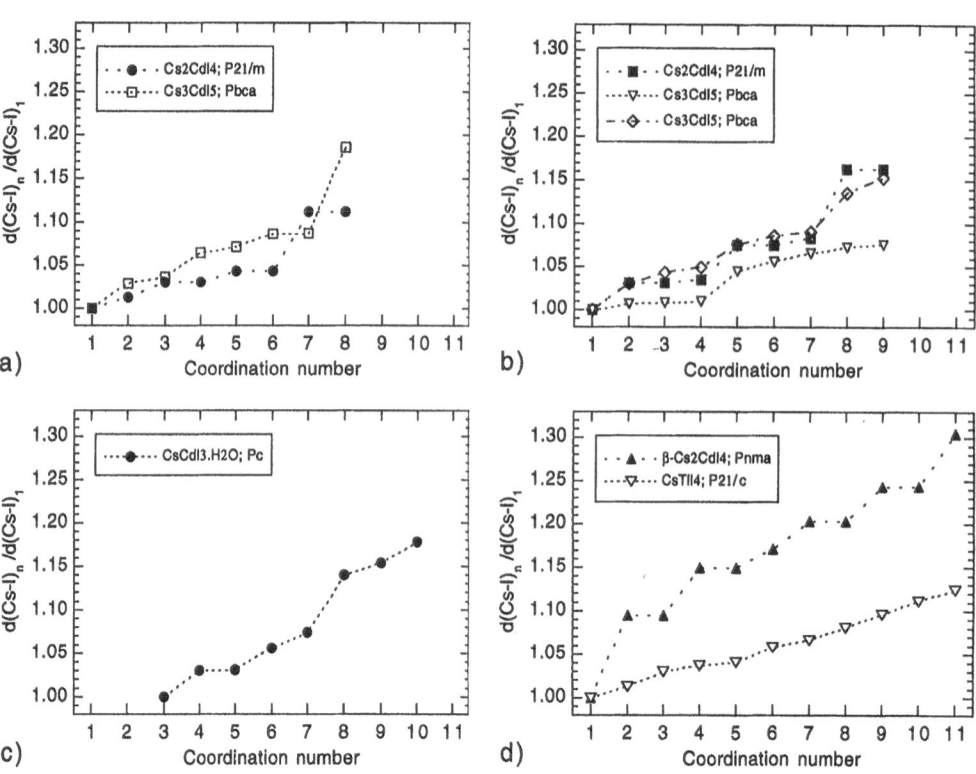

geometrical method. The geometrical method also shows that there are structural weaknesses in the chemical description. An inhomogeneous coordination could indicate a structural anomaly, explained from the effects of lone pairs, polarization, etc.

Acknowledgements We would like to thank Prof. Emil Mackovicky and Dr. Tonci Balić Žunić for providing us with the IVTON program. B.S.E. Kerri Oikarinen is also gratefully acknowledged for linguistic comments on the manuscript.

References

1. Sjövall R (1997) Thesis, Lund University, Sweden
2. Balić Žunić T, Vicković I (1994) J Appl Crystallogr 29:305
3. Balić Žunić T, Mackovicky E (1996) Acta Crystallogr B52:78
4. Philippot E, Ribes M, Maurin M (1971) Rev Chimie Minér 8:99
5. Lidin S, Larsson AK (1991) Acta Chem Scand 45:856
6. Klug HP, Alexander L (1944) J Amer Chem Soc 66:1056
7. Sjövall R, Svensson C (1988) Acta Crystallogr C44:207
8. Figgis BN, Gerloch M, Mason R (1964) Acta Crystallogr 17:506
9. Mackovicky E (1996) Private communication
10. Hansen S (1993) J Solid State Chem 105:247
11. Sjövall R (1989) Acta Crystallogr C45:667
12. Sjövall R, Svensson C (1997) Z Kristallogr 212:732
13. Touchard V, Louër M, Auffredic JP, Louër D (1987) Rev Chimie Minér 24:414
14. Thiele G, Rotter HW, Zimmermann K (1986) Z Naturforsch 41B:269
15. Sjövall R, Svensson C, Lidin S (1996) Z Kristallogr 211:234
16. Shannon RD (1976) Acta Crystallogr A32:751

Progr Colloid Polym Sci (1998) 108:209
© Steinkopff Verlag 1998

AUTHOR INDEX

Progr Colloid Polym Sci (1998) 108:210
© Steinkopff Verlag 1998